D0869152

Cambridge Planetary Science Series

Editors: W. I. Axford, G. E. Hunt, R. Greeley

Wind as a geological process

on Earth, Mars, Venus and Titan

RONALD GREELEY
Department of Geology and Center for Meteorite Studies, Arizona State University
JAMES D. IVERSEN
Department of Aerospace Engineering, Iowa State University

Wind as a geological process
on Earth, Mars, Venus and Titan

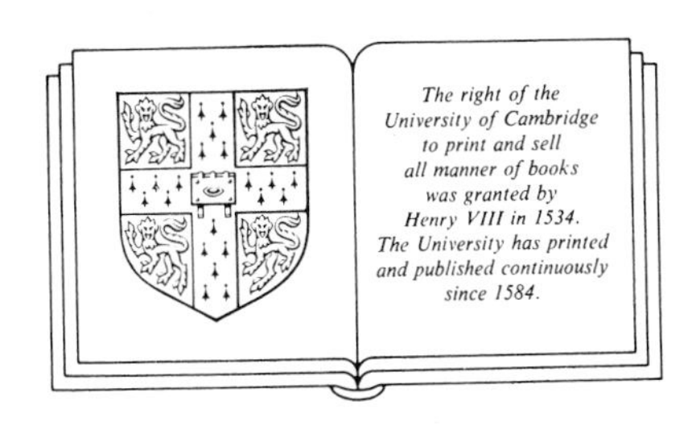

CAMBRIDGE UNIVERSITY PRESS
Cambridge
New York New Rochelle Melbourne Sydney

Published by the Press Syndicate of the University of Cambridge
The Pitt Building, Trumpington Street, Cambridge CB2 1RP
32 East 57th Street, New York, NY 10022, USA
10 Stamford Road, Oakleigh, Melbourne 3166, Australia

First published 1985
First paperback edition 1987

Printed in Great Britain at The Alden Press, Oxford

Library of Congress catalogue card number: 83-18878

British Library cataloguing in publication data

Greeley, Ronald
Wind as a geological process on Earth,
Mars, Venus and Titan. – (Cambridge planetary science series; 4)
1. Wind erosion
I. Title II. Iversen, James D.
551.3′7′2 QE597

ISBN 0 521 24385 8 hard covers
ISBN 0 521 35962 7 paperback

AP

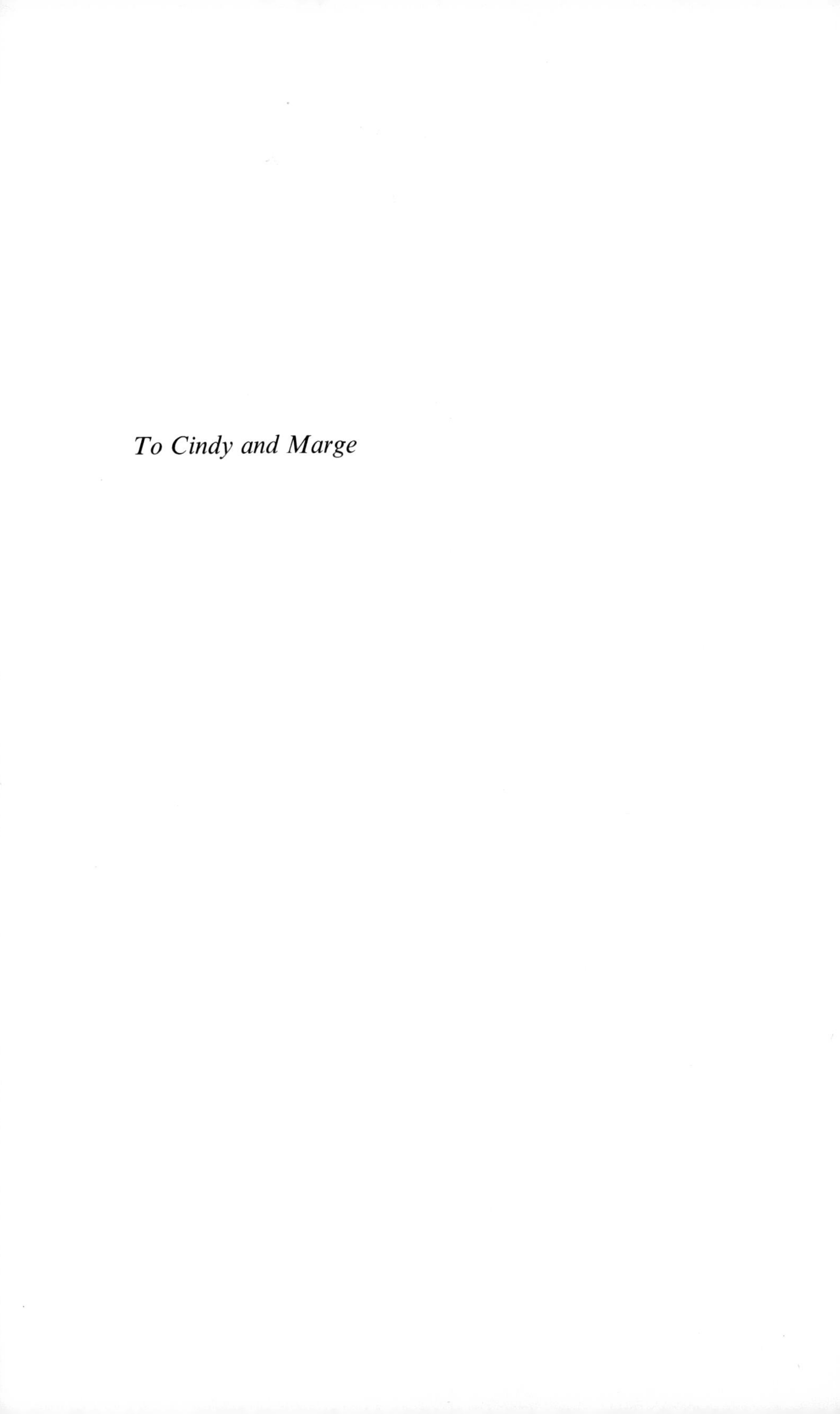

To Cindy and Marge

CONTENTS

PREFACE

Ralph Bagnold – an engineer by training, a military man by profession, and in many ways a geologist at heart – melded his interests into an elegant study of aeolian processes that has spanned many decades. In 1941 Bagnold published the first edition of his book, *The Physics of Blown Sand and Desert Dunes*. Often referred to simply as 'Bagnold's classic book', it is indeed a classic in every sense of the word. The fact that nearly every subsequent paper dealing with aeolian processes refers to the Bagnold book bears testimony that the basic principles described by him are essentially correct and have withstood the test of time.

Our book deals with aeolian processes in the planetary context. It is not our intent to 'replace' Bagnold's book or the research it represents. We learned that was neither required nor possible early in our own research program! Instead, we have built upon the solid foundation laid by Bagnold, testing the relationships defined by him through different approaches, and extrapolating the results to other planetary environments by attempting to predict how aeolian processes operate on Mars, Venus and, perhaps, Titan, the largest of the saturnian satellites.

We begin with an introduction to aeolian processes and a general overview of aeolian activity on the planets. We then discuss, in Chapter 2, the requirements for aeolian activity – a dynamic planetary atmosphere and a supply of particles capable of being moved by the wind – and describe in Chapter 3 the physical processes involved in particle movement by the wind. In Chapters 4 and 5 we describe wind-eroded and wind-deposited features and landforms. Next we consider interaction between the wind and topography and then close with a chapter on windblown dust (fine-grained material carried aloft in suspension).

Insofar as is practical, we have integrated non-Earthly aspects of aeolian activity into the appropriate chapter sections. Typically, we begin a section with a discussion of Earth (our 'ground truth'), extend the discussion to Mars, and then close the section with speculations for Venus and Titan.

Our intention is that this book be used as reference and text for upper division or graduate courses in comparative planetology. Perhaps more than any other field, planetology requires a multidisciplinary approach to combine talents from the geological sciences, engineering, chemistry, and physics. One of the biggest difficulties in comparative planetology is communication among the various disciplines. Consequently, we have attempted to write this book in such a way that it can be understood by anyone with a science or engineering background. Our own somewhat disparate backgrounds, in geological sciences and in engineering, have often forced us to reevaluate our own and each other's viewpoints, and we hope those experiences have helped us achieve our objectives. Terms and commonly used jargon are defined where first used; an expanded glossary is also included for reference.

RG and JDI
1983

Acknowledgments

Writing a book is a substantial project. Such a project can be undertaken only with the assistance and encouragement of friends, family, and colleagues. Among the many individuals who helped in this effort, we thank P. Thomas (Cornell University), A. Peterfreund (Brown University), and M. Malin and S. Williams (both of Arizona State University), who read the entire manuscript and provided helpful discussions for improvement, as well as providing reviews for separate chapters. Critical reviews of individual chapters were also provided by R. Sharp (California Institute of Technology), D. Gillette (National Center for Atmosphere Research), S. Idso (US Department of Agriculture), L. Lyles and colleagues (Wind Erosion Research Laboratory), H. Tsoar (Ben Gurion University of the Negev), J. Veverka (Cornell University), and G. Takle (Iowa State University).

We acknowledge, with gratitude, photographic work provided by D. Ball, assisted by J. Riggio and Joo-Keong Lim, typing of countless drafts by M. Schmelzer, D. Keller, T. Gautesen-Borg, C. Mathes, D. Reil, and T. Krock, drafting of figures by the Technical Graphics section at Iowa State University, proofreading by C. Freeley, and the assistance in locating various planetary images by J. Swan and L. Carroll of the US Geological Survey, and L. Jaramillo of Arizona State University.

Finally, we thank Steve Dwornik and Joe Boyce, Discipline Scientists for Planetary Geology of the National Aeronautics and Space Administration, for support of our research on planetary aeolian processes.

1

Wind as a geological process

1.1 Introduction

Beginning with the first tentative probes into space in the mid-1960s, the geological exploration of the solar system has revealed a remarkable diversity in the planets and their satellites. Each planetary body displays combinations of surface features that reflect unique geological histories and environments. Yet, when the surfaces of the terrestrial planets and satellites are analyzed in detail, we find that many of them have experienced similar geological processes in their evolution.

The discipline of *comparative planetary geology* has as its goal the definition of the fundamental processes that have shaped and modified the planets, satellites and other 'solid surface' bodies in the solar system. For simplification, we shall refer to all such objects simply as *planets*. The giant gaseous planets, such as Jupiter and Saturn, are excluded from study because they apparently lack solid surfaces and thus are not appropriate for geological analyses. The goal of planetary geology is achieved by determining the present state of planets, by deriving information of their past state(s) – or geological histories – and by comparing the planets to one another.

Comparative planetary geology has shown that nearly all of the planets have been subjected to major geological processes, including impact cratering, volcanism, tectonism (crustal deformation), and gradation. Gradation involves the weathering, erosion, and deposition of crustal materials through the actions of various agents, such as wind and water. This book deals with wind, or *aeolian*, processes (Fig. 1.1). Aeolian is defined (Gary *et al.*, 1972) as 'pertaining to the wind; especially said of rocks, soils, and deposits (such as loess, dune sand, and some volcanic tuffs) whose constituents were transported (blown) and laid down by atmospheric currents, or of landforms produced or eroded by the wind, or of sedimentary structures (such as ripple marks) made by the wind, or of geologic processes (such as erosion and deposition) accomplished by the wind'. Thus, any planet or satellite having a dynamic atmosphere and a solid surface is subject to aeolian, or wind, processes. A survey of the solar system shows that Earth, Mars, Venus, and possibly Titan, meet these criteria (Table 1.1). These planets afford the opportunity to study a basic geological process – aeolian activity – in a comparative sense, with each planet being a vast, natural laboratory which has strikingly different environments. Because terrestrial processes and features have been studied for many years, Earth is the primary data base for interpreting aeolian processes on the planets. However, because surface processes are much more complicated on Earth – primarily because of the presence of liquid water and vegetation – some aspects of aeolian processes that are difficult to assess on Earth are easier to understand on the other planets. For example, on Mars the lack of competition from other processes during the

Fig. 1.1. View of the great dust storm of December, 1977, in the Central Valley of California, showing dust originating near the base of the mountains to the left and rising to several hundred meters. Dust storms have direct cultural and geological effects and are part of the general aeolian regime. (Copyright 1978, UNIFO Enterprises, San Francisco, California.)

Table 1.1. *Relevant properties of planetary objects potentially subject to aeolian processes*

	Venus	Earth	Mars	Titan
Mass (Earth = 1)	0.815	1	0.108	0.02
Mass (kg)	48.7×10^{23}	59.8×10^{23}	6.43×10^{23}	1.34×10^{23}
Density (water = 1)	5.26	5.52	3.96	1.90
Surface gravitational acceleration (m/sec^2)	8.88	9.81	3.73	1.36
Diameter (km)	12 104	12 756	6787	5140
Surface area (km^2)	4.6×10^8	5.1×10^8	1.4×10^8	8.3×10^7
Atmosphere (main components)	CO_2	N_2, O_2	CO_2	N_2
Atmospheric pressure at surface (mb)	9×10^4	10^3	7.5	$\approx 1.6 \times 10^3$
Mean temperature at surface (°C)	480	22	−23	−200
Liquid water on surface	no	yes	no	no
Orbital radius (AU)	0.72	1.00	1.52	9.53
Orbital period (yr)	0.62	1.00	1.88	29.6[a]
Orbital eccentricity	0.007	0.017	0.093	0.056[a]
Obliquity (°)	< 3	23.5	25.1	—
Axial rotation rate (days)	243 (retrograde)	1	1.027	—
Solar flux (erg/(cm^2 sec))	2.61×10^6	1.37×10^6	0.59×10^6	0.15×10^5

[a] Values for the planet Saturn.

last aeon or more permits the cumulative effect of aeolian processes to be relatively better observed than on Earth.

In this chapter we discuss the general approach for investigating aeolian processes in the planetary context and consider the relevance of aeolian processes to other geological problems. Finally, we provide an overview of aeolian activity on Earth, Mars, Venus, and Titan.

1.2 Approach to the problem

Aeolian processes incorporate elements of geology, meteorology, physics, and, to some degree, chemistry. A unified study of these processes therefore requires a multidisciplinary approach. The approach commonly employed is not only multidisciplinary but combines field studies with laboratory simulations and theory.

Let us take the determination of the *threshold curve* as an example of this multidisciplinary, combined approach. The threshold curve (Fig. 1.2; Appendix A) relates the minimum wind speeds required to set particles of different sizes into motion and is probably the most important relationship within the various aspects of aeolian processes. The threshold curve for Earth was first derived by R. A. Bagnold (1941), a British Army engineer who spent considerable time in the deserts of Egypt. Bagnold conducted a series of wind tunnel experiments in which he varied particle size to determine the minimum wind speeds necessary to set the particles into motion. Wind velocities were expressed in terms of the shear stress exerted on the sand surface by the wind, which is a function of the wind velocity profile. He then field tested the results under natural conditions, making careful measurements of wind velocity profiles and various particle characteristics.

Using the results from his laboratory experiments and field studies, Bagnold then derived the mathematical expressions for the movement of sand by the wind, in terms of the fundamental physics that are involved. Thus, he was concerned with a *geological* problem (windblown sand) that required knowledge of *meteorology* (wind velocities above the sand

Fig. 1.2. Comparison of the threshold friction speed versus particle diameter for Mars, Earth, and Venus. (From Iversen *et al.*, 1976*b*.)

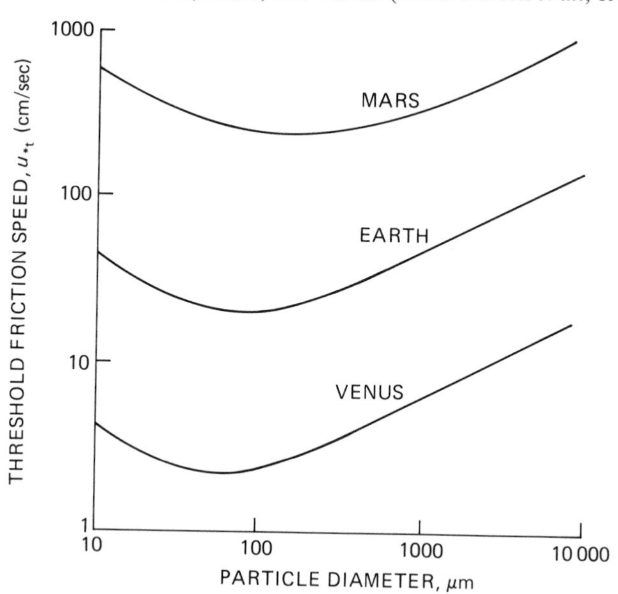

surface), and *physics* (expression of the movement of the sand). His approach was to combine engineering practices utilizing wind tunnels with geological and meteorological field methods to derive mathematical models of the general problem.

This same approach is used to study aeolian processes on the planets. In the mid-1950s, planetologists first began to consider the possibility that Mars may experience aeolian processes. Dean McLaughlin (1901–65), a professor at the University of Michigan, used his combined training and interests in astronomy and geology to analyze the distribution of albedo markings on Mars. From his analyses, he concluded (McLaughlin, 1954*a,b*; reviewed by Veverka & Sagan, 1974) that many of the markings were aeolian and derived a map of deduced wind directions, which is remarkably similar to wind patterns based on recent spacecraft data. Later, Sagan & Pollack (1969) adopted the basic expressions for wind threshold speeds derived by Bagnold, substituted the appropriate values for the martian environment (gravity and atmospheric density), and derived a threshold curve for sand movement on Mars. Although there were a great many uncertainties in their extrapolation, because knowledge of the martian environment was extremely limited, it was the first attempt to apply terrestrial aeolian parameters to an extraterrestrial problem.

Knowledge of the martian environment expanded with each of the United States' missions to Mars – Mariner 4 (1965), Mariners 6 and 7 (1969), Mariner 9 (1971–72), and the Viking mission (1976–79) – as well as the Soviet Orbiters, Mars 3 (1971) and Mars 5 (1974). Concurrent with incoming spacecraft data, an understanding of the dynamics of carbon dioxide at low atmospheric densities and knowledge of particle motion in the martian environment were gained through various laboratory simulations (Hertzler *et al.*, 1967; Iversen *et al.*, 1973; Greeley *et al.*, 1976, 1980*a*). These simulations culminated in a series of wind tunnel experiments in which atmospheric composition and density were duplicated for Mars. But because some martian parameters, such as the lower gravity, could not be simulated in the experiments, certain parameters had to be analyzed theoretically in order to derive the final threshold curve for Mars, shown in Fig. 1.2.

The ultimate test for this approach to planetary problems is a measurement made on the planet concerned. In the case described for Mars, measurements of winds obtained through the meteorology experiment on board the Viking Landers, and observations of dust storm activity, show that the threshold curves are essentially correct (Sagan *et al.*, 1977).

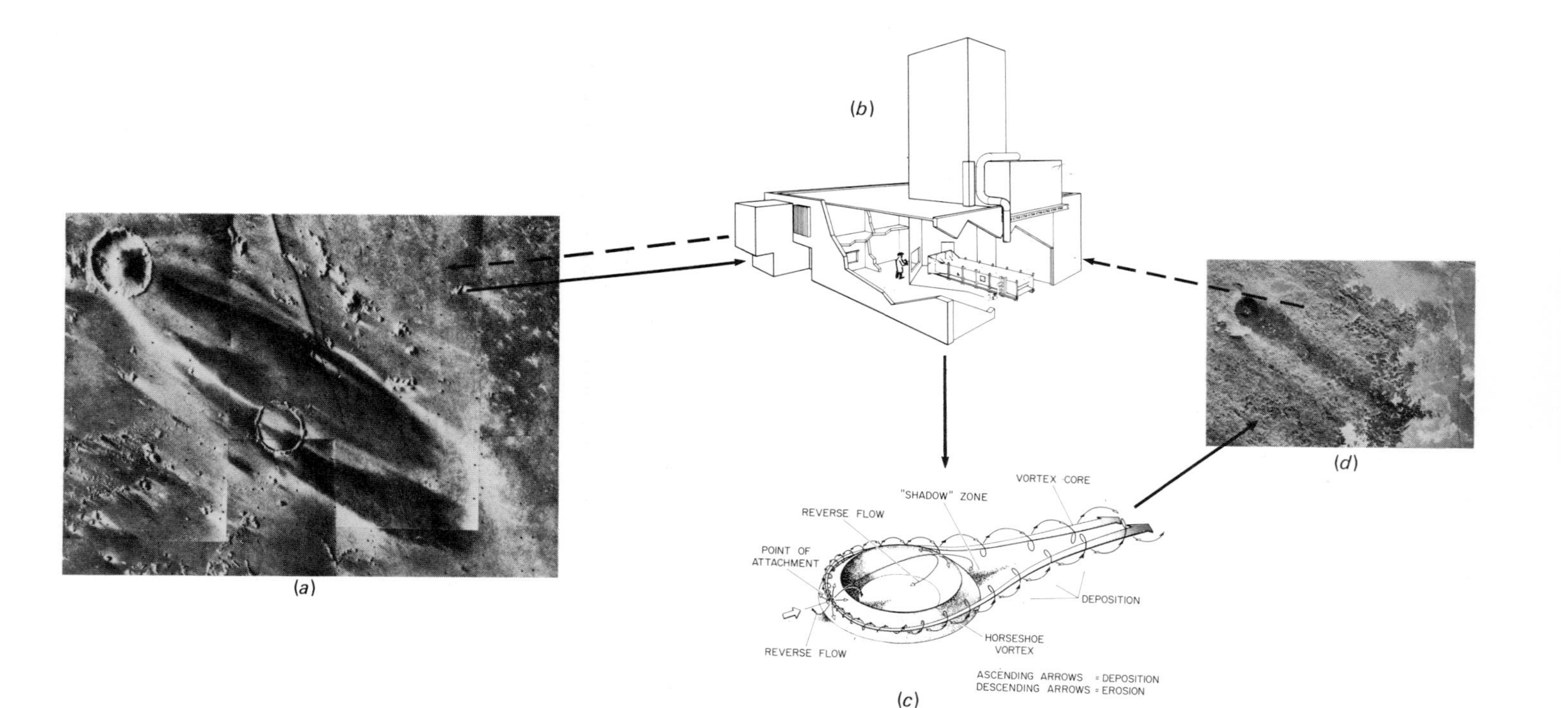

Fig. 1.3. Illustration of the three-fold approach to problems in planetary geology, combining spacecraft data analysis with laboratory simulations and field studies: (*a*) *definition of the planetary problem* (formation of dark streaks associated with wind swept craters on Mars), (*b*) *laboratory simulation* 'Earth-case' (winds blown across the model of a crater), (*c*) *derivation of the model* (air flow patterns and zones of wind erosion and deposition are determined from the wind tunnel tests), (*d*) *field study* (measurements of the air flow and geological studies of the natural site at Amboy lava field, California). The results from the field study are used to verify–calibrate–modify the wind tunnel simulations; then, once confidence in the methodology is obtained for the 'Earth-case', the wind tunnel tests are run under conditions simulating the martian environment as closely as possible. Extrapolation to the planetary case usually requires the use of theory with the simulations because some parameters, such as gravity, cannot be duplicated in experiments conducted on Earth. In the example shown here, dark crater streaks on Mars were found to be erosional features resulting from the vortices shed from the rims of the craters. (After Greeley *et al.*, 1974*b*.)

We can outline a general procedure for studying aeolian processes in the planetary context, using the example shown in Fig. 1.3 (Greeley, 1982):

(1) identification of the general problem and isolation of specific factors for study;
(2) investigation of the problem under laboratory conditions simulating the 'Earth case' where various parameters can be controlled;
(3) field testing of the laboratory results under natural conditions to verify that the simulations were done correctly;
(4) correction, modification, and/or calibration of the laboratory simulations to take the field results into account;
(5) laboratory experiments for the extraterrestrial case to duplicate or simulate, as nearly as possible, the planetary environment involved;
(6) extrapolation to the planetary case using a combination of the laboratory results and theory for parameters, which cannot be duplicated, such as gravity differences;
(7) field testing of the extrapolation via spacecraft observations and application of the results to the solution of the identified problem.

Although we are a long way from carrying out this approach in the study of all aspects of aeolian processes for Earth, Mars, Venus, and Titan, the results presented here draw upon this general approach as much as possible. As one might expect in defining the various problems, we commonly find that many aspects of aeolian processes are not well understood, even for Earth, let alone for other planetary environments. Consequently, a benefit of the approach outlined here is not only to provide a logical means for solving extraterrestrial problems, but to contribute toward solving problems dealing with aeolian processes on Earth as well.

1.3 Significance of aeolian processes

It is estimated that more than 500×10^6 metric tons of dust (particles $\leqslant 20$ μm) are transported annually by the wind on Earth (Peterson & Junge, 1971). Dust storms reduce visibility on highways and are responsible for loss of life and property due to many accidents each year. Atmospheric dust, whether raised by winds or injected into the atmosphere by volcanic processes, can also have a significant effect on temperature. Such effects have been documented on Earth, both locally and globally, and have been observed on Mars, as shown in Fig. 1.4. Thus, aeolian processes can have a direct effect on changing the climates of the planets.

Windblown sand (particles 0.0625–2 mm) causes numerous problems, primarily through abrasion of man-made objects and encroachment on cultivated lands and developed areas. For example, special precautions must be taken to prevent erosion of structures by windblown sand in some regions. As shown in Fig. 1.5, the lower parts of power poles in sandy regions must be sheathed in metal to prevent their being worn away by sands driven by the wind.

Any process that is capable of eroding and transporting vast quantities of material is important in the geological context. Much of the present landscape in desert regions results from aeolian processes. Vast areas are blanketed with sheets of windblown silt and clay, called *loess* (Fig. 1.6). It is estimated that one-tenth of Earth's land surface is covered with loess and loess-like deposits in thicknesses of 1–100 m (Pécsi, 1968). Loess soils constitute some of Earth's richest farmlands.

The geological column shows ample evidence of aeolian processes throughout Earth's history, as reviewed by Reineck & Singh (1980). Glennie (1970) discusses ancient aeolian sediments and provides a list of factors to enable the recognition of windblown deposits. Thick sand

Fig. 1.4. Effect of atmospheric dust on atmospheric temperatures on Mars (modeled). (*a*) Two modeled temperature profiles (morning, 0600, and afternoon, 1600) as a function of the height above the surface for clear atmospheric conditions. (*b*) Modeled temperature profiles during the global dust storm. Model values are similar to measurements made by Mariner 9 during the dust storm of 1971–72 (shaded area). (Pollack, 1979, after Gierasch & Goody, 1973.)

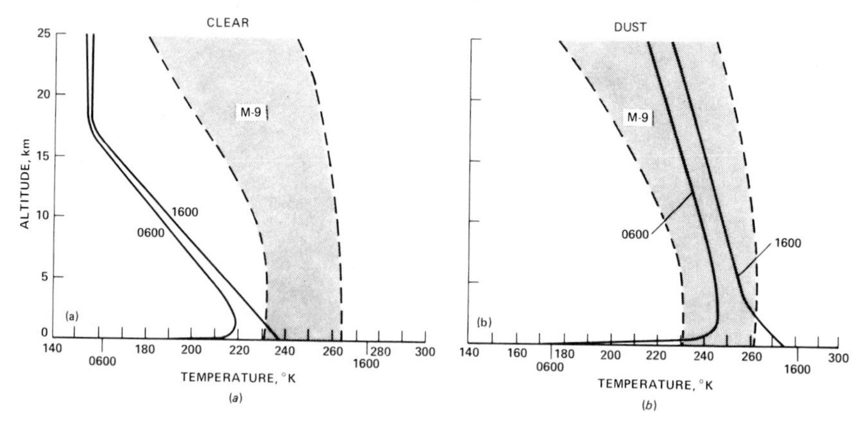

deposits (Fig. 1.7) occur in many areas and represent ancient sand 'seas'. For example, the Permian Age Coconino Sandstone is tens to hundreds of meters in thickness and covers thousands of square kilometers of the Colorado Plateau. Enormous cross-beds and other sedimentary structures attest to its aeolian origin. In some cases, windblown sediments are important reservoirs of water and petroleum. Understanding sedimentary structures within these aeolian deposits and knowledge of their environments of deposition can help in fully realizing their potential as supplies of water and oil.

1.3.1 *Relevance to Earth*

The understanding of aeolian processes is essential to the control of such processes at those places on Earth where control is important. The necessity for control can be generally classed into three groups – environmental, agricultural, and transportation – although there are significant problems in other endeavors as well. Environmental problems have to do with the effects of dust on health, visibility, and climate, as well as on engineering considerations such as abrasion by windblown grains. Agricultural problems involve soil erosion by wind and the effects on plants of blowing soil and sand. The effects on transportation include the protection from, or removal of, blown snow or sand on highways, railroads, and airport runways.

Desertification is a term coined for the conversion of land to deserts. Although the causes of desertification are controversial, thousands of square kilometers are converted to deserts annually. Whether primarily man-caused or resulting from natural cycles on Earth, aeolian processes play a significant role in desertification. For example, during periods of drought, topsoil dries out and is easily removed by the wind, converting arable land to desert. By understanding the relationships between surface roughness (e.g., windbreaks) and threshold wind speeds, it is possible to retard topsoil erosion and to slow down, or halt, desertification in some areas.

The desertification problem is enormous. Although more than one-third of Earth's land is arid or semiarid, somewhat less than one-half of this area is so dry that it cannot support human life. Over 600 million people live in dry areas, and about 80 million of these live on lands that are nearly useless because of soil erosion and encroachment of sand dunes or other effects of desertification (Eckholm & Brown, 1977). Desertification of arid lands, much of which is the result of human activity, is evident on all inhabited continents of Earth, but the largest areas of severe desertification are in

Fig. 1.5 (a)

(b)

Africa and Asia (Dregne, 1980). Agricultural land damaged by wind erosion in the US varies from 400–6000 km^2/yr (Kimberlin *et al.*, 1977).

Desertification control measures can be many and varied and depend not only on the locality, but also on knowledge of ecological management and the physics of aeolian erosion. If desert pavements or other covers (such as vegetation) can be maintained, wind erosion of soil can be reduced. Overgrazing, unwise cultivation practices, and the use of off-road vehicles for various purposes weaken and destroy natural or artificial covers (Secretariat of the United Nations, 1977; Dregne, 1980; Péwé, 1981).

Control of aeolian phenomena in the area of transportation is somewhat easier because the human equation concerning the existence and quality of life is not involved to the same extent as in the desertification problem. Technical problems can still be quite difficult, however, and, in the case of snow, removal rather than control is often the technology applied, even though control methods would usually be more economical and energy efficient. Methods of control of drifting sand and snow are illustrated in Figs. 1.8 and 1.9. The control of windblown material, whether sand, dust, or snow, is a worldwide problem that is exemplified by laboratory and field test research in Asia (Dyunin & Komarov, 1954; Zhonglong & Yuan, 1980), Europe (Jensen, 1959; Norem, 1979; Iversen & Jensen, 1981), and the US (Iversen, 1980*a*,*b*, 1981; Ring *et al.*, 1979; Tabler, 1980*a*,*b*).

1.3.2 *Relevance to planetary science*

Loess deposits cover extensive parts of Earth's surface (Fig. 1.6). Even where relatively young and well exposed on the surface, loess deposits are extremely difficult to identify by remote-sensing methods. Yet, identification of such deposits could be important in understanding planetary surfaces. For example, substantial parts of Mars appear to be mantled with material interpreted to be aeolian sediments. However, other processes, such as volcanism, could lead to similar-appearing terrain. Thus,

Fig. 1.5. (*a*) Power pole in the Coachella Valley, near Palm Springs, California, sheathed with metal to retard abrasion by windblown sand. The metal has been eroded between 8–24 cm above the ground, a reflection of the zone of maximum abrasion by windblown sand. Prevailing wind is from the left. (Photograph by R. Greeley, December, 1982.) (*b*) Fence post abraded to a depth of more than 10 cm during the 20 December 1977 sand and dust storm in southern California. Scour grooves in the soil are up to 12 cm deep and total deflation was ≈ 20 cm. Note that the fence wire protected parts of the post from abrasion. (From Wilshire *et al.*, 1981; copyright 1981, Geological Society of America.)

Fig. 1.6. Map of Earth showing arid and semiarid regions, polar deserts, the distribution of loess deposits, and the principal directions of dust transport. Major deserts include (1) Great Basin, (2) Sonoran, (3) Chihuahuan, (4) Peruvian, (5) Atacama, (6) Monte, (7) Patagonian, (8) Sahara, (9) Somali-Chabli, (10) Namib, (11) Kalahari, (12) Karroo, (13) Arabian, (14) Rab'al Khali, (15) Turkestan, (16) Iranian, (17) Thar, (18) Taklimakan, (19) Gobi, (20) Great Sandy, (21) Simpson, (22) Gibson, (23) Great Victoria, and (24) Sturt. (Sources: Meigs, 1953; Flint, 1971; Péwé, 1981.)

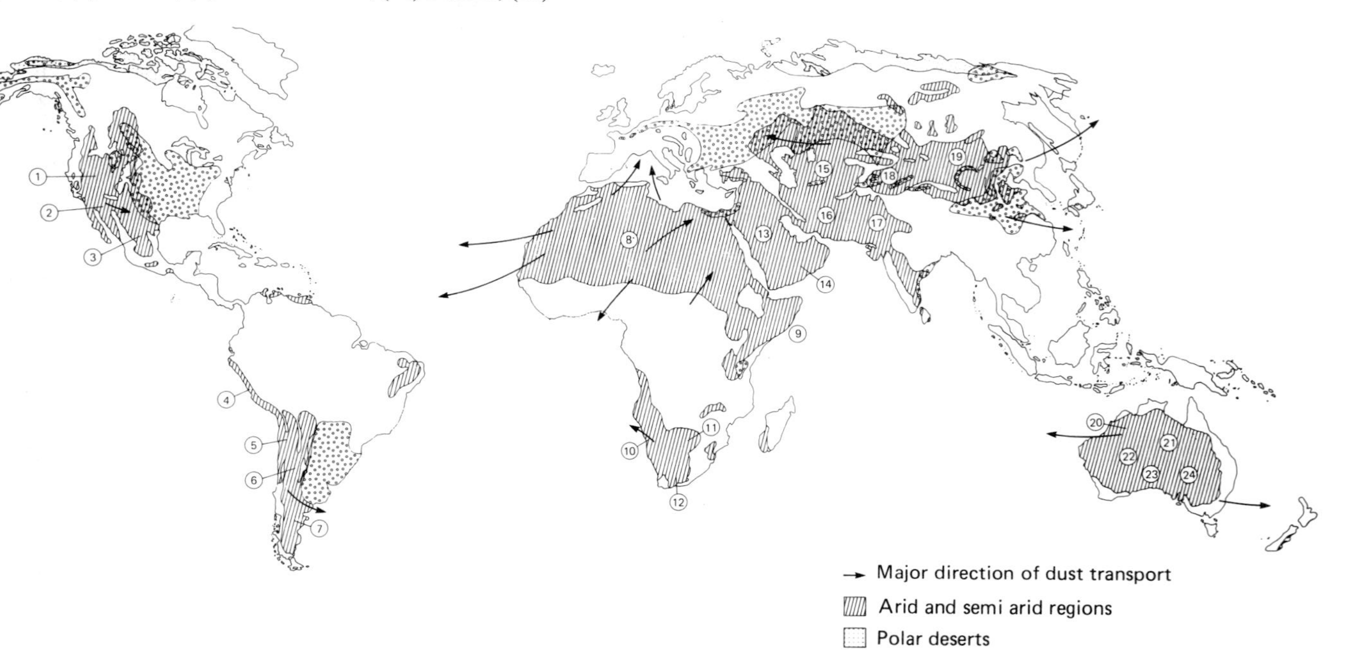

there is need for a definitive means to identify fine-grained aeolian sediments by remote sensing.

The aeolian activity is a direct link between the atmosphere and lithosphere, and the identification of aeolian landforms on planetary surfaces can provide clues to atmospheric processes. For example, identification of yardangs, certain types of dunes, and wind 'streaks' associated with craters (Fig. 1.3) enabled patterns of near-surface winds to be derived for Mars (Thomas & Veverka, 1979), which have been used to formulate global wind circulation models.

Small-scale aeolian features include *ventifacts* (rocks that are sculpted by wind abrasion) and aeolian sedimentary structures. These features can be observed only directly on the ground or inferred from remote-sensing data. Ventifacts can provide information about local wind directions and the length of time that a surface has been exposed. Identification of ventifacts is also relevant to other aspects of planetology. For example, rocks (Fig. 1.10) at the Viking Lander sites (Mutch *et al.*, 1976), which show pitted surfaces, have been interpreted as vesicular igneous rocks and are

Fig. 1.7. Cross-bedding in the Permian Toroweap Formation exposed in Walnut Canyon near Flagstaff, Arizona; each set of beds is about 4 m thick. (Photograph by R. Greeley, 1982.)

Fig. 1.8. Snow fence system (3.7 m high) along Interstate Highway 80 in Wyoming, USA. (Photograph by J. D. Iversen, 1979.)

Fig. 1.9. Sand dune control in northern Denmark. Sand is trapped by a system of staggered poles and tree branches. (Photograph by J. D. Iversen, 1981.)

part of the basis for identifying the surrounding plains as volcanic. Alternatively, the pitting in the rocks could result from aeolian processes and might not necessarily be igneous in origin.

Impact crater frequency distributions are widely used in planetary sciences as a means to obtain relative dates for different surfaces. The older a surface, the more impact craters it should have. On planets subjected to aeolian processes, the degradation of craters by erosion or burial by aeolian sediments can drastically alter the crater record and invalidate crater-derived ages. Thus, knowledge of rates of aeolian erosion and deposition for a wide range of planetary environments is required in order to assess the possible effects on the impact record.

Aeolian processes can both mix and sort sediments. When subjected to winds, deposits consisting of a wide range of particle sizes, such as river sediments or glacial deposits, may have coarse particles left behind, thus leading to 'lag-deposit' surfaces (Fig. 1.11). Conversely, windblown dust derived from a wide range of rocks may become compositionally 'homogenized' in dust storms and settle on widespread surfaces, as can

Fig. 1.10. View of the martian surface from the Viking Lander 1, showing a 20 cm rock (right side) that is coarsely pitted. Pits may be volcanic vesicles, or abrasion features resulting from wind motion, or both. Note the 'scoured' zones around both rocks. (VL image 11A037.)

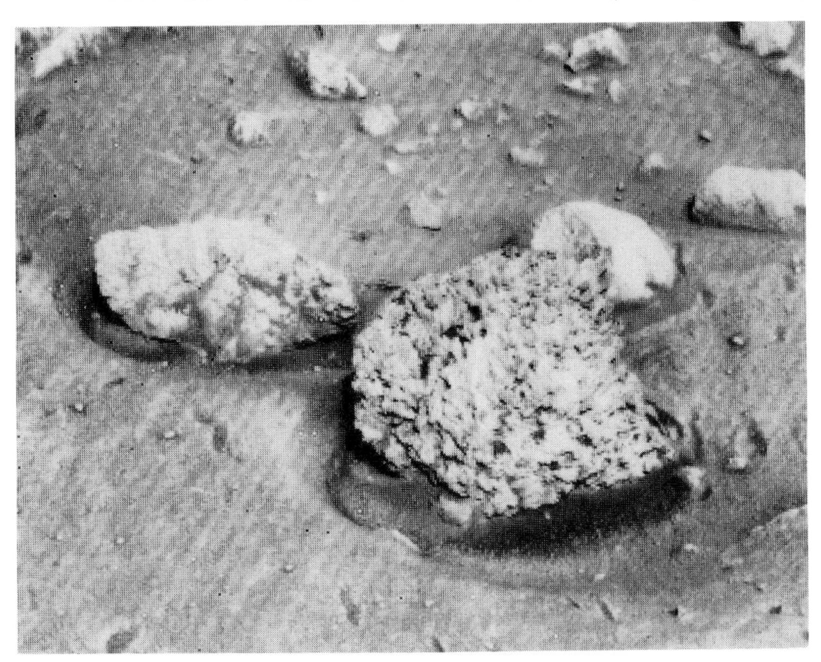

occur on Mars. Remote sensing of either of these cases (wind-sorted or wind-mixed sediments) could lead to erroneous conclusions about the surface composition of the areas observed.

1.4 Aeolian features on the planets

In this section we briefly discuss the movement of particles on Earth by the wind and present the erosional and depositional features that result from aeolian processes. We then discuss Mars, Venus, and Titan, giving their general geology and an introduction to their aeolian features.

1.4.1 *Earth*

On Earth, aeolian processes occur primarily in regions (Fig. 1.6) where there is an abundance of small particles, where vegetation is absent or minimal, and where winds are strong and frequent. Thus, most aeolian activity is found in desert regions (hot and cold deserts) and in coastal areas.

Fig. 1.11. Desert pavement surface at Amboy, California; the pavement consists of basalt fragments underlain by silt and clay. The rectangular area in the middle of the photograph is a 3 m square test plot in which all the surface fragments were cleared away, exposing the silt and clay. Development of desert pavement is probably a combination of wind deflation and other processes, such as swelling of clays, to push the rock fragments to the surface. (Photograph by S. H. Williams, March, 1980.)

Particles are commonly transported by the wind in one of three modes (Fig. 1.12): *suspension* (mostly silt and clay particles, i.e., smaller than about 60 μm), *saltation* (mostly sand-size particles, 60–2000 μm in diameter), and *traction* (rolling, sliding, pushing of particles along the surface). However, saltation is probably the critical mode of transport, as can be explained by examining the threshold curve (Fig. 1.2) which shows that the grain size most easily moved by the wind is about 100 μm in diameter (fine sand). Particles smaller than this size are more difficult to move, primarily because of cohesion and various aerodynamic effects (discussed in more detail in Chapter 3). However, once these small grains are set into motion, they are easily carried by the wind, usually in suspension. Particles larger than about 100 μm become progressively more difficult to move by the wind, simply because they are more massive. Thus, as wind strength increases over a surface of particles, the first grains to move are fine sands. Observations of these grains show that they first begin to quiver, then roll along a distance of a few grain diameters or fly into the

Fig. 1.12. Diagram showing the three principal modes of aeolian transport of grains: surface shear stress exerted by the wind causes grain (A) to lift off the surface, carries it downwind back to the surface where it bounces (B) back into flight; this motion is termed *saltation*; grain at (C) hits a large rock – possibly causes some erosion – and elastically rebounds to a relatively high saltation trajectory; grain at (D) strikes the surface and 'triggers' other grains into saltation; grain at (E) strikes the surface containing very fine particles (too fine to be moved by the wind alone in this case; see threshold curve, Fig. 1.2) and sprays them into the wind where they are carried by turbulence in *suspension*; grain at (F) strikes larger grain and pushes it downwind a short distance in a mode of transport termed *impact creep*, or *traction*.

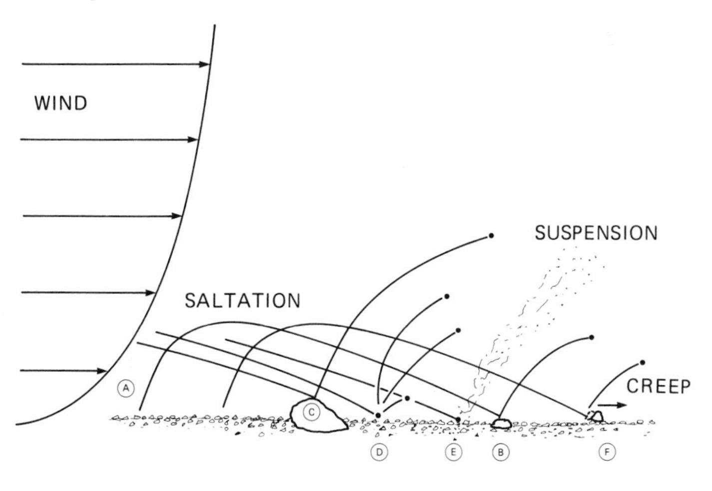

air at a fairly steep angle. After reaching some maximum height, they are carried along by the wind, falling back to the surface where they impact and bounce back into the air. This movement is *saltation* (from the Latin *salto*, meaning to dance, leap, or spring), and the path described by the grain is termed the *saltation trajectory*.

When the grain returns to the surface during saltation, it may cease motion, either because the wind is not strong enough to keep it moving or because it becomes trapped among rocks or other particles. Alternatively, the saltating grain may plough into a bed of other grains, setting them into motion and causing a cascading effect. Saltation impact is one of the primary mechanisms for raising 'dust', and other small grains, by winds that would otherwise be too gentle to initiate particle movement. Particles too large to be moved solely by wind forces can also be transported along the surface by saltation impact, a mechanism referred to as *impact creep*, a type of traction.

Aeolian deposits

Wind is an extremely effective agent for sorting material by sizes during transport, a result of the energetics described above. Thus, an important characteristic of aeolian deposits is their strongly unimodal grain-size distribution.

By far the most extensive aeolian sediments consist of silt-size particles, or *loess* (Fig. 1.6). Deposits of sand-size particles can assume many forms, including isolated dunes, dune fields, and sand seas, which may cover thousands of square kilometers. The exact form that sand deposits take appears to be a function of many complex factors including sand grain diameter, the wind regime, and local setting, as discussed in Chapter 7.

Aeolian erosion

Removal of loose particles from the ground is termed *deflation*, and the resulting landforms are deflation features, or *blowouts*. Wind is also capable of eroding rocks composed of slightly indurated particles by 'plucking' mechanisms (McCauley & Grolier, 1976), and it has been suggested that wind alone can abrade crystalline rocks (Whitney, 1979). Winds that carry sand and silt are extremely effective agents of erosion, leading to 'sand blasting' of surfaces exposed to the wind. Ventifacts are common in most deserts of the world and have been identified on Mars (McCauley *et al.*, 1979).

Wind erosion on the scale of landforms also occurs. *Yardangs* are elongated hills sculpted by the wind. They resemble inverted boat hulls and can be tens of meters high and many kilometers long. Although most yardangs develop in easily eroded material, such as volcanic ash, they have also been reported in crystalline rocks.

An extensive terminology has developed to describe the myriad of aeolian features on Earth, and the same terms are usually applied to features seen on the other planets. But it must be recognized that many of the specialized terms for terrestrial features carry explicit or implied meanings for origin or material properties – features that usually cannot be determined on extraterrestrial planets. Thus, planetary usage is restricted to those terms that are very broad in their meaning.

1.4.2 *Mars*

Mars has been observed telescopically from Earth for more than 100 years. Global-scale red and yellow patterns were seen to shift with the martian seasons and are particularly active in times of southern hemisphere summer. The patterns have been attributed to a wide range of processes, including biological activity. Some of the earliest interpretations of these patterns as aeolian were those of Dean McLaughlin (1954*a*,*b*).

Spacecraft observations confirmed speculations that Mars experiences aeolian processes. When the US Mariner 9 Orbiter and the Soviet missions, Mars 2 and 3, arrived at Mars in 1971, the planet was being swept by a global dust storm. Even before the spacecraft arrived, however, the shifting yellow patterns were being tracked telescopically from Earth. Thus, the directors of spacecraft operations for both the US and Soviet missions were aware of the dust storm well in advance of the encounter (a term used for the arrival of spacecraft at a planet). Evidently, the Soviet spacecraft could not be reprogrammed to adjust to the dust storms and were thus failures. Fortunately, the mission sequence for Mariner 9 was able to be altered so that the spacecraft could be put into a holding orbit to ride out the dust storm. During this phase of the mission, various measurements were made and images taken of the dust clouds (Fig. 1.13). These observations stimulated the initial formulation of models to explain dust storm activity on Mars. By imaging the same regions on the planet throughout the mission, surface features and cloud patterns could be tracked to see if they changed with time.

As the dust storm cleared in early 1972, the Mariner 9 cameras began to reveal the true geological complexity of Mars (Fig. 1.14), including abundant features attributable to aeolian processes. By far the most abundant aeolian features are various albedo patterns on the surface, which appear, disappear, or change their size, shape, and position. Termed

Fig. 1.13. Mariner 9 arrived at Mars during a major global dust storm in 1971; dust completely obscured the surface and formed thick clouds to heights at least 20 km above the average surface elevation. This 4-frame Mariner 9 mosaic (*a*) shows the dust clouds and four prominent 'spots' (the Tharsis Montes; see Fig. 1.14); it was not until the dust settled that the incredible diversity at the martian surface was revealed. The spots turned out to be huge volcanoes, such as the 600 km in diameter Olympus Mons, shown here (*b*) as an inset (VO image 649A28.)

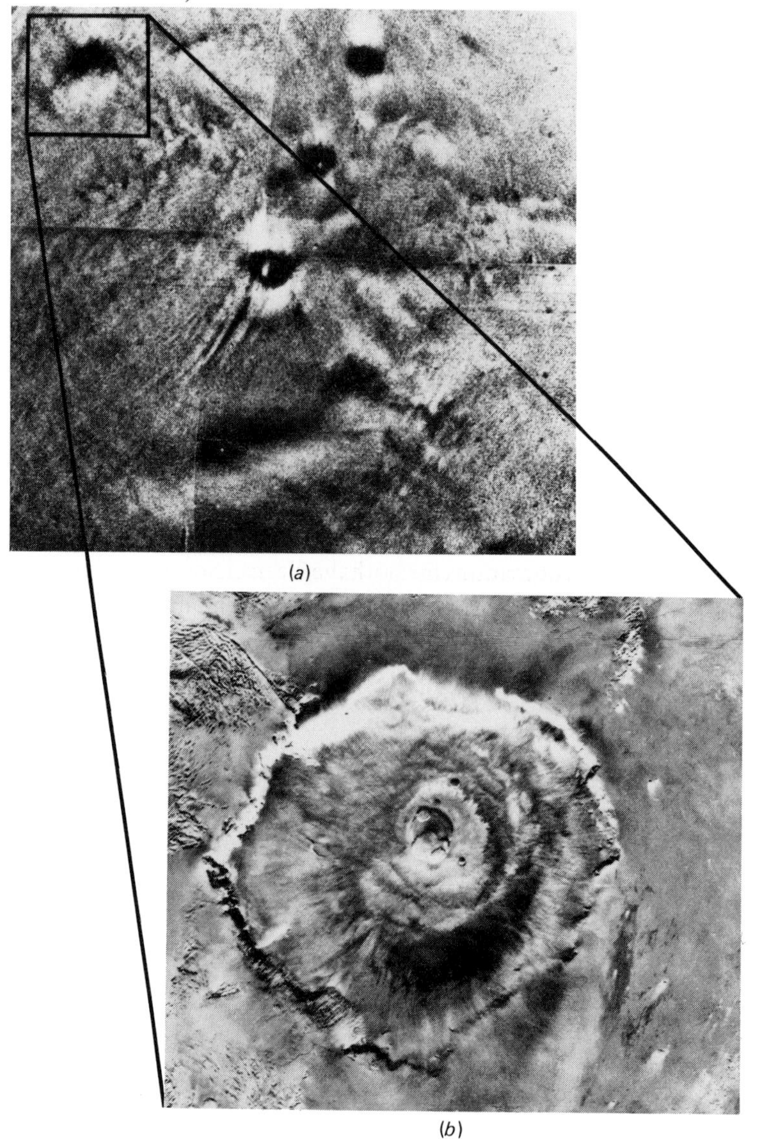

variable features (Sagan *et al.*, 1972), most appear to be associated with topographical structures, such as craters (Fig. 1.3).

Mariner 9 showed many other aeolian features (McCauley, 1973), including dunes, yardangs, and various pits and grooves, considered to result from deflation. Substantial parts of Mars also appeared to be mantled by deposits considered to be aeolian in origin (Soderblom *et al.*, 1973*a*), perhaps similar to loess deposits on Earth.

The Viking mission (1976–81), consisting of two orbiting spacecraft and two landers, added greatly to the catalog of martian aeolian features and contributed to the better understanding of aeolian processes on Mars. The Orbiters were equipped with: (1) cameras capable of obtaining pictures with resolutions as good as 10 m, (2) an instrument using infrared radiation to map the thermal inertias of surface materials, and (3) a second infrared system to map the water content of the atmosphere. The Viking Landers were capable of measuring various chemical and physical properties of surface materials near the two landing sites, and of monitoring wind speeds, wind directions, atmospheric temperatures, and surface pressures.

In addition the Landers provided our first pictures from the surface of Mars (Fig. 1.10), showing in exquisite detail various pitted rocks and 'drifts' of sediments presumed to be deposited and sculpted by the wind.

More than 100 000 images of Mars have been obtained, ranging from global views to close-ups of the surface. The pictures cover periods when the atmosphere was very clear, as well as periods of heavy dust storm activity. Meteorological measurements have been made for nearly four years, along with thermal inertia mapping at both global scales and at high-resolution scales for local areas. These data provide a great wealth of information for the synthesis of aeolian processes on Mars.

Dunes

One of the most impressive discoveries of the Mariner 9 and Viking missions is the existence of an enormous sand sea in the north circumpolar region of Mars. The field covers more than 7×10^5 km^2, larger than Rub al Khali in Arabia – the largest active erg on Earth. All of the dunes are either of the transverse or barchan type (Fig. 1.15). Mapping the orientations of the dunes and coupling the results with other indicators of wind directions have enabled regional maps of the wind circulation pattern to be derived (Tsoar *et al.*, 1979; Breed *et al.*, 1979). Two major wind directions are suggested: off-pole winds that become easterly due to Coriolis forces during summer and on-pole winds that become westerly during winter.

Fig. 1.14. (*a*), (*b*) Shaded airbrush relief map of Mars based on Mariner 9 images, showing principal named regions and sites of the two Viking Landers (VL 1, VL 2). (Base map courtesy of the US Geological Survey.)

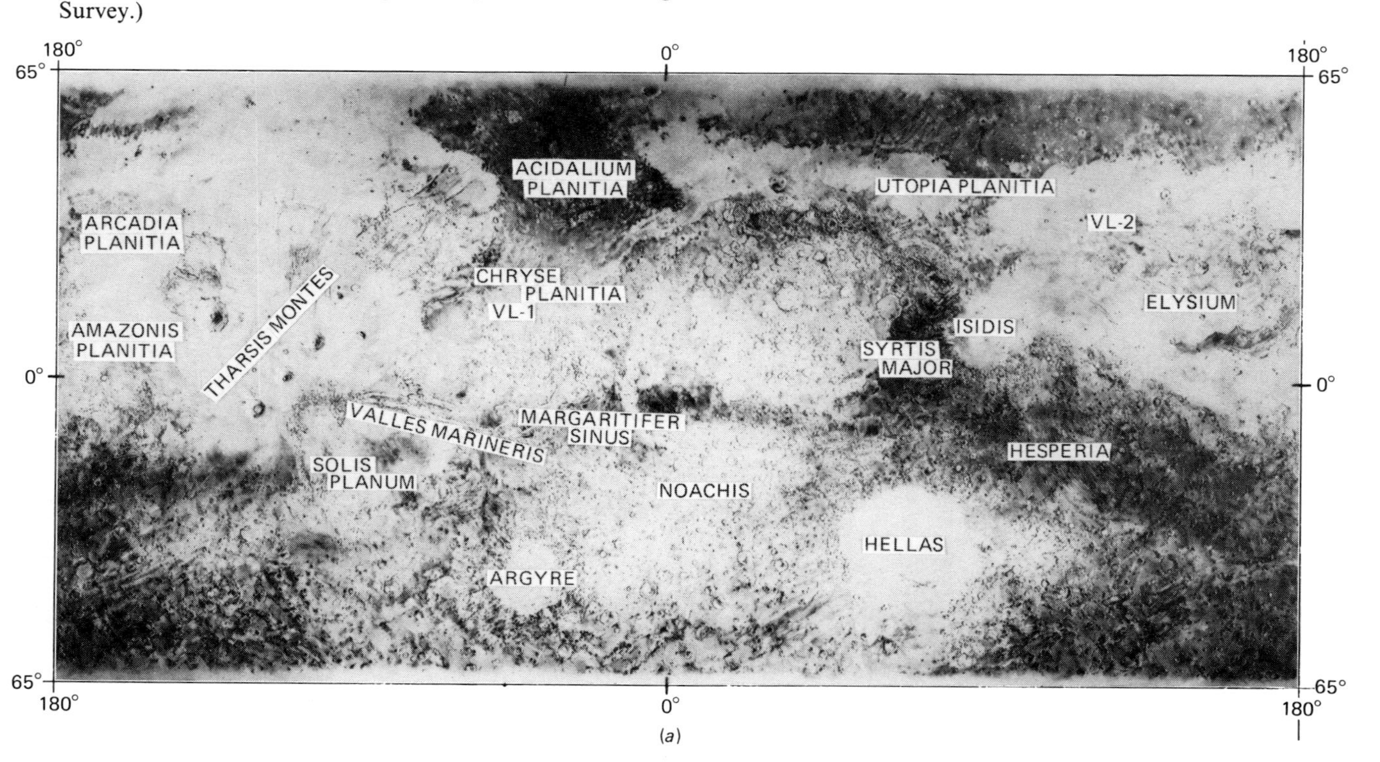

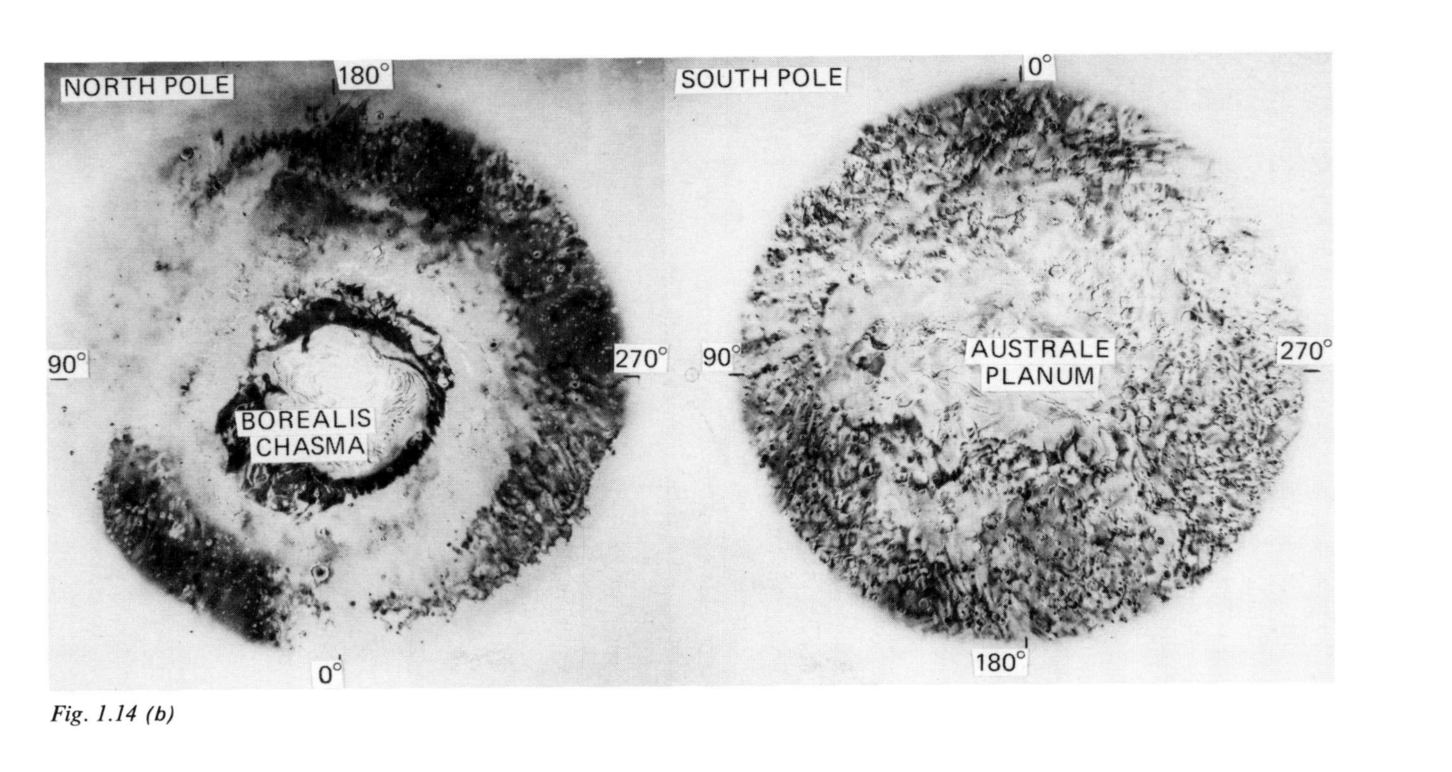
NORTH POLE
180°
90°
270°
BOREALIS
CHASMA
0°
SOUTH POLE
0°
90°
AUSTRALE
PLANUM
270°
180°

Fig. 1.14 (b)

The low albedo (i.e., dark) appearance of the dunes suggests a composition other than quartz, an observation consistent with the apparent lack of quartz on Mars (Smalley & Krinsley, 1979). Because basaltic lavas are very common over much of Mars, including the smooth plains south of the dune field, it is suggested that the north polar dunes are composed of windblown basaltic particles.

Although the greatest concentration of dunes occurs in the north polar region, dunes have been observed in nearly all regions of Mars. In many localities, the dunes appear to be deposits of sand trapped by local topographical features, e.g., inside impact craters or within small valleys (Fig. 1.16).

Yardangs

Yardangs have been found in several areas on Mars. Most yardangs occur in equatorial regions, notably in the Amazonis region, the Aeolis region, the Ares Valles, and Iapygia. Some of the largest features are interpreted to be early stage yardangs (Ward, 1979); they are ≈50 km long, 1 km wide, 200 m high, and appear to have developed from the erosion of mesas. From studies of terrestrial yardangs, Ward concludes that the martian features formed relatively recently in the geological history of Mars. He considers most yardangs to be composed of friable rocks, such as ignimbrites (many of the yardang localities are near known volcanoes), or

Fig. 1.15. Viking Orbiter view of the north polar region of Mars showing isolated dunes and dune complexes; the image covers an area ≈30 km wide. (VO frame 544B07.)

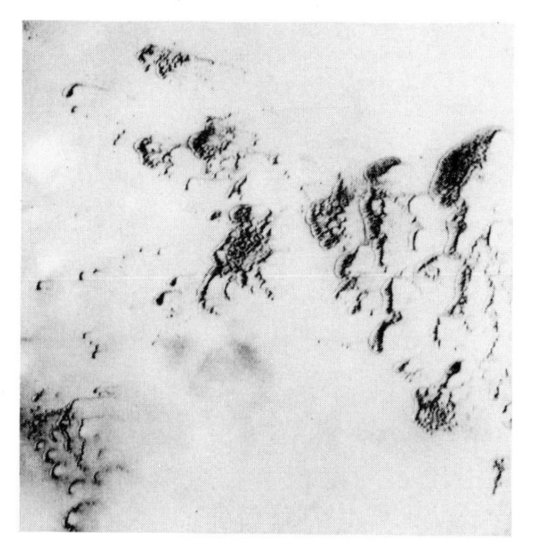

indurated regolith (*regolith* in this sense being fragmental debris generated by impact cratering).

On Earth, most yardangs develop by erosion of grains that are loosened by weathering processes involving liquid water; Ward suggests that on Mars – in the absence of liquid water – exfoliation, salt weathering, or freeze–thaw processes may loosen the grains, but that the net weathering rate would be slower than on Earth.

Fig. 1.16. Dune field on the floor of an ancient impact crater in the region south of Sinus Sabaeus on Mars, centered at 46°S, 339°W. Craters and other topographical depressions are natural traps for windblown sediments, and the crater shown here is typical of many that have been photographed from orbit. The crater is ≈150 km across. (VO frame 94A42.)

Variable features

Variable features occur as either light or dark forms, although 'mixed' forms are also found in which both light and dark streaks occur in association with the same topographical structure. The most common forms are streaks associated with craters (Fig. 1.3). The origin of crater streaks is a matter for debate, with several models having been proposed. However, nearly all investigators agree that streaks are caused by the deposition or erosion of relatively thin ($\approx$cm) deposits of particles that shift in response to winds.

Several different models of light- and dark-streak formation can be postulated as functions of particle sizes, particle compositions, and various wind characteristics, such as wind velocities and turbulence (Greeley *et al.*, 1974*a*, *b*, 1978). Most models of streak formation must take into account the flow patterns generated by winds blowing over and around craters. Wind tunnel simulations and limited field studies show that a horseshoe vortex (Fig. 1.3) wraps around the crater rim and creates an erosive zone in the wake of the crater and a depositional zone in the immediate lee of the crater rim. The size and shape of zones of erosion and deposition are functions of crater geometry, wind speeds, time, and other parameters (Iversen *et al.*, 1975*a*, 1976*c*).

Some elongated bright streaks associated with craters in the south polar region appear to be accumulations of carbon dioxide frost. Based on their form and seasonal behavior, these bright streaks have been interpreted to be accumulations of wind-transported frost in the lee of craters, similar to the formation of bright crater streaks involving dust and sand particles (Thomas *et al.*, 1979). Because frost particles have a fairly low density, threshold wind speeds are lower than for sand, and frost streaks, therefore, are more active.

Dust storms

Earth-based and spacecraft observations show that major dust storms on Mars (Fig. 1.17) typically begin in the southern hemisphere during spring or early summer, close to perihelion (Gierasch, 1974; Briggs *et al.*, 1979). In some years, no major storms may develop; in other years, several major storms may occur. Major dust storms originate in areas of both the southern (Hellespontus, Noachis, and Solis Planum), and northern hemisphere (Isidis Planitia).

Global dust storms appear to go through three phases. In phase I, numerous local dust storms occur in the southern hemisphere, associated with zones of strong surface thermal contrasts and in association with the sublimating south polar cap (Peterfreund & Kieffer, 1979). These local storms contribute dust to the global atmosphere, increasing diurnal and thermal tides. This increases the likelihood of a global storm because of the energy absorbed by the atmosphere, due to the suspended dust. Phase II is the expansion of a local storm into a global event; typically taking three to seven days, this is rapid for a global event. Phase III marks the decay of the storm and lasts from 50 to 100 days. The first areas to clear are the poles and topographically high regions, such as the summits of the shield volcanoes.

Mariner 9 and Viking results show that the average particle size in the atmosphere is less than 2 μm, or about the same as the particles carried over the Atlantic by major Saharan dust storms. The dust on Mars was found to be well mixed in the atmosphere to heights of 30–40 km and had the effect of raising the atmospheric temperature by as much as 50 K (Fig. 1.4).

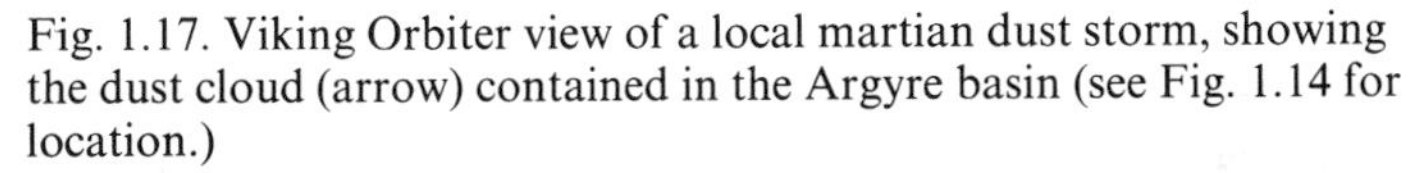

Fig. 1.17. Viking Orbiter view of a local martian dust storm, showing the dust cloud (arrow) contained in the Argyre basin (see Fig. 1.14 for location.)

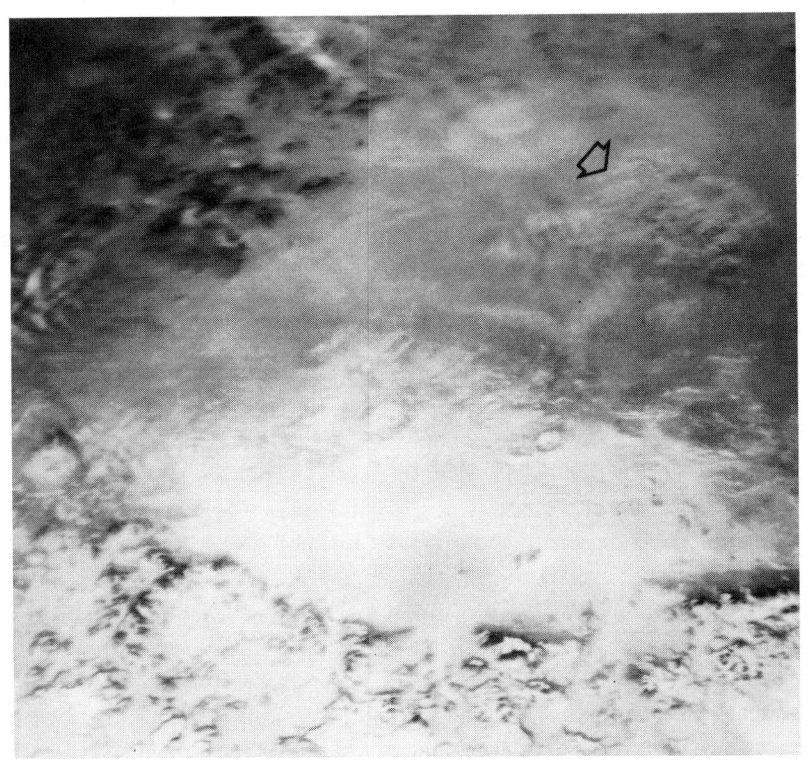

Calculation of dust deposition from the storms suggest significant mantling of the surface of Mars over geological time.

In summary, aeolian processes appear to be the dominant process currently active on the surface of Mars and have played an important role in the geological past.

1.4.3 *Venus*

Of all the terrestrial planets, Venus remains the least understood despite the fact that it is the so-called sister planet of Earth, due to its similar size and density. Thus far in solar system exploration, most spacecraft missions have been directed toward gaining information on the venusian atmosphere.

The Pioneer–Venus mission (1978–80) involved an orbiting spacecraft, plus small probes that penetrated the atmosphere to measure the temperature, pressure, and composition. Radar altimetry measurements obtained via the Orbiter, supplemented with earth-based observations, have allowed the general topography of Venus to be determined (Fig. 1.18). Most of its surface consists of rolling plains, with only about 15% of the surface deviating more than 1 km from the median elevation (McGill *et al.*, 1983). Topographical data have been combined with earth-based radar images to enable preliminary interpretation of venusian geology. Interpretations include large shield volcanoes, impact craters, and tectonic rifts (Malin & Saunders, 1977; McGill *et al.*, 1983), although such interpretations are rather speculative because of limitations of the data.

The Soviets have landed several spacecraft on the surface of Venus, which have survived the 480 °C temperatures and 90 b atmospheric pressure long enough to return pictures from four sites, some wind measurements, and data on surface compositions. Surface materials analyzed by Venera 8, 9, 10, 13, and 14 are chemically similar to common igneous rocks on Earth (Barsukov *et al.*, 1981). Gamma-ray detectors measured the abundance of K, U, and Th. Although the interpretations of the rock types are model dependent, the enriched potassium abundance at the Venera 8 site corresponds best to the rock type syenite. Depleted potassium abundance at the Venera 9 and 10 sites are interpreted to represent basaltic rocks.

The atmosphere of Venus is composed primarily of carbon dioxide with minor amounts of hydrochloric, hydrofluoric, and sulfuric acids. With a surface pressure of about 90 b, it has the highest atmospheric density of all the terrestrial planets (Table 1.1). Venus is completely enveloped in a perpetual shroud of cloud that hides the surface from view. Repetitive

Fig. 1.18. Rendition of the surface of Venus showing topographical relief obtained from Pioneer–Venus radar altimeter data. Venera landing sites are south of Beta Regio. (Courtesy of M. Malin, Arizona State University.)

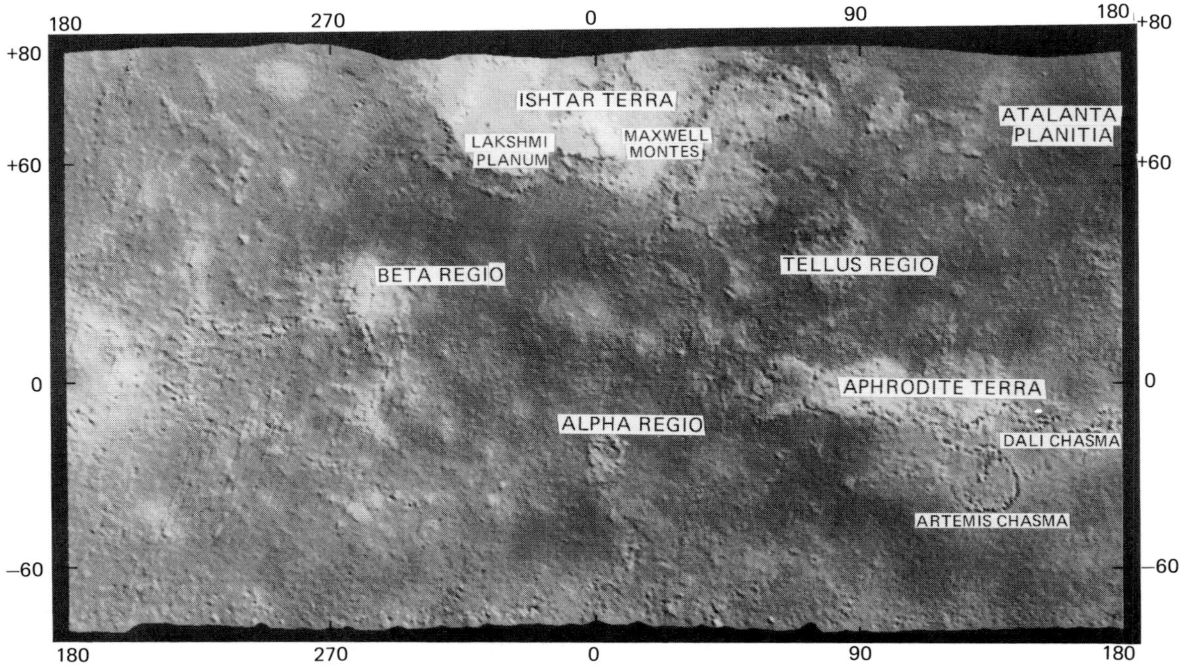

pictures of the cloud tops, obtained over a period of eight days during the flyby of Mariner 10 in 1974, showed circulation patterns and allowed wind speeds to be determined for the upper atmosphere (Murray *et al.*, 1974). Although speeds of about 100 m/sec were obtained for the upper clouds in the equatorial zone, when extrapolations were made to the surface the winds were estimated to be very sluggish.

The Soviet Landers, Venera 9 and 10, measured wind speeds near the surface for two sites on Venus of 0.5–1 m/sec at the height of the wind sensors (1–2 m above surface). More recent measurements of wind speeds, obtained by the Pioneer–Venus atmospheric probes, have been extrapolated to the surface and yield values of 1–2 m/sec (Counselman *et al.*, 1979). These values are well within the range predicted for particle threshold (Fig. 1.2), based on a combination of theory and extrapolations of wind tunnel experiments (Iversen *et al.*, 1976*b*; Greeley *et al.*, 1984).

Venera images of the surface of Venus (Fig. 1.19) show rock fragments of several centimeters, and larger, set in a mass of the fine (< 1 cm) material interpreted to be sand size or smaller (Florensky *et al.*, 1977; 1983). This bimodal-size distribution is indicative of fluid transport and, because liquid water cannot exist in the extremely high temperatures on Venus, it is assumed that the fluid involved is the atmosphere, or wind. Thus, it is likely that aeolian processes are active at present on Venus (Hess, 1975; Sagan, 1975) and have probably been active in the geological past.

1.4.4 *Titan*

Titan is the largest satellite in the saturnian system and is the only known satellite in the solar system to have a substantial atmosphere. Although earth-based observations indicated the presence of an atmosphere, little was known about its composition, density, and temperature. The Voyager 1 spacecraft flew past Titan in the fall of 1981, made several critical measurements, and obtained images of the atmosphere. The atmosphere was found to be composed predominantly of nitrogen with small amounts of methane. Methane abundance increases toward the surface, where it is just under 10% of the total mass of the atmosphere at that level. Atmospheric pressure at the surface is estimated to be 1.6 b, and the temperature is about 93 K. Thus, all three phases of methane could be present on Titan – solid methane ice, bordering liquid methane oceans, all covered by gaseous methane clouds.

Images of Titan show a thick haze that completely hides the surface (Fig. 1.20). High-resolution images show several discrete haze layers up to 55 km above the more opaque atmospheric haze (Stone & Miner, 1981). A dark 'hood' over the northern polar cap and a bright southern hemisphere suggest some sort of seasonal effect.

The diameter of Titan is estimated to be 5140 km, or larger than that of the planet Mercury (Tyler *et al.*, 1981). When combined with estimates of its mass, Titan's density is about 1.9 ± 0.06 gm/cm^3, which suggests an equal mix of ice and rock.

Whether the atmosphere is dynamic – some 'structure' is observed in the form of haze layers and hemispheric differences in the clouds – and whether granular particles exist are unanswered questions. Dunes composed of methane ice particles and ice grains being blown in the dense, extremely

Fig. 1.19. Venera 9, 10, 13, and 14 images showing the surface of Venus at four sites. The spacecraft is seen in the middle of each image and the horizon is visible in the upper corners. The surface is quite variable from site to site, but in all areas, rocks are seen set in a groundmass of particles, presumed to be in a size range capable of wind transport. See Fig. 1.18 for the general landing site area.

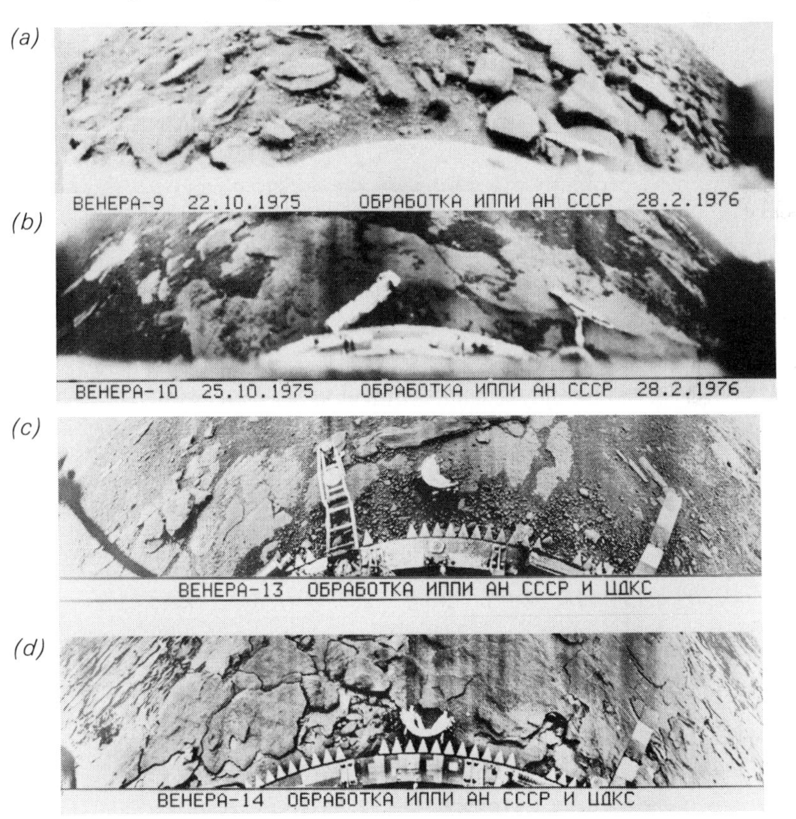

cold nitrogen atmosphere border on the realm of science fiction but remain a possibility.

Thus, we see a wide range of known, probable, and possible features on Earth, Mars, Venus, and Titan, which can be attributed to aeolian processes. In the chapters which follow, we will discuss various aspects of aeolian processes and the resulting landforms and other features related to winds – all within a planetary context.

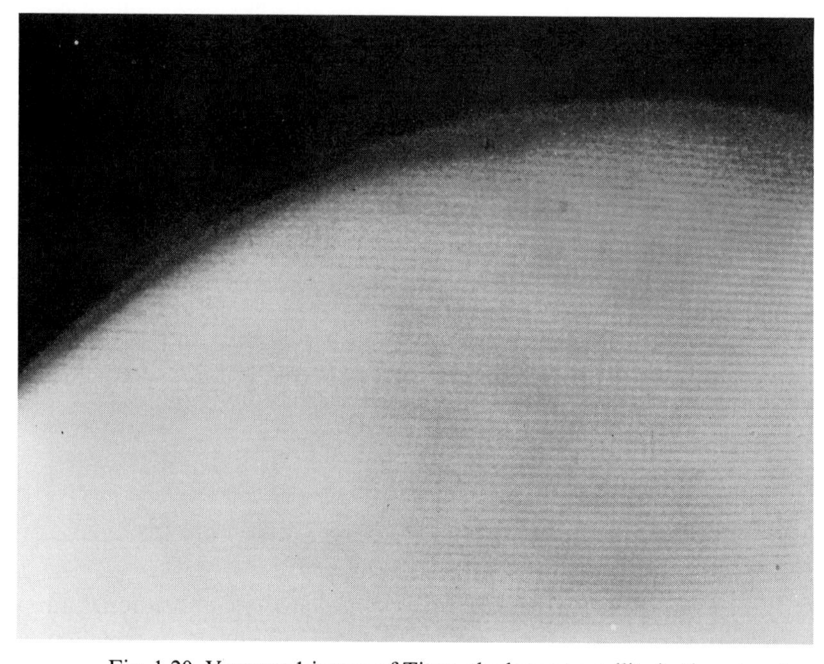

Fig. 1.20. Voyager 1 image of Titan, the largest satellite in the saturnian system and the only satellite known to have a substantial atmosphere. This view shows a haze layer that merges with a darker 'hood' or cloud layer over the north pole. The atmosphere was found by Voyager to consist predominantly of nitrogen, plus small amounts of methane. Estimates of the surface temperatures on Titan suggest that all three phases (gas, liquid, solid) of methane may exist. (Jet Propulsion Laboratory, photograph P-23108 C.)

2

The aeolian environment

2.1 Introduction

Aeolian processes basically involve the interaction of the atmosphere – the external gaseous part of some planets – with the lithosphere, or the solid surfaces of the planets. The zones where this interaction occurs constitute the aeolian environment. In order to understand variations in style and vigor of aeolian activity among the planets, we must first understand the nature of the aeolian environment. In this chapter we review the origins and characteristics of the atmospheres of Earth, Mars, Venus, and Titan and discuss the motions of the atmospheres relevant to aeolian transport. We then discuss the possible origins and characteristics of windblown particles.

2.2 Properties of atmospheres

For simplicity it is useful to consider a one-dimensional atmosphere, or one in which the various properties are functions only of distance above the surface. Because atmospheric properties are controlled partly by solar energy – which is a function of position on the planet's surface – such a one-dimensional atmosphere is not really possible. Nonetheless, this approach is useful for considerations of an 'average' planetary atmosphere, recognizing that there can be diurnal, latitudinal, and seasonal variations.

2.2.1 *Hydrostatic equilibrium*

In our idealized model, the pressure gradient in a vertical column of air above a planetary surface is given by

$$\mathrm{d}p/\mathrm{d}r = -\rho g \tag{2.1}$$

where p is the pressure, ρ is atmospheric density, g is gravitational acceleration, and r is the spherical radius from the planet's center. For a thermally perfect gas, the relationship (called the equation of state) between pressure, density and temperature is

$$p = \rho R_g T \tag{2.2}$$

where T is temperature and R_g is the specific gas constant appropriate for the composition of the atmosphere. For an atmosphere of uniform temperature (isothermal), the density according to Eq. (2.2) is simply proportional to the pressure, and Eq. (2.1) can be integrated to yield

$$p = p_o \exp(-z/H_s) \tag{2.3}$$

where z is height above the surface and $H_s = R_g T/g$ is the so-called *scale height*. The scale height is thus the altitude in an isothermal atmosphere for which the pressure has decreased from its surface value by a factor 1/e (where e is the base of natural logarithms, 1/e = 0.368).

2.2.2 *Adiabatic lapse rate*

Because the atmospheric pressure decreases with elevation, a parcel of air which moves quickly up or down will experience an *adiabatic* expansion or compression (i.e., with negligible heat exchange with the surroundings) with a corresponding change in temperature. For a neutrally stable atmospheric layer – one with an adiabatic temperature gradient – the temperature variation with altitude is

$$T = T_o - gz/C_p \tag{2.4}$$

Using Eq. (2.4), Eq. (2.1) can be integrated to yield the pressure variation with height for a neutral (adiabatic) atmosphere

$$p = p_o (1 - gz/C_p T_o)^{C_p/R_g} \tag{2.5}$$

where T_o is the surface temperature and C_p is the specific heat capacity at constant pressure. Meteorologists define a parameter called *potential temperature*, θ, such that

$$\theta = T + gz/C_p \tag{2.6}$$

so that θ is a constant for a neutral atmosphere (i.e., $\partial\theta/\partial z = 0$). The lapse rate is the rate of change of temperature with change in height. For a neutral atmosphere the lapse rate $\partial T/\partial z$ is $-g/C_p$ and thus $\partial\theta/\partial z = 0$.

Temperature decrease with altitude in a static atmosphere can be either less than or greater than the adiabatic lapse rate. If the magnitude of the lapse rate is greater than adiabatic ($\partial\theta/\partial z < 0$), as a parcel of air moves upward its temperature will decrease according to Eq. (2.4) (adiabatically) and it will have a temperature higher than its surroundings, but since it has the same pressure as the surroundings its density will be less, according to Eq. (2.2). Thus the parcel of air is buoyant and will go still farther upward. This corresponds to an unstable atmospheric layer, one in which vertical transfer of mass and momentum by turbulence is enhanced by the unstable temperature gradient. Conversely, if the lapse rate is less than adiabatic ($\partial\theta/\partial z > 0$), the upward-moving air parcel would be colder than its surroundings and would sink back to its former position. Such a lapse rate corresponds to a stable layer of air in which turbulent exchange is inhibited. Stable, neutral (adiabatic), and unstable temperature profiles are illustrated in Fig. 2.1.

2.3 Comparisons of planets and their atmospheres

The two planets closest to Earth in geological and atmospheric characteristics are Mars and Venus. There are, however, some important differences in the properties of the atmospheres among the three planets, primarily as a result of differences in distance from the sun and – in the case of Mars – planetary size. Titan is markedly different from the inner planets (Table 2.1).

The primary difference between Earth and Venus is the distance from the sun and the resultant solar flux, which probably accounts for the surprising differences in the atmospheric compositions, pressures, and temperatures. Geological history also is a consideration; it has been suggested that Earth's atmosphere might be very much like that of Venus

Fig. 2.1. Hypothetical constant-slope temperature profiles for unstable, stable, and neutral atmospheres. Potential temperature is defined by Eq. (2.6).

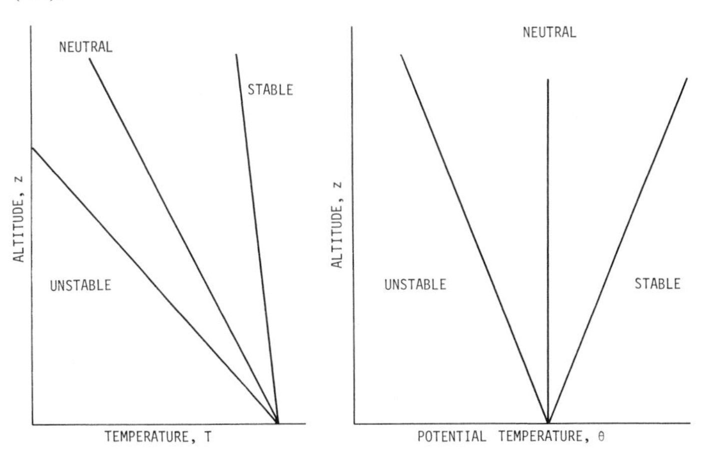

Table 2.1. *Atmospheric properties of Venus, Earth, Mars, and Titan*

	Venus	Earth	Mars	Titan
Surface pressure (mb)	90 000	1013	7	1600
Surface temperature (K)	750	288	218	93
Scale height (km)	14.9	8.4	10.6	21
Adiabatic lapse rate (deg/km)	10.7	9.8	4.5	1.3
Atmospheric mass (kg)	$4.66(10)^{20}$	$5.27(10)^{18}$	$2.72(10)^{16}$	$9.76(10)^{18}$
Atmosphere–planet mass ratio	$9.6(10)^{-5}$	$8.8(10)^{-7}$	$4.23(10)^{-8}$	$7.3(10)^{-5}$
Composition (%)	96 CO_2 3.5 N_2	77 N_2 21 O_2 1 H_2O	95 CO_2 2.7 N_2 1.6 Ar	90 N_2 10 CH_4

were it not for the fact that most of the CO_2 on Earth is presently stored in carbonate rocks such as limestone. Mars is significantly smaller than Earth and Venus and has a correspondingly smaller gravitational acceleration, while Titan is still smaller and much farther from the sun.

The differences in planetary size and mass among the planets, their evolutionary histories, and in their distance from the sun have led to considerable differences in their atmospheres. Not only are the compositions different, but the atmospheric surface pressures span a wide range (Table 2.1). The differences in atmospheric pressure and temperature among Venus, Earth, and Mars are emphasized in Figs. 2.2 and 2.3.

A unique feature of the surface pressure on Mars has been discovered by the Viking Landers. The pressure on the surface exhibits a seasonal variation with a maximum deviation of about 15% from the mean (Fig. 2.4). The pressure minimum corresponds to maximum extent of solid carbon dioxide within the south polar cap, condensed from the atmosphere, at the end of southern hemisphere winter. The secondary minimum at northern hemisphere vernal equinox is due to maximum accumulation in the northern cap (which is mostly ice). The asymmetry between the northern and southern hemispheres is due primarily to the orbital eccentricity and axial tilt (Table 1.1). Higher-frequency, smaller-amplitude variations in surface pressure are due to traveling storms. The large pressure jump at the Lander 2 site shortly after winter solstice was due to a planetwide dust storm (Leovy, 1979).

The striking differences in atmospheric properties between Venus, Earth, and Mars are both mystifying and intriguing. The differences and their effects on surface features, revealed by space probes, have led to new

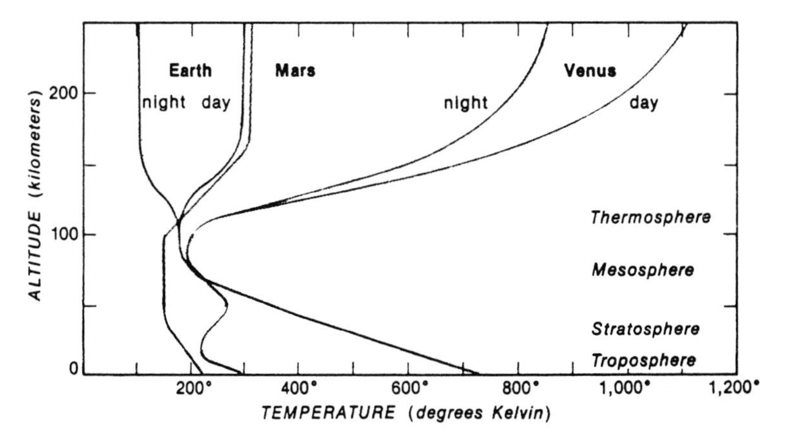

Fig. 2.2. Temperature versus altitude for Venus, Earth, and Mars (from Pollack, 1981). The names are those given to regions within Earth's atmosphere.

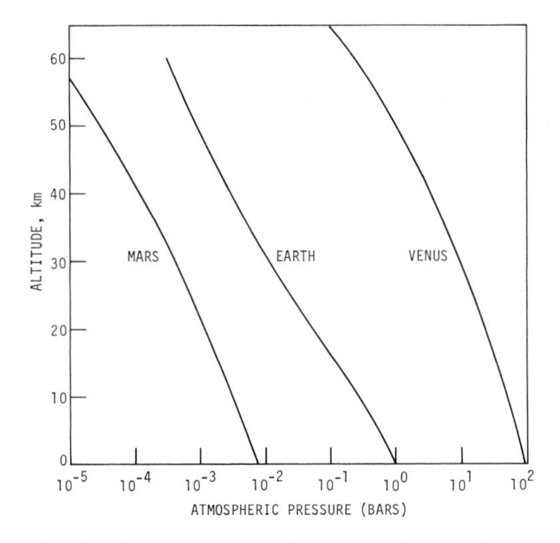

Fig. 2.3. Pressure versus altitude for Venus, Earth, and Mars. Data from Seiff & Kirk (1977), Seiff *et al.* (1979), and Government Printing Office (1962). The standard altitude for Earth is an average used for aeronautical purposes.

investigations of the effects that atmospheric density has on aeolian processes and have increased our understanding of these phenomena on Earth, as well as on the other planets experiencing aeolian activity.

2.3.1 *Origin of atmospheres*

Many factors affect the atmospheric environment of a planet, including its position in the solar system, the orbital characteristics, and the history of its volatiles and deposits. These and other factors interact to make the atmospheric history complicated and difficult to understand (Pollack, 1979, 1981; Pollack & Yung, 1980).

Sources of the constituents that make up the atmospheres include the solar nebula (the stellar cloud from which the planets formed), solar wind, colliding bodies (meteors, comets), and outgassing of volatiles from the interior after planetary formation. Changes in composition and other properties of the atmospheres, after formation, may have occurred because of escape of gases (particularly lighter ones) to space and the transfer of volatiles to surface reservoirs, due to condensation, absorption, and chemical weathering.

Earth is unique among the planets in having a surface covered mostly by liquid water. At one time in the distant past, however, Mars may also have had reservoirs of liquid water on the surface. The Viking Orbiter images of the martian surface show valleys and channels with streamlined 'islands' and gully patterns, which are considered by most investigators to have been

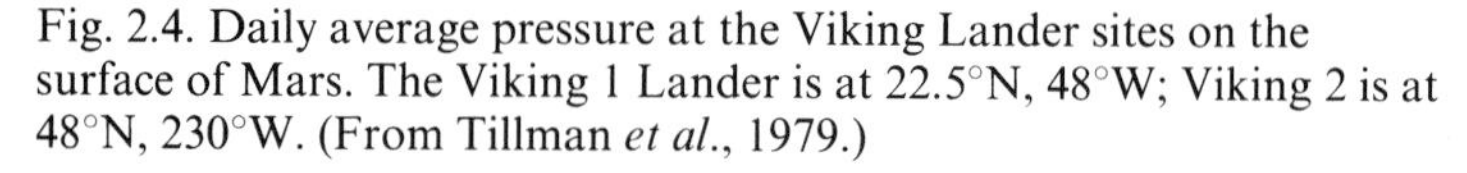
Fig. 2.4. Daily average pressure at the Viking Lander sites on the surface of Mars. The Viking 1 Lander is at 22.5°N, 48°W; Viking 2 is at 48°N, 230°W. (From Tillman *et al.*, 1979.)

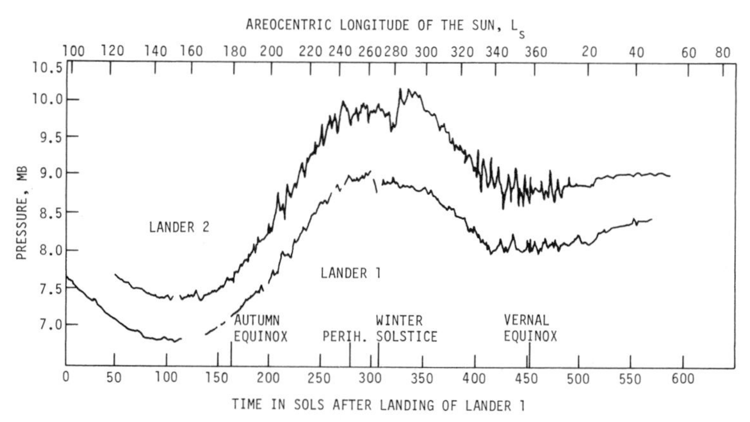

formed by liquid water (Carr, 1981; Baker, 1982). Thus, Mars may have had a denser atmosphere in the past and may have been much warmer.

In the geological past, Venus also may have had substantial liquid water on its surface. As shown by Donahue *et al.* (1982), the discovery of a hundred-fold enrichment in deuterium on Venus, by the Pioneer–Venus mission, suggests sufficient water in the past to have formed Earth-like oceans. Depending upon solar flux and atmospheric heating, they estimate that a billion years would be required for Venus to lose a 'terrestrial' ocean of water.

The atmosphere on Venus is fundamentally different from Earth in that it is hot near the surface and cold at the top, just the opposite of Earth. The present high surface temperature on Venus (Table 2.1) is thought to result from the so-called runaway *greenhouse effect* caused by its proximity to the sun. The greenhouse effect is created when the atmosphere is more transparent to incident radiation from the sun than it is to the thermal radiation emitted from the planet's surface. Water vapor and carbon dioxide released from the planetary surface can lead to the increased surface temperature due to the greenhouse effect (Goody & Walker, 1972). Because Venus is closer to the sun, its initial surface temperature was higher than for Earth or Mars, and the surface temperature increased more rapidly than did water vapor content, thereby creating a perpetual subsaturation, even though water content was increasing. The water vapor thus stayed in the atmosphere, and the surface temperature continued to increase to its present high equilibrium level. On Earth and Mars, however, the starting temperatures were lower, and the greenhouse effect did not increase temperature rapidly enough to maintain subsaturation. The resulting saturation and precipitation led to condensation, deposition, and further precipitation, and thus limited the amount of water the atmospheres could contain. On Venus, the hydrogen from dissociated water vapor has escaped to space, and the oxygen has presumably combined with surface materials, leaving an atmosphere of mostly carbon dioxide.

2.4 The atmospheric boundary layer

Wind – the motion of the atmosphere relative to the surface – is caused by variations in solar heating from one point to another, and by variations in the way in which solar radiation is reflected and absorbed. The general circulation of the atmosphere, including jet streams and atmospheric 'fronts', is relatively independent of the surface except for very large features such as mountain ranges and oceans. In contrast, winds close to the surface can be greatly affected by small-scale features such as trees and boulders. That part of the atmosphere which is significantly affected by the direct influence of the surface is called the *atmospheric boundary layer*. The boundary layer is typically of the order of 1 km thick. Because the characteristics of the boundary layer govern the transport of sand and dust from and to the surface, as well as within the atmosphere, we shall discuss the boundary layer in some detail. Typical wind speed profiles for the lowest 100 m within the boundary layer for Earth, Mars, and Venus are shown in Fig. 2.5.

2.4.1 *Turbulence*

Most atmospheric flow, at least on Earth and Mars, is *turbulent*, including flow within the atmospheric boundary layer. It is difficult to define turbulence precisely, but its principal characteristic is that the motion is 'irregular'. We must, therefore, rely on statistical methods to define a time-averaged turbulent flow.

For either a very viscous fluid or very slow flow, the motion can be *laminar*. In laminar flow, the exchange of momentum from one flowing layer to another is caused only by molecular activity. However, as the speed increases, or the viscosity decreases, this type of smooth flow becomes unstable, and exchange of momentum starts to take place by lateral motion of parcels of the gas, i.e., turbulence, rather than by molecular motion alone. Thus the interchange of momentum is greatly facilitated with the onset of turbulence and the tangential stress exerted by wind blowing over a surface, called the *shear stress*, is much greater than for purely laminar flow.

Fig. 2.5. Logarithmic wind speed profiles for Earth, Mars, and Venus, corresponding to static threshold for 100 μm sand particles. Values for the surface friction speed are 2.45, 21, and 150 cm/sec for Venus, Earth, and Mars, respectively, for surface pressures of 90, 1.013, and 0.007 b.

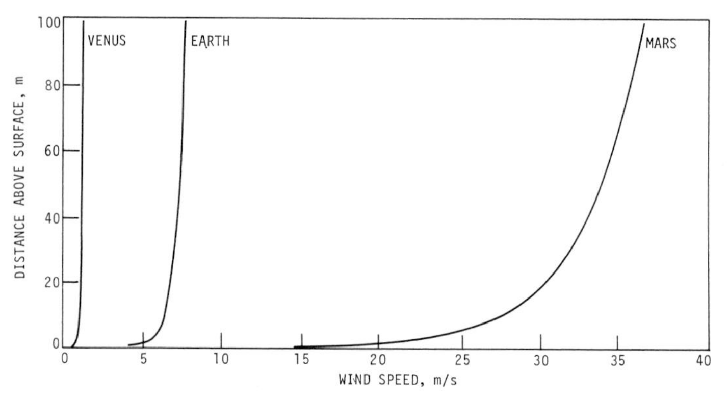

The features of a turbulent flow are characteristic of the flow and not of the fluid medium. Thus, the gross features of turbulent motion generated from flow around a specific object, such as a rock, look the same whether the fluid motion is taking place in water or in air. The parameter which indicates whether the flow speed is sufficiently high, or viscosity sufficiently low, for the two flows to be similar is called the *Reynolds number*. Generally, if the Reynolds numbers for the two flows are equal, or if each is larger than some critical value, the two flows are said to be dynamically similar.

The Reynolds number, R, is defined as

$$R = \rho\, UL/\mu \tag{2.7}$$

where ρ is the fluid density, U is a characteristic wind speed, L a characteristic length, and μ the coefficient of absolute viscosity, which is a function primarily of fluid composition and temperature. The Reynolds number is essentially a measure of the ratio of inertial forces to the forces caused by viscosity (molecular motion). Therefore, if the Reynolds number is large, inertial forces predominate and the flow is turbulent.

The Reynolds number can be useful for making planetary comparisons. Suppose that there are three identically shaped topographical features situated on flat plains on Mars, Earth, and Venus, the surface roughnesses (e.g., boulders) are correctly scaled, and that the critical Reynolds number for turbulent flow is exceeded on all three planets. It would then be expected that all three flow patterns around the topographical features would be the same, even though the atmospheric properties are quite different. Thus, it is possible to interpret aeolian processes on Venus or Mars in terms of processes on Earth, so long as the Reynolds-number similarities criteria are satisfied. Care must be taken, however, because even though the flow patterns might be similar, other differences may arise because of factors such as particle speed.

2.4.2 *Turbulent boundary layer*

For laminar flow over a surface, the shear stress within the fluid is proportional to the rate of strain. For such a flow over a horizontal surface, the relationship between shear stress and the vertical velocity gradient is approximately

$$\tau = \mu \frac{\partial U}{\partial z} \tag{2.8}$$

where τ is the surface shear stress, μ the viscosity coefficient, U the fluid velocity, and z the distance above the surface.

If a surface is aerodynamically smooth (defined below), laminar flow develops in a very thin layer adjacent to the surface, even for flows having large Reynolds numbers (in which most of the boundary layer is turbulent). The velocity within the laminar sublayer varies approximately linearly with distance above the surface, so that

$$U = \tau z/\mu = u_*^2 z/\nu, \; u_* z/\nu \leqslant 10 \tag{2.9}$$

The *surface friction speed, u_**, is the square root of the ratio of shear stress to fluid density ($u_* = (\tau/\rho)^{1/2}$), and the kinematic viscosity, ν, is the ratio of absolute viscosity, μ, to fluid density ($\nu = \mu/\rho$). The laminar sublayer is very thin, ranging from less than 1 mm for winds blowing over quiescent sand surfaces on Earth and Venus to less than 9 mm on Mars (White, 1981*a*).

Above the laminar sublayer, the shear stress is sustained by turbulent exchange, and the coefficient of proportionality in Eq. (2.8) becomes a function of the wind velocity profile and distance above the surface. In other words, the coefficient is a function of the flow rather than of the fluid. Above the sublayer, the wind speed becomes a logarithmic function of height. For an aerodynamically smooth surface, the time-averaged wind speed above the laminar sublayer is written:

$$\frac{U}{u_*} = \frac{1}{0.4} \ln \frac{9u_* z}{\nu} \tag{2.10}$$

This equation is valid for values of surface friction Reynolds number, $R_* = u_* D_p/\nu \leqslant 5$, when the surface is said to be aerodynamically smooth and consists only of a smooth, quiescent sand surface of mean particle diameter, D_p.

If $R_* \geqslant 70$, the sand particles become sufficiently larger than the thickness of the laminar sublayer, the laminar sublayer ceases to exist and the surface is said to be aerodynamically rough. The velocity profile then becomes independent of viscosity, such that

$$\frac{U}{u_*} = \frac{1}{0.4} \ln \frac{z}{z_o} \tag{2.11}$$

where z_o is the equivalent roughness height. For a quiescent sand surface, z_o is about 1/30 the sand particle diameter. For most rough surfaces, whether they consist of pebbles, rocks, or trees, the equivalent roughness height under neutral conditions must be determined by measurement of the wind speed for at least two heights above the surface. Typical values of the roughness height range from 1 mm, or less, over the open sea to 50 cm for a forest to 5 m for a metropolitan city center. The roughness height is a function not only of the shape of the roughness elements, but also of the average distance between them (Fig 2.6).

It must be recognized that wind velocities, as represented by Eq. (2.11), are time-averaged values. For most flows, an instantaneous measurement of the wind speed profile is not a monotonic function of height. Rather, wind speeds must be averaged over at least 30 min to obtain a reasonable profile because the turbulent fluctuations of largest scale have characteristic durations of that order of time. For example, we measured wind velocities in the Mojave Desert (Fig. 2.7; Iversen & Greeley, 1978) for flow over a lava field. Each measurement point represents the time-averaged wind speeds for at least 1 hr. The different curves are each for a different day or time period within a day, and so represent different wind strengths. The logarithmic velocity profiles plot as straight lines on the semilogarithmic coordinate system, as shown.

The wind speed data represented in Fig. 2.7 are for flow over a fairly rough surface (hummocky lava with local relief exceeding 8 m) with an equivalent roughness height, z_0, of about 1 cm. Data were also recorded at the same time on another meteorological tower about 3 km upwind and situated in the midst of a smooth, alluvial plain. The roughness height for the alluvial plain was measured to be about 2 mm or about one-fifth as rough as the lava flow area. Thus, not only does the roughness height affect

Fig. 2.6. The aerodynamic roughness height, z_0, is a function not only of the roughness element shape and Reynolds number but also of the distance between elements. For closely packed, nearly spherical elements such as sand particles, the roughness height is $\approx 1/30$ the diameter. The maximum roughness of $\approx 1/8$ the diameter occurs when the center-to-center distance is about twice the diameter. Further increase in spacing distance decreases the roughness height.

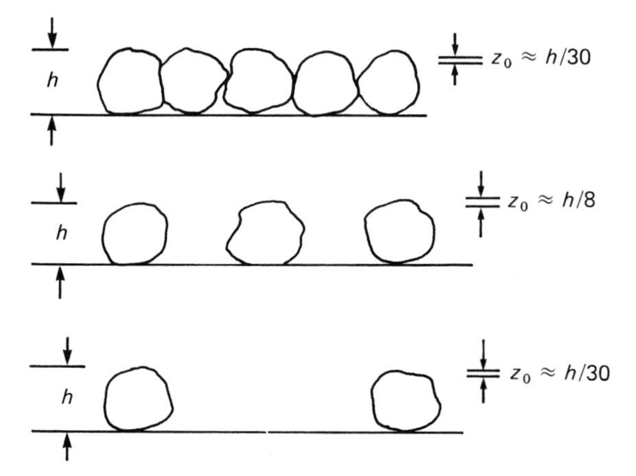

the velocity profile, as illustrated by Eq. (2.11), but an increase in roughness also results in an increase in surface shear stress. This is illustrated in Fig. 2.8, which shows that, for equivalent time periods, the friction speed measured for the rougher surface was, on the average, about 20% higher than for the smoother surface (equivalently, the surface shear stress is then about 44% higher).

Abrupt changes in surface roughness can have a significant effect on wind flow. Let us consider a wind blowing across a smooth, flat plain which passes over a rougher surface such as a lava flow, then onto a smooth plain on the other side of the lava flow. As shown in Fig. 2.9, the wind velocity profile takes some distance to reestablish an equilibrium profile after crossing the lava flow, and an 'internal' boundary layer with increasing thickness develops downwind of the leading edge of the change in roughness (see Blom & Wartena, 1969). In effect, there is an 'overshoot' in the surface shear stress which develops at the leading edge. Thus, for a change from a smooth to rough surface, there is an increase in surface shear

Fig. 2.7. Logarithmic wind speed profiles over a rough, flat plain. Data for each set of 5 points represent wind speeds averaged over at least 1 hour during strong winds. The data were taken during a two-month period. (From Iversen & Greeley, 1978.)

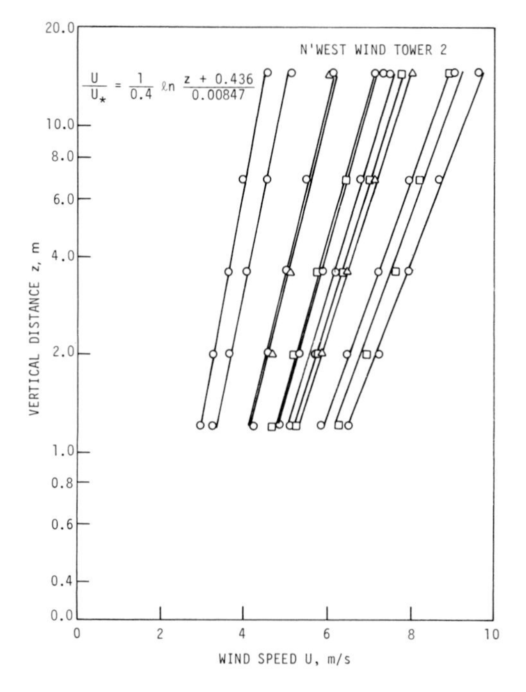

stress that is higher than the downwind equilibrium value (stress levels 1–2 in Fig. 2.9). Because the rougher surface eventually causes a decrease in wind speed gradient near the surface, the temporarily high gradient from the upwind smoother surface causes a temporary increase in surface shear stress. The ability of the wind to move loose particles is therefore greater at the leading edge of the rough surface than it is further downstream; we would expect erosion to occur in this area. For the opposite change – rough to smooth – the stress level at the leading edge is less than the downwind equilibrium value, creating a 'shelter' effect, or a type of 'wind shadow' at the leading edge of the smooth surface; we would expect net deposition in this area.

The variation in atmospheric density from Venus to Earth to Mars causes significant differences in the values of wind speeds necessary to initiate particle motion on their surfaces. The wind profiles near the surface of the three planets are illustrated in Fig. 2.5. Although the predicted value of wind speed at threshold on Mars is more than 27 times greater than on Venus (Fig. 1.2), the threshold value of dynamic pressure (one-half the density times velocity squared) does not vary that much from planet to

Fig. 2.8. Ratio of friction speeds for two meteorological towers for data taken concurrently. Tower 2 was centrally located in a rough (lava flow) flat plain. Tower 1 was 3 km upwind in a smoother (alluvial) flat plain. (From Iversen & Greeley, 1978.)

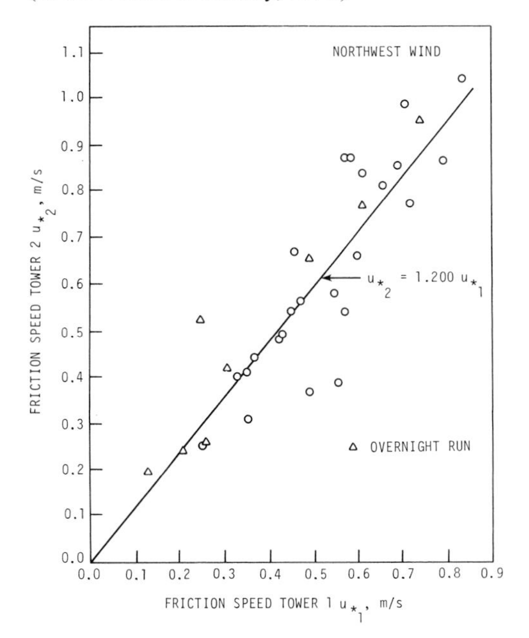

planet, ranging from about 0.06 mb on Mars to 0.33 mb on Venus at 10 m height.

2.4.3 *Effect of non-neutral stability*

The logarithmic wind speed profile, expressed by Eq. (2.11), is valid only for that part of the atmospheric boundary layer near the surface. If the lapse rate is not adiabatic, then the logarithmic curve will approximately hold for a somewhat shorter height above the surface. A parameter which characterizes the stability of an atmospheric layer is the gradient Richardson number

$$R_i = g(\partial\theta/\partial z)/[(\theta)(\partial U/\partial z)^2] \tag{2.12}$$

The Richardson number is positive for a stable layer and negative for an unstable one. The Richardson number, and therefore an estimate of the stability, can be obtained by measuring the temperature and velocity variations with altitude.

For the neutrally stable logarithmic layer, the dimensionless wind shear $0.4z(\partial u/\partial z)/u_*$ is unity. For the stable layer the wind shear factor is greater than 1, and for an unstable layer it is less than 1. Data for an area of planar homogeneity, i.e., a flat plain with a uniformly rough surface, are shown in

Fig. 2.9. Schematic of wind speed profiles (upper) and surface shear stress distribution (lower) associated with discontinuous surface roughness changes.

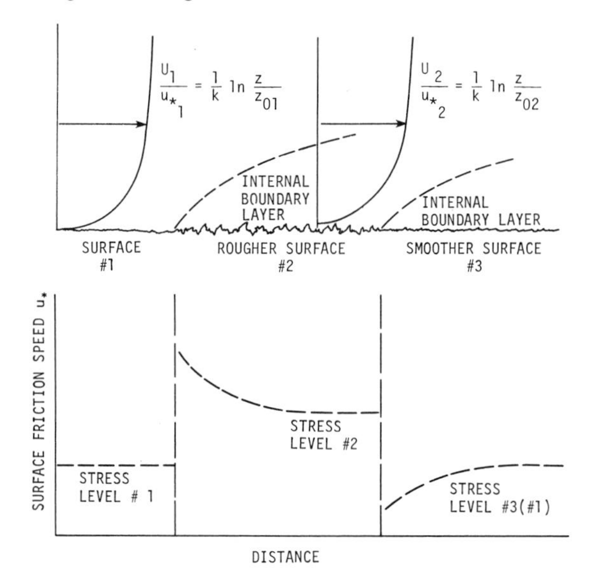

Fig. 2.10. The corresponding velocity profiles are illustrated in Fig. 2.11 for all three types of stability, i.e., stable, neutral, and unstable. For the non-neutral profiles, the curves are approximately logarithmic for $z/L^* \leqslant 0.1$. If the winds are strong enough to exceed the threshold for movement of sand and dust, turbulent mixing is usually enough for thermal mixing to result in an adiabatic, or neutral layer. The height of the logarithmic layer for very strong winds can be of the order of 100 m.

2.4.4 *Pressure gradient and Coriolis forces*

If defined with respect to an absolute inertial frame of reference, a moving parcel of air follows a straight line when not subjected to any perturbing forces. To an observer located on the surface of a rotating planet, however, the path of the parcel of air appears curved because of the

Fig. 2.10. Dimensionless wind shear data for non-neutral atmosphere boundary layers. The data to the left of the vertical line are for unstable layers and to the right are for stable layers (From Businger *et al.*, 1971.)

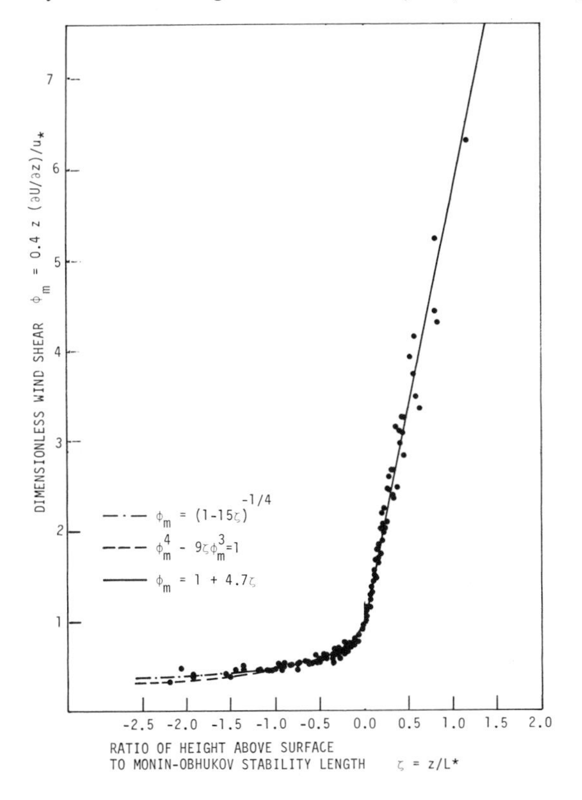

rotation of the planet. The apparent force which seems to be acting on the air parcel is called the *Coriolis force*.

The equations of motion for a uniform (i.e., with planar homogeneity) planetary boundary layer written in a coordinate system on the surface are

$$\begin{aligned} -2\omega V \sin\phi_o &= -\frac{1}{\rho}\frac{\partial p}{\partial x} + \frac{\partial \tau_x}{\rho \partial z} \\ 2\omega U \sin\phi_o &= \frac{1}{\rho}\frac{\partial p}{\partial y} + \frac{1}{\rho}\frac{\partial \tau_y}{\partial z} \end{aligned} \qquad (2.13)$$

For these equations, the orthogonal x and y coordinates are parallel to the surface and z is perpendicular. The wind speeds are U and V in the x and y directions respectively and, because of the assumption of planar homogeneity, are functions only of vertical height, z. The rate of rotation around the planetary axis is ω, and the latitude is ϕ. The pressure gradients in the x and y directions are $\partial p/\partial x$ and $\partial p/\partial y$, and the shear stresses due to turbulent

Fig. 2.11. Wind speed profiles near the surface for stable, neutral and unstable atmospheric boundary layers. z_o is the surface roughness height and L^* is the Monin–Obhukov stability length.

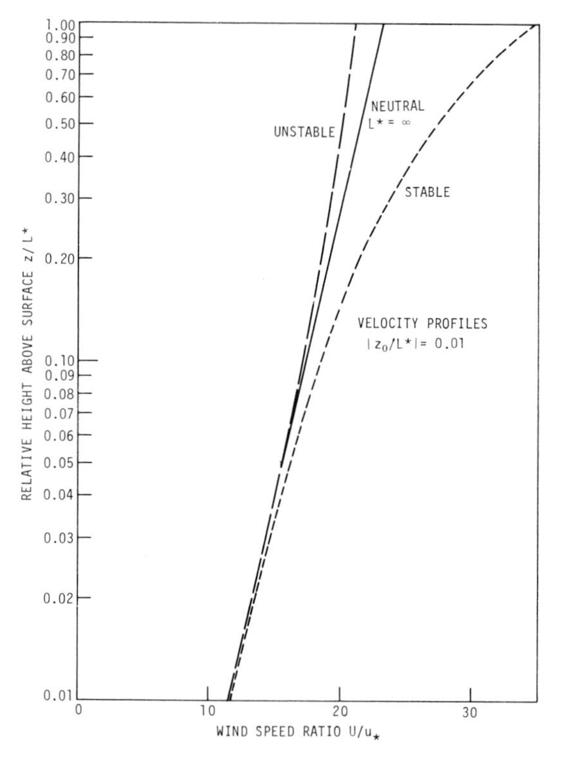

friction are τ_x and τ_y. The factors on the left side of Eq. (2.13) are the Coriolis terms, which appear because the equations are written in a rotating, non-inertial coordinate system.

Above the boundary layer, the shear stress gradients are negligible, and the so-called geostrophic wind speed components are thus

$$U_g = -\frac{1}{\rho}\frac{\partial p}{\partial y}\Big/(2\,\omega\,\sin\phi_o)$$
$$V_g = \frac{1}{\rho}\frac{\partial p}{\partial x}\Big/(2\,\omega\,\sin\phi_o) \qquad (2.14)$$

The consequences of Eq. (2.14) are shown in Fig. 2.12. If the pressure gradient components are both positive, the x component of wind speed U is negative and V is positive, and the resultant wind vector is perpendicular to the resultant pressure gradient, i.e., the wind speed is parallel to the *isobars* (lines of constant pressure). Fig. 2.12 also shows that the wind speed vector is rotated 90° to the left of the pressure gradient vector, which in turn points in the direction of increasing pressure. Thus, the wind rotates anticyclonically (in a clockwise direction), around a high-pressure area. The reverse is true for a cyclone, i.e., a low-pressure area with winds rotating in a counterclockwise direction. These conditions are true only in the northern hemispheres for Earth and Mars, where the latitude, ϕ, is positive. The directions of rotation are reversed in the southern hemisphere because of

Fig. 2.12. Vectors of geostrophic wind corresponding to positive, horizontal pressure gradients in the northern hemisphere.

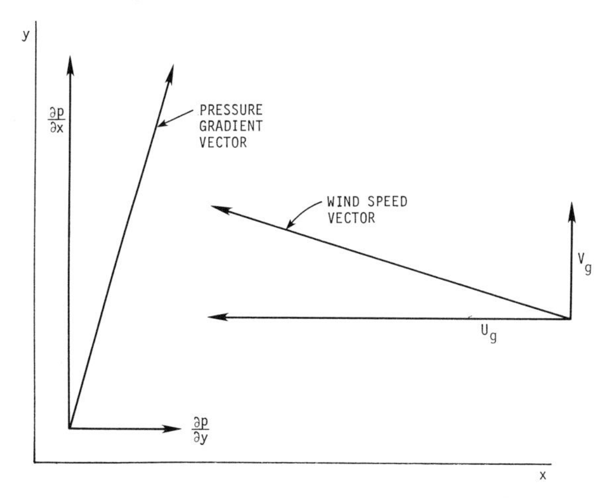

the negative value of $\sin\phi_o$ in Eq. (2.14). The rotation rate on Venus is so slow that the Coriolis forces are negligible.

2.4.5 *Friction forces – the Ekman spiral*

With the help of Eq. (2.14) for the geostrophic wind speed components, Eq. (2.13) can be rewritten as

$$
\begin{aligned}
-2\,\omega\,\sin\phi_o(V - V_g) &= \frac{\partial \tau_x}{\partial z} \\
2\,\omega\,\sin\phi_o(U - U_g) &= \frac{\partial \tau_y}{\partial z}
\end{aligned}
\qquad (2.15)
$$

The existence of the shear stress gradients within the boundary layer causes a rotation of the wind speed within it with respect to the geostrophic wind direction. This is easier to see by rotating the x–y coordinate system until the geostrophic wind is in the positive x direction (Fig. 2.13). Since the shear stress derivatives are negative (shear stress is maximum at the surface) and in the coordinate system of Fig. 2.11, V_g is zero, the y component of wind speed, V, is positive within the boundary layer, and the x component, U, is also positive but less in magnitude than the geostrophic wind, U_g. The wind vector profile through the boundary layer – as seen from above (Fig. 2.13) – resembles a spiral curve and is thus called the *Ekman spiral*, after Ekman's (1902) original solution. The wind speed vectors are shown in a three-dimensional view in Fig. 2.14. Again in the southern hemisphere the latitude, ϕ_o, is negative, and the patterns would be mirror images of those in Figs. 2.13 and 2.14.

In general, there are relationships among the variables, geostrophic

Fig. 2.13. Plan-view of the Ekman spiral boundary-layer profile in the northern hemisphere.

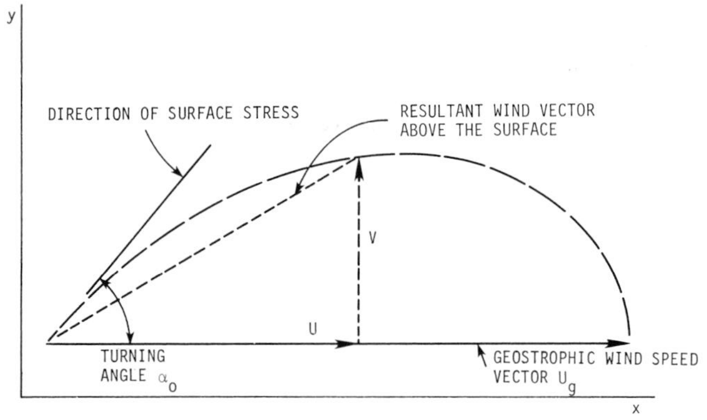

wind, U_g, boundary-layer height, δ, surface friction speed, u_*, surface roughness length, z_o, Coriolis parameter, $f = 2\omega \sin\phi_o$, and the angle, α_o, between the direction of surface shear stress (and wind just above the surface), and geostrophic wind. For a neutral layer, the relationships become

$$\begin{aligned} \frac{u_*}{U_g} &= F_1 (u_*/fz_o) \\ \alpha_o &= F_2 (u_*/fz_o) \\ \frac{\delta f}{u_*} &= \text{constant} \end{aligned} \tag{2.16}$$

The parameter, u_*/fz_o, is one form of the *Rossby number*. For a non-neutral layer, the ratio of surface friction speed to geostrophic speed, u_*/U_g, Ekman turning angle, α_o, and the boundary-layer thickness parameter also become functions of the roughness–stability parameter, z_o/L^*, where L^* is the Monin–Obhukov length (Monin & Yaglom, 1971).

The Rossby number varies by several orders of magnitude, depending on wind strength, latitude, and surface roughness. Measured values of the ratio of friction speed to geostrophic wind speed on Earth range from

Fig. 2.14. Three-dimensional view of the Ekman spiral in the northern hemisphere.

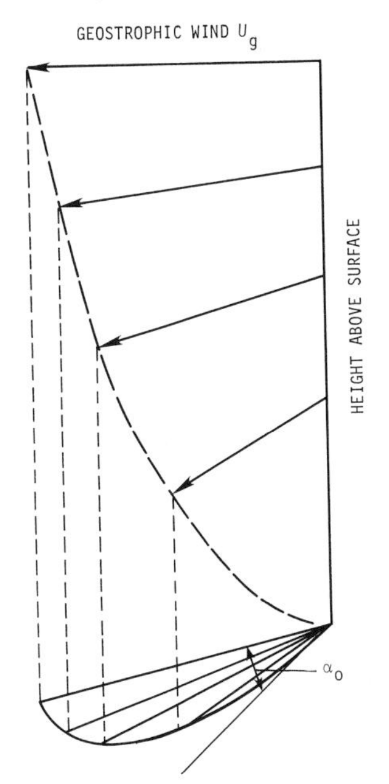

about 0.02 to 0.04 for neutral layers, with smaller values for stable layers and larger values for unstable layers, particularly at small Rossby numbers. The turning angle, α_o, is usually between 10 and 35° and increases with increase in the ratio of friction speed to geostrophic speed (Csanady, 1972). Tennekes (1973) derives the following expressions (for neutral layers):

$$\ln\left(\frac{U_g u_*}{u_* f z_o}\right) = 2 - \ln\left(\frac{u_*}{U_g}\right) + \left(\frac{0.1225}{u_*^2} - 25\right)^{1/2}$$
$$\delta f / u_* = 0.3 \qquad (2.17)$$
$$\alpha_o = \sin^{-1}(100 u_* / 7 U_g)$$

The threshold characteristics used to obtain the wind speed profiles in Fig. 2.5 can be substituted into Eq. (2.17). The results are displayed in Table 2.2. The geostrophic wind speed on Mars is about six times as high as the value for Earth and the boundary layer is about seven times as thick. Pollack *et al.* (1976) have also calculated geostrophic speeds and turning angles for threshold conditions on Mars, with similar results.

2.4.6 *Diffusion*

Diffusion of contaminant gases, or of tiny particles by atmospheric turbulence, takes place at a rate many orders of magnitude greater than can be accounted for by molecular diffusion alone. Thus, the rate at which a 'plume' from a point source (e.g., a smokestack) broadens as it extends from the source in a downwind direction depends upon the level of turbulence. That level of turbulence, in turn, depends upon the strength of the wind and the roughness of the surface, as well as the relative stability of the atmospheric boundary layer.

The diurnal variation in solar radiation on Earth and Mars sets up a cycle of temperature variation in the planetary boundary layer, which strongly affects the type of wind profile and the resulting levels of turbulence and turbulent diffusion. Before sunrise, the air layer near the surface is stable because the ground has cooled through thermal radiation. After sunrise on cloudless days, the land surface heats up more rapidly than the air above. The air adjacent to the surface warms, and the air layer becomes unstable. This layer, which increases in thickness as the ground becomes warmer, is capped by an inversion, i.e., by the stable layer above it. The elevation of the inversion is often noticeable from above, as dust and haze are trapped below it, since the stable layer of air above the inversion inhibits the production of turbulence and, therefore, turbulent diffusion.

Table 2.2. *Ekman boundary layer characteristics at minimum threshold for Earth and Mars*

Planet	Earth	Mars
Latitude, ϕ_o	40°N	40°N
Friction speed, u_* (m/sec)	0.21	1.5
Rossby no., u_*/fz_o	$4.84(10)^7$	$6.96(10)^6$
Geostrophic wind, U_g (m/sec)	9.89	62.7
Boundary-layer thickness, δ (m)	674	4940
Turning angle, α_o	17.7°	20.0°

Strong winds, such as are necessary for the initiation of motion of dust and sand, produce turbulence and so tend to break up the alternating layers of stably and unstably stratified air. Mixing of warm and cold parcels of air is enhanced by the mechanically produced turbulence, and the temperature gradient thus tends toward the adiabatic or neutrally stable lapse rate.

Dust, composed of tiny particles of diameter 20 μm, or less, once diffused by turbulence to considerable heights above the surface, can be carried by the wind for long distances. For example, it has been estimated that dust from the Sahara Desert in Africa contributes up to 200 million metric tons per year to tropospheric aerosols, perhaps half of the total mineral dust for Earth (Morales, 1979). Atmospheric transport of Saharan dust has been documented for distances of up to several thousand kilometers. The dust in the atmosphere is important for its effect on climate and on regions where erosion and deposition of the dust occur. Atmospheric dust can affect temperature (Fig. 1.4) and climate by altering radiation and optical properties of the atmosphere.

2.5 **Windblown particles**

What constitutes aeolian particles and where do they come from? Examination of the wind threshold curve (Fig. 1.2) and consideration of average wind strengths on the planets give some indication of the sizes of particles that we might expect to be moved by the wind. Particles most readily moved by the wind fall within the size range called 'sand'. Many definitions have been proposed for sand (e.g., Pettijohn *et al.*, 1972) but, in its common usage, *sand* is loose, granular material in the size range 62.5–2000 μm (2 mm) in diameter, regardless of its composition or origin.

Windblown sand is transported primarily in saltation, traction, and impact creep.

The physical properties of windblown snow particles cover a wide range. Depending upon temperature, snow can range from very dry to very wet, where the particles cohere and freeze together. Average sizes of falling particles range from 0.06–5 mm. However, when blown by strong winds, snow crystals are broken and abraded into particles with rounded corners, having average diameters in the sand-size range, i.e., 150–200 μm. Particles which are much smaller than that quickly sublime and disappear. Thus, for dry snow, the particles behave much like sand.

Particles smaller than sand require stronger threshold winds and are therefore usually set into motion by saltation impact of sand or are raised by dust devils. The sedimentologist divides these fine particles into silt (39–62.5 μm in diameter) and clay (< 39 μm in diameter). In the aeolian context, windblown silt and clay are commonly called 'dust' which is transported in suspension. Dust tends to occur in two size ranges (Péwé, 1981 and Fig. 2.15), but the subdivisions do not coincide with the sizes for silt and clay. Coarse dust is about 5–50 μm in diameter and is usually transported by dust devils or local dust storms for distances seldom in excess of 100 km. Fine dust is typically 2–10 μm in diameter and is tropospherically sorted material, frequently transported as aerosols. Because fine dust is so small, it remains in suspension until it is brought down by rain. Fine dust particles may also clump together by various

Fig. 2.15. Cumulative size–frequency distributions for various particles: volcanic particle data (Heimaey scoria, Iceland, and Ukinrek maars, Alaska) courtesy of S. Self; sands and loess data from Krumbein & Sloss (1963); impacted basalt data from Gault *et al.* (1963); dust data from Péwé (1981): dust 1 represents samples from dust devils, local dust storms, loess, and volcanic ash – all were transported distances of (generally) less than 100 km; dust 2 represents a sample of desert dust carried from the Sahara to Barbados, West Indies, and is typical of global-scale dust storms which are tropospherically sorted.

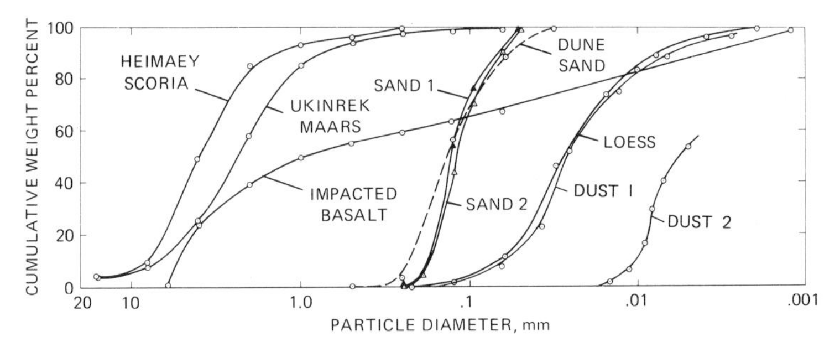

mechanisms to form larger aggregates which then settle out of the atmosphere.

Particles larger (>2 mm in diameter) than sand are typically moved in the aeolian regime by impact creep, in which they are rolled and pushed along the ground by the wind, or moved by the impact of smaller, saltating grains; occasionally larger grains may also move in saltation by storm-strength winds. Grains 2–4 mm in size are called *granules*; particles 4–16 mm across are termed *pebbles*, again without connotation as to composition or origin. These large particles may form wind ripples, discussed in Chapter 6. Rock fragments larger than pebbles are seldom moved very far by the wind but are nonetheless affected by aeolian processes. They may be abraded, or sculpted, by the wind to form ventifacts, discussed in Chapter 4. And, because wind is an excellent fluid for sorting sediments of mixed sizes, sand and dust are commonly winnowed away, leaving coarser material behind. This coarse material can form a 'lag' deposit, contributing to the development of *desert pavement* (Fig. 1.11). Desert pavement plays an important role in enhancing the transport of sand by saltation because of the increased elastic rebound of grains bouncing along the coarse, pebble surface.

An accounting of the abundances of different particle sizes of all origins shows that several size ranges seem underrepresented in sedimentary deposits (Pettijohn *et al.*, 1972). For example, the relative abundance of grains 1–8 mm in diameter is low. This could result from various mechanical instabilities of grains in that size range, which cause them to be relatively 'short lived' before breaking into smaller grains. This leads us to consider the processes involved in the formation of windblown materials and the sources of the materials from which they are formed.

2.6 Processes of particle formation

The 'rock cycle' relates the main rock types and materials found on planetary surfaces to the processes involved in their formation and their subsequent modification. The ultimate source of all material is either magma, generated in planetary interiors, or cosmic material, such as meteorites, introduced from external sources. Windblown particles are included in the sedimentary part of the cycle and are derived from the breakdown of various rocks and minerals. Fig. 2.16, modified from Pettijohn *et al.* (1972), shows the five major processes involved in the formation of aeolian sediments: weathering, cataclastic processes, volcanism, precipitation–biological activity, and aggregation. A review of weathering in different planetary environments is given by Gooding (1983).

2.6.1 *Weathering*

Weathering refers to the chemical and physical breakdown of materials through exposure to the atmosphere, hydrosphere, and biosphere. In general, weathering leads to smaller particles, usually of chemically stable compositions for the local environment. Physical weathering predominates in cold climates and involves breaking materials into smaller pieces. Chemical weathering involves chemical reactions and predominates in most climates, especially where moisture is present. Given the common rocks and minerals in typical environments on Earth, the end products of weathering are quartz grains of sand and silt size, small flakes of muscovite mica, clay minerals, such as montmorillonite, and various soluble carbonates and iron oxides.

Salt weathering is a process which may be locally important in the production of particles (Wellman & Wilson, 1965; Cooke & Warren, 1973). As noted by Cooke & Smalley (1968), 'salt' refers to a wide variety of compounds that form crystals from supersaturated solutions. Although salt weathering may occur in any one of several modes, pressures exerted by salt crystal growth in pores and crevices appear to be an extremely effective means for shattering rocks. For example, Goudie *et al.* (1979) used an experimental apparatus in which sand was subjected to cycles of salt weathering under controlled conditions and found that silt-size particles are readily produced from the larger grains (Fig. 2.17).

After considerations of the weathering environment on Mars, Malin (1974) concluded that salt weathering is probably an active process in the generation of particles and may have been important for much of martian geological history. Analyses of fine particles on Mars at the Viking Lander sites suggests that they are composed of clay minerals such as nontronite and are probably the weathering products of mafic rocks such as basalt (Clark & Baird, 1979). The presence of chlorine and other salt-forming

Fig. 2.16. Diagram showing the five principal processes leading to the formation of particles in the size range appropriate for aeolian transport. (Modified from Pettijohn *et al.*, 1972.)

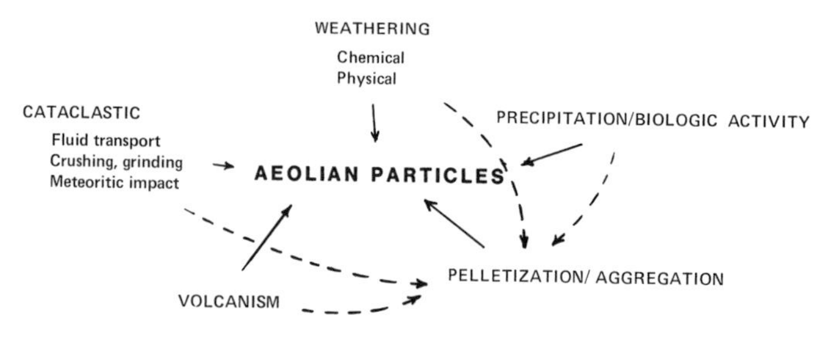

Fig. 2.17. Scanning electron micrograph of an aeolian quartz grain subjected to salt weathering in the laboratory, showing cracks developed after 40 cycles of weathering with Na_2SO_4, demonstrating the potential effectiveness of producing dust from sand-size grains. (From Goudie *et al.*, 1979; copyright Academic Press, 1979.)

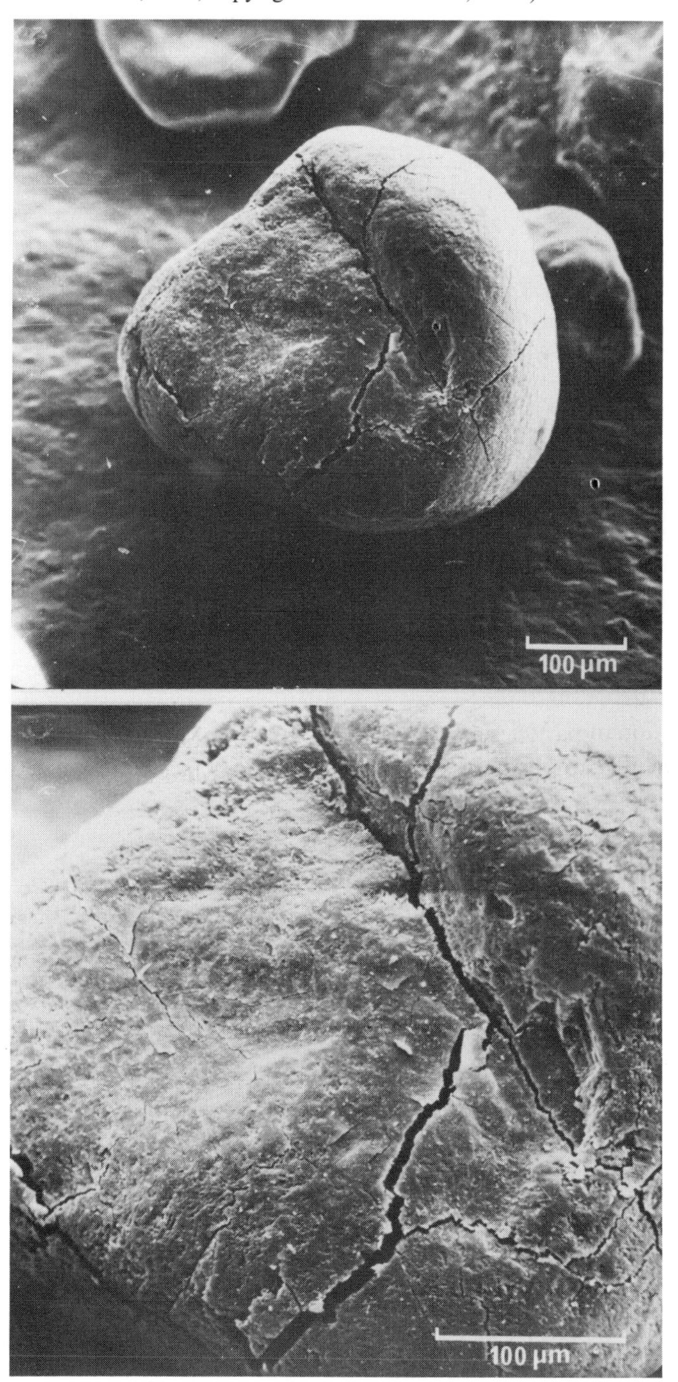

elements support the idea that salt weathering is important in the production of fine grains on Mars.

Relatively little direct information concerning weathering is available for Venus. As discussed in Chapter 1, interpretations of the Soviet Venera results include the presence of both granitic and basaltic rocks. Considerations of these rock types and their weathering products in the highly chemically reactive environment of Venus have been made by Nozette & Lewis (1982) and by McGill *et al.* (1982). They conclude that at high elevations, such as Maxwell Montes (Fig. 1.18), magnesite, enstatite, quartz, tremolite, andalusite, dolomite, and either sulfides or sulfates might be produced, and they suggest that these areas are probable sources for aeolian materials on Venus.

2.6.2 *Cataclastic processes*

Cataclastic processes involve those events that break, fragment, or crush rocks and particles. Although some geologists include these processes as part of physical weathering, we shall follow the usage of Pettijohn *et al.* (1972) and consider cataclastic events separately, which include: (1) breakage resulting from collision of particles in fluid (wind or water) transport, (2) fragmentation resulting from impact cratering, and (3) crushing and grinding by mass movements (e.g., landslides, rock fails), tectonic processes (e.g., faulting), and glacial processes.

Kuenen (1960) has shown that *attrition* – the reduction of particles in mass – is 100–1000 times more effective in air than in water, partly because the collisions of particles tend to be 'cushioned' in the denser fluids and are less energetic. Attrition of windblown grains on Mars may be very rapid; the high wind speeds required for threshold and the subsequent high particle velocity results in energetic collisions. Sagan and colleagues (1977) have proposed the term 'kamikaze' (from the Japanese referring to divine wind) grains to describe the particles subject to rapid break-up, through collisions and impacts with rocks, of grains on Mars.

Fragmentation resulting from impact cratering is not a significant process currently operating because of the infrequent occurrence of impact cratering, but it was certainly important in the past when rates of cratering were higher. Although most particles initially formed by this process undoubtedly have been recycled many times on Earth, many particles in the current aeolian regime on Mars may have been formed by impact cratering and have survived through time in the less active geological environment of Mars.

Results from laboratory experiments and analyses of the predominantly impact-generated soils on the Moon provide clues to the sizes of particles generated by impact. Impact cratering results from the near-instantaneous transfer of energy from the impacting object (e.g., meteoroid) to the target surface. Because impact velocities are extremely high – commonly in excess of 10–15 km/sec – substantial kinetic energy is focused on the target. As a result, parts of the target are vaporized, melted, and fragmented, then excavated to form a crater. Laboratory experiments (Gault *et al.*, 1963) show that small-scale impacts into solid basalt blocks produce particles in the sizes appropriate for aeolian transport (Fig. 2.15). Similar experiments (Stöffler *et al.*, 1975) involving targets of loose, well-sorted sand show that the particles are broken into small sizes, and that the mass becomes inefficiently sorted. Analyses of lunar soil samples returned by the Apollo astronauts confirm these laboratory results and show that, after more than three aeons of impact cratering on all scales, a fragmental surface layer containing soils that average 0.03–1.2 mm in diameter has been produced (Lindsay, 1976).

Thus, we may expect a significant contribution of particles through impact cratering to the aeolian regime in the early histories of the planets. An analysis of the Viking Lander 2 sites shows a block-size distribution very similar to that seen on the rim of the impact crater Tycho on the Moon (Binder *et al.*, 1977), suggesting that rock fragments on the martian surface still reflect particles formed from impact processes dominant hundreds of millions of years ago.

2.6.3 *Volcanism*

Volcanic processes generate fragmental particles (Fig. 2.15), termed *pyroclastics* or *tephra*, which range in size from very fine ($\approx$microns) to blocks the size of small houses. Depending upon the violence of eruption, volcanic dust (ash) is frequently injected directly into the atmosphere and can be carried by winds over the entire globe, as we discuss in Chapter 7. Sands of volcanic origin include both pyroclastic materials and reworked (weathered) volcanic rocks. Because of the apparent prevalence of volcanism on Mars (Greeley & Spudis, 1981), it has been suggested that much of the material inferred to mantle the surface could be windblown volcanic ash. Considerations of magmatic processes on Mars (Wilson & Head, 1981) show that explosive eruptions should have occurred. Magma and lava that come in contact with water or ice commonly fragment violently into fine-grained material. Many black sand beaches, such as those in Hawaii, have formed in this manner. Occasionally, active volcanic fields may be blanketed with pyroclastics that have been formed by lava–water interaction. Because subsurface ice and water probably existed on Mars during volcanic activity, it has been suggested that large quantities of ash and other fine particles may have been generated by lava–water interactions similar to those on Earth (Hodges & Moore, 1979; Frey *et al.*, 1979) and introduced into the aeolian regime.

Because of the high atmospheric surface pressures on Venus, explosive volcanism and the production of pyroclastics may be suppressed (Head & Wilson, 1981).

2.6.4 *Precipitation and biological activity*

Chemical and biochemical precipitation may lead to sand-size particles, such as those found in the formation of carbonate sands. Biota contribute to particle formation through mechanical breaking of sediments (e.g., expanding tree roots during growth), generation of fecal pellets, and/or contribution of skeletal remains.

2.6.5 *Aggregation*

With the exception of precipitation, all of the processes described above involve the generation of small particles from big ones. In this section we consider sand-size aggregates that are built from smaller particles. One way these 'composite' grains can form is indirectly from deposits in dry lakes (playas). These deposits, typically, are water transported silt and clay; as the lake dries up, the deposits crack, offering jagged, irregular edges to the wind. Winds peel off flakes and chunks of dried mud and abrade the material as it is bounced along, quickly forming sand-size pellets. In most respects, the particles behave as sand and are capable of forming ripples (King, 1916) and dunes, as reported in many areas on Earth (Coffey, 1909; Huffman & Price, 1949; Butler, 1974). Extensive deposits of this type of particle have been described in Australia where it is referred to as *parna*. In most playa deposits, the binding of silt and clay is enhanced by salt cementation.

Aggregation of fine grains may also result from electrostatic charges. Laboratory experiments (Greeley, 1979) show that in a cloud of windblown grains, both the size and magnitude of electrostatic charges are functions of collisional impact velocity, grain size (Fig. 2.18), grain composition, and atmospheric pressure. Electrical charges result from one or more processes including friction, contact electrification, and grain breakage; they have been measured in clouds of aeolian material in natural conditions on Earth (Boning, 1927), and aggregation resulting from them has been proposed as a mechanism for settling dust from the atmosphere (Beavers, 1957).

The present environment on Mars appears to be well suited for the generation of electrostatic charges (Mills, 1977), and it has been suggested that electrostatic aggregates may be an important part of the aeolian regime (Greeley, 1979). Laboratory simulations of windblown particles in a martian environment show that sand grains quickly break into fine dust (Fig. 2.19; the kamikaze effect described earlier), but that with time the grains cling together as aggregates held together by electrostatic charges (Fig. 2.20).

2.7 Sand and dust sources

Let us now consider the sources of the particles most commonly involved in aeolian processes: sand, and dust. Sand has caught the

Fig. 2.18. Electrical current generated by windblown particles of glass, shells, and quartz, as a function of particle size. Both the sign and magnitude of the current vary widely with size and composition, leading to the possibility of aggregation of grains by electrostatic charges. (From Greeley, 1979; copyright, American Geophysical Union.)

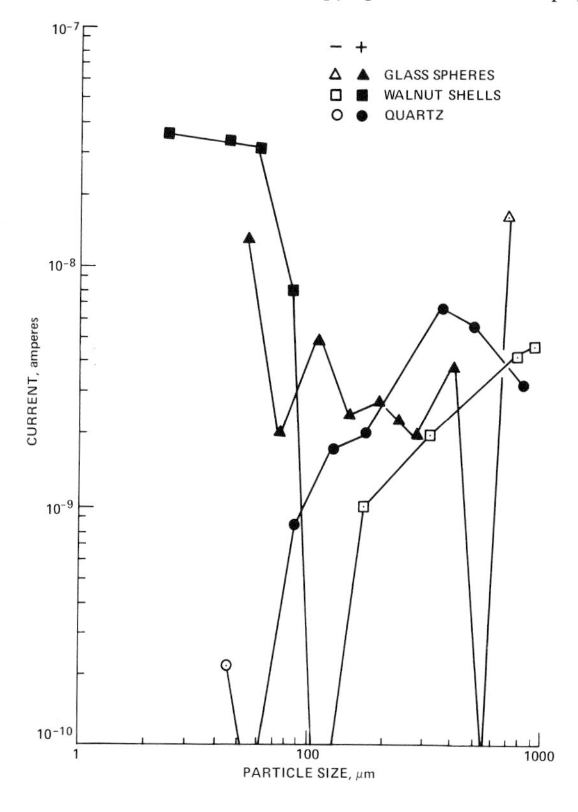

imagination of both layman and specialist for many years, as evidenced by numerous papers and frequent references to sand and sand related topics. Reviews of the origin and characteristics of sand are provided by Kuenen (1960) and Krinsley & Smalley (1972); the book by Pettijohn *et al.* (1972) provides a detailed account of sand and sandstone.

Most particles of sand size on Earth are composed of quartz, although some aeolian sands may be of other compositions, such as the gypsum sands of New Mexico. Quartz is extremely durable and resists attack by chemical processes. Experiments by Kuenen (1960) have shown that, as particles are reduced in size, the rate of attrition progressively decreases, until a diameter of about 0.1 mm is reached, at which size further attrition is virtually halted. Thus on Earth, once sand grains are formed, most particles remain as sand within the 'sediment' part of the rock cycle.

Most quartz sand is derived from the chemical weathering of granitic rocks, composed primarily of feldspar and quartz crystals in sizes typically ranging from about 1 mm to 1 cm across. Feldspar is less resistant to chemical reactions than quartz, and as it weathers chemically, the quartz grains are released from the rock. In some granitic areas, notably in deserts, the crumbled mass of feldspar and quartz grains and partly decomposed granite (referred to colloquially as 'rotten granite') forms a granular surface termed *grus*, which is an important source for windblown sands.

Smalley (1966) has developed another model to explain the release of

Fig. 2.19. Cumulative size–frequency distribution of particles subjected to attrition in martian simulation. Average particle size prior to attrition was 100 μm; after $\approx$20 min the average size was $\approx$15 μm, but as the experiment continued, the average particle size increased. Examination of the particles showed that aggregation had occurred, forming particles such as are shown in Figure 2.20 (From Greeley, 1979; copyright, American Geophysical Union.)

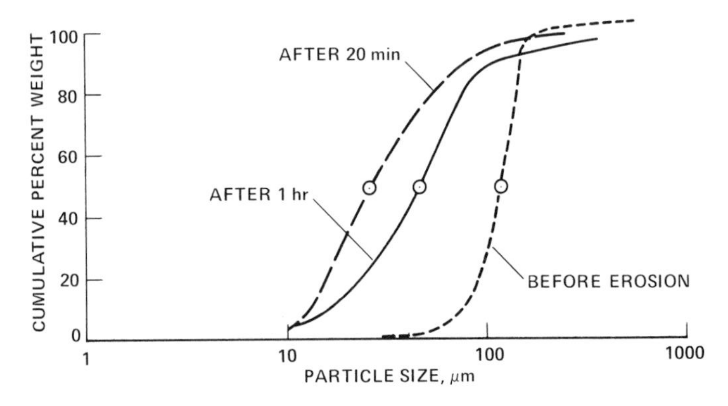

quartz from granite by purely mechanical processes. As magma cools, quartz is generally the last mineral to crystallize. In addition, there are two types of temperature-dependent quartz, each with a different crystal structure. As the crystallizing magma cools through 573 °C, there is a transition from 'high'-temperature to 'low'–temperature quartz. This transition involves a change in crystal structure, which results in a decrease in the volume of the crystal. Because the quartz crystals are constrained by other grains (mostly feldspar) which formed at higher temperatures, stresses develop within the 'low' quartz. These stresses remain until the rock is exposed to weathering whence the stresses are relieved through fracturing and shattering of the crystals, leading to the formation of quartz grains of sand size.

Initially, the quartz grains are irregular and angular. As they are weathered and abraded in transport they become more rounded, as shown

Fig. 2.20. Scanning electron micrograph (SEM) of a large ($\approx 500\ \mu m$ in diameter) aggregate of fine grains produced in a simulated martian environment. Grains are bonded primarily by electrostatic charges. (Photograph courtesy of D. K. Krinsley.)

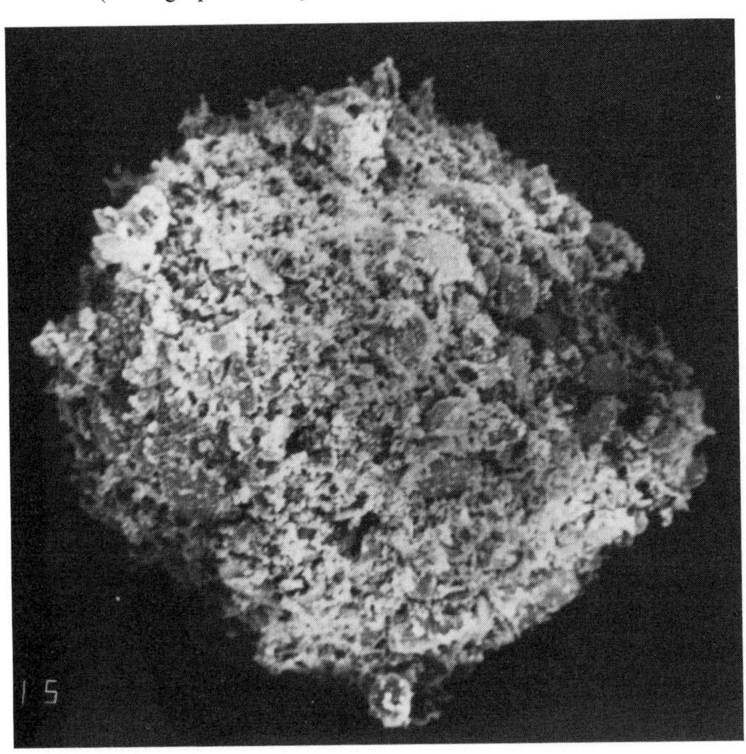

in Fig. 2.21; sharp edges and corners are frequently chipped and flaked off from the grains during transport, contributing to the production of silt-size grains. This finer material is typically winnowed away by suspension in the wind, contributing toward the maintenance of the sand-size sorting of the mass left behind. The quartz particles released by chemical weathering tend to retain the general form of the original crystal, i.e., an elongate grain remains elongate but the edges become rounded.

Once sand is produced, regardless of origin (Fig. 2.16), it may become part of the aeolian regime directly, or it may be subjected to a host of processes. On Earth, because of the long life of quartz sand and the prevalence of processes involving liquid water, almost all aeolian sand has experienced some part of the water cycle. In most cases, even desert sands were first transported by water, to be picked up later by the wind. The Nubian sands of the eastern Sahara, however, pass directly from weather-

Fig. 2.21. Scanning electron micrograph of a typical aeolian sand grain showing well-rounded shape and dimpled surface giving it a frosted appearance. Large depressions probably result from chipping during strong wind storms. The grain is from the Sabha sand sea, south-central Libya. (From Krinsley & Doornkamp, 1973; copyright, Cambridge University Press.)

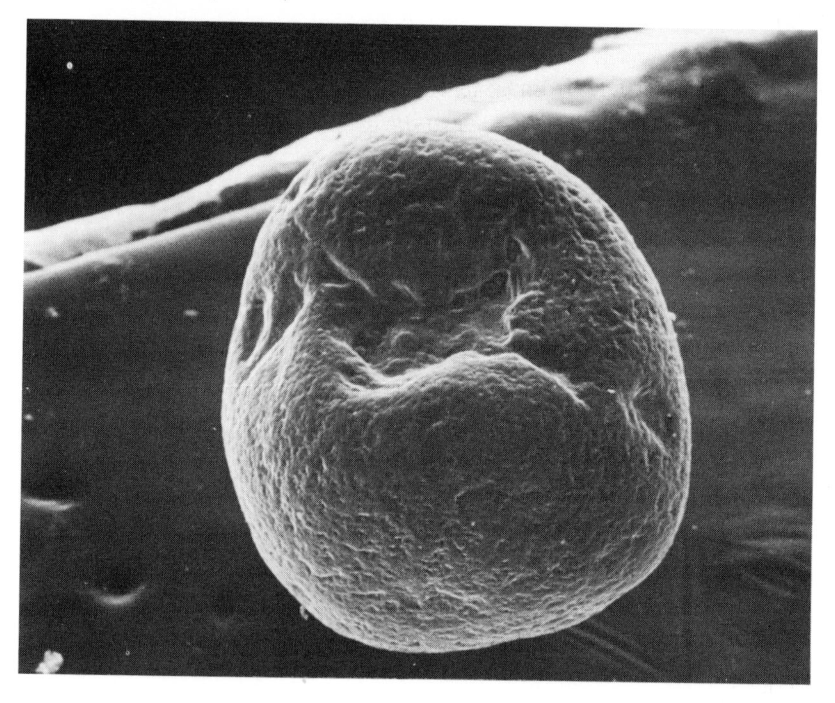

ing of bedrock into the aeolian regime without involving the hydrological cycle (Mabbutt, 1977).

Ultrafine markings on quartz sand grains, as revealed by scanning electron microscopy (SEM), can indicate the processes to which the grains have been subjected. As documented by Krinsley & Doornkamp (1973), each major mode of transportation – wind, rivers, glaciers – leaves distinctive textures on the surfaces of quartz grains. Aeolian sands are usually 'frosted', with the frosting being a dull opaque surface, in contrast to typical river sands, which are polished and shiny. The frosting is attributed both to abrasion and to chemical precipitation resulting from desert dew (Kuenen & Perdok, 1962). Krinsley & Smalley (1972) examined quartz grains abraded in laboratory wind machines, grains collected from glacial regions and grains from temperate coastal dunes, and found that their surfaces were characterized by sharp, somewhat regular, plates arranged parallel to each other. Grains collected from hot deserts also showed plates, but the plates appeared rounded and worn, not sharp. They attribute this appearance to the result of local solution and reprecipitation at essentially the same site on the grain.

What are the sources for silt-size particles which eventually may become windblown dust? Most silt is of quartz composition and, like sand, we can trace its ultimate origin to granitic rocks. Several authors have addressed the question of silt production; there appear to be two general modes of formation. Silt may be produced by the gradual wearing down in size of sand by abrasion, or it may form by the fracturing of sand grains and the chipping of fine particles from larger fragments, as discussed in Section 2.6. Overwhelming evidence argues against the gradual wearing down of grains to silt (Rogers *et al.*, 1963). The lack of a continuous distribution of sizes from large to small, noted earlier, is contrary to the hypothesis of gradual wearing of large to small grains. (If that were the case, then there ought to be a continuum of sizes.) In addition, abrasion experiments show that once grains are reduced to about 1 μm, further reduction in size is extremely retarded. And, finally, the observation of the angularity of silt grains versus the rounded edges of sand grains – all in the same deposit – suggests that the silt has been produced from chipping of the edges of the sand.

Although other rocks and minerals, and other processes, contribute to particle generation, we can conclude that on Earth most windblown sand and dust are quartz grains weathered from granites. Once released, the grains may be further reduced in size by abrasion, salt weathering, or crushing by glaciers, mass movements, etc. Can we draw the same conclusion for the other planets subject to aeolian processes? Probably not, for at least two reasons: first, quartz-containing rocks may not be present and, second, the different surface environments probably result in different – or at least differently expressed – weathering processes.

Krinsley *et al.* (1979) and Smalley & Krinsley (1979) considered aeolian abrasion and sedimentation on Mars in comparison with Earth and concluded that Mars is probably quite different. With the apparent absence of granite on Mars, they conclude that quartz sand is unlikely to be present. There are, however, several lines of evidence to point toward the existence of sand-size particles on Mars. The numerous dunes (Figs. 1.15 and 1.16) indicate that sand grains are present; there is nothing inherently different about Mars that would cause the physics of particle movement and dune formation to be substantially different from that on Earth. In addition, estimates of grain sizes derived from measurements of thermal inertias from the Viking Orbiters suggest that sand-size grains are present in a great many areas on Mars (Peterfreund, 1981). And, finally, observations at the Viking Lander sites suggest that sand-size grains are present (Moore *et al.*, 1977, 1979; and others).

Although the presence of sand on Mars can be established, it is much more difficult to state the composition and characteristics of the particles. Most investigators feel that the grains could be comminuted basalt; certainly the darkness of the north polar dunes could support this contention (Tsoar *et al.*, 1979).

It is relatively easy to summarize the knowledge of the characteristics of aeolian particles on Venus because so little is known. As reviewed by McGill *et al.* (1983) and Gooding (1983), weathering is probably dominated by chemical processes, although thermally driven exfoliation may also be important. Abrasion in fluid transport is probably non-existent – there is no liquid water, so grains are not subjected to any part of the water cycle at present, and wind speeds appear to be barely above threshold – thus, the collisional energetics would be at a minimum.

Explosive volcanism can lead to the production of sand and dust on Earth; however, if explosive volcanism has occurred on Venus, Head & Wilson (1981) predict that the explosive velocities would be reduced by a factor of between 2 and 4 in comparison to Earth, primarily because of the higher atmospheric pressure at the surface. Thus, fragmentation of volcanic rocks would also be less than on Earth. Cataclystic processes associated with impact cratering undoubtedly generated particles early in the history of Venus, but it is not possible to determine whether or not the particles would have survived to the present in a form amenable to aeolian processes.

3

Physics of particle motion

3.1 Introduction

Many images of Mars show sand dunes and wind-related surface features that are clear evidence of atmospheric processes. These remarkable pictures have sparked new interest in understanding the complex phenomena associated with windblown particles, which in turn has led to increased knowledge and understanding of aeolian processes on Earth. In this chapter we discuss some of the basic physics of particle motion, particularly in regard to the effects of widely different atmospheric densities on the terrestrial planets. Most of these phenomena, such as the characteristics which determine onset of motion and the transport rate of material in motion, are not yet completely understood because of their physical complexity and the difficulty of observation.

Much of the early research on the basic physics of windblown sand was performed by R. A. Bagnold. As an officer in the British Army, he led expeditions across the sand seas of Egypt in the 1930s. His observations provided the foundation for understanding sand motion and the formation of sand dunes and ripples. Later, in England, he built a wind tunnel in order to make quantitative measurements of threshold wind speeds and mass transport rates. His findings are reported in his delightfully lucid cornerstone book (Bagnold, 1941) and in many papers.

A parallel effort on the problems of soil erosion has been carried out by the US Department of Agriculture at a wind tunnel laboratory in Manhattan, Kansas. Most of the understanding of the physics of soil movement by wind, especially as related to agricultural applications, was derived by W. S. Chepil who reported his work in a large number of papers during a 25-year period beginning about 1940. The description of the physics of blowing snow has evolved from the work of many researchers around the world, including those in the Soviet Union (notably A. K. Dyunin) and at the US Army Corps of Engineers laboratory in New Hampshire.

3.2 Classification of motion

From wind tunnel studies and field observations in Egypt, Bagnold defined three modes of particle movement by the wind: suspension, saltation, and creep or surface traction (Fig. 1.12). In order to define the boundaries between these modes of transportation, it is necessary to introduce the concepts of terminal speed, U_F and threshold friction speed, u_{*_t}. The forces on a particle falling through quiescent air are primarily those of weight in the direction of motion and of aerodynamic drag opposing the motion, as illustrated in Fig. 3.1. A particle dropped from rest will accelerate due to the gravitational field until the drag force increases to equal the weight (for small particles or for a fluid of high viscosity, the drag is proportional to the speed of the particle; for large particles or low viscosity, the drag increases with the square of particle speed). The expression for terminal speed, found by equating weight and drag force, is

$$U_F = (4 \rho_p g D_p / 3 \rho C_D)^{1/2} \tag{3.1}$$

where ρ_p is particle density, D_p is particle diameter, ρ is atmospheric density, and the drag coefficient, C_D, is a function of the Reynolds number, $U_F D_p / \nu$. For Reynolds numbers of 0.1 and less, C_D for a sphere is equal to 24 divided by the Reynolds number (the so-called Stokes flow). If the gas is

Fig. 3.1. Sketch illustrates the primary forces acting on the spherical particle on its downward flight through the atmosphere, i.e., weight acting downward in the direction of motion and aerodynamic drag opposing the motion. When these two forces are equal and opposite, the particle is moving at its terminal speed downward, relative to the field.

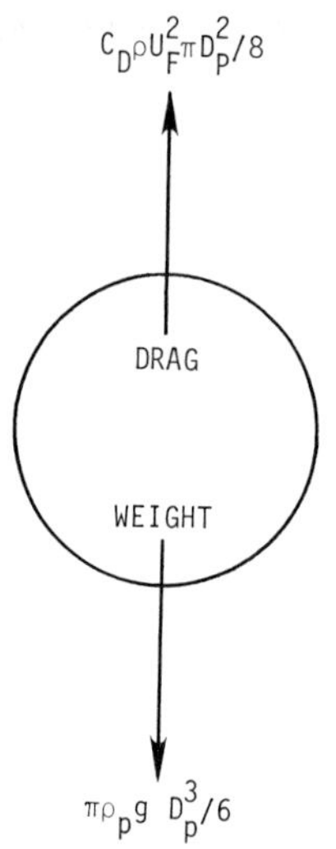

of very low density so that the ratio of molecular mean-free-path-to-particle diameter is of the order unity, the drag coefficient is also a function of that ratio, called the *Knudsen number*.

The friction speed, as mentioned in Chapter 2, is not a true wind speed but is equal to the square root of the ratio surface shear stress, τ, to atmospheric density, ρ. The value of the vertical component of fluctuating turbulent speed near the surface is approximately equal to the surface friction speed, u_*. Thus, if the terminal speed, U_F, is less than the vertical component of turbulence, the turbulent eddies are capable of transporting the particles upward. This type of motion is called *suspension*.

However, if the terminal speed is significantly larger than the vertical component of turbulence (or friction speed, u_*), the particle trajectory is relatively unaffected by turbulence, the trajectory path is a much smoother curve and the trajectory height is of the order of centimeters, much less than for fine particles in suspension, which can reach heights of the order of kilometers. The larger particles travelling in low, smooth trajectories are said to be moving in *saltation*.

Finally, if particles are too large or heavy to be lifted from the surface by the wind, but not too large to be rolled or pushed along the surface by the impact of grains in saltation, the motion is called *creep*.

From this discussion, one can see that suspension grades into saltation. However, a boundary (albeit somewhat indistinct) between suspension and saltation can be made by setting $U_F/u_* = 1$ (Iversen *et al.*, 1976*b*). Below a certain value of u_*, particles of a given diameter are not capable of being set in motion by the wind. The lowest value of u_* at which particles begin to move is called the *static threshold friction speed*, u_{*_t}. Thus the boundary, U_F/u_*, can be written

$$\frac{U_F}{u_*} = \frac{U_F}{u_{*_t}} \Bigg/ \frac{u_*}{u_{*_t}} = 1 \tag{3.2}$$

The suspension–saltation boundary is thus a function of the ratio of friction speed to that at threshold, u_*/u_{*_t}, and the ratio of terminal speed to threshold friction speed, U_F/u_{*_t}. The latter term is a function of the particle properties and the fluid medium and is plotted as a function of particle diameter for the planets Venus, Earth, and Mars, in Fig. 3.2. The values of particle diameter for the saltation–suspension boundary at threshold are 30 μm, 52 μm, and 210 μm for Venus, Earth, and Mars respectively.

Bagnold found the very finest dune sands to have an average diameter of about 80 μm or, more typically, larger, while percentages of particles less than 50 μm are very small. It is probable that most of the finer particles go into suspension and are carried far downwind of a sand dune area. The dune sands on Mars probably consist of larger particles than the average size on Earth, since particles less than 200 μm appear to be susceptible to suspension. If there are sand dunes on Venus, however, they would most likely contain particles down to the size of about 30 μm.

3.3 Threshold of motion

Watching grains of sand through a telescope at the onset of their motion by the wind is a fascinating experience. As the wind speed increases to just under the threshold value, the particle begins to oscillate; as the speed increases, the particle takes off suddenly in a vertical, or nearly vertical, direction.

Fig. 3.2. Ratio of terminal speed, U_F, to threshold friction speed, u_{*t}, for Venus, Earth, and Mars, as a function of particle diameter. The values assumed for the calculations for Venus, Earth, and Mars respectively; density ratio $\rho_p/\rho = 41, 2160, 240\,000$; kinematic viscosity $= 0.0044, 0.146$, and 11.19 cm^2/sec; gravitational acceleration $= 877, 981, 375$ cm/sec^2.

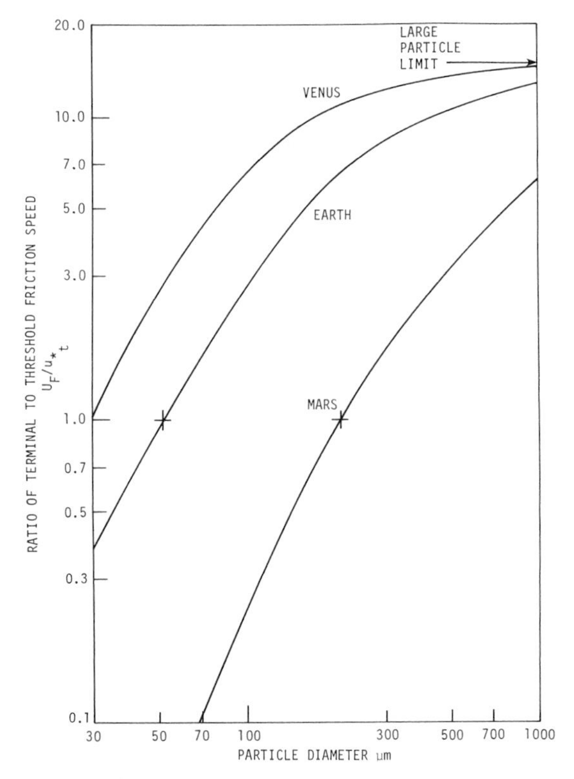

Bagnold defined two levels of threshold, one for 'static' conditions and one for actively saltating grains. In the latter case, if particles are introduced from upstream, continuous movement of particles from the initially quiescent surface begins at a lower wind speed. This is called the *impact threshold*, which is about 80% of the static value. The static threshold – defined as the wind speed at which continuous motion starts without impact from upwind (Eq. (3.2)) – is the threshold condition discussed in most of the rest of this chapter.

3.3.1 *Particle forces*

Fig. 3.3 depicts a loose grain (drawn as a perfect sphere) at rest on top of a bed of similar particles. The forces on the particle include drag and lift forces, D and L, moment, M, weight, W_t, and interparticle force (cohesion), I_p. The lift and drag forces and moment are aerodynamic in origin, caused by the viscous fluid flowing over the particle. Despite many years of study, these forces have not been determined accurately because of the difficulty of measurement. The various cohesive forces are understood even less.

At threshold, the particle forces are assumed to be in equilibrium about point p. Any slight increase in wind speed will cause the particle to lift from the surface. The moment arms, a, b, and c, are shown in Fig. 3.3. Thus the

Fig. 3.3. Schematic of an erodible spherical particle resting on other like particles. Forces include aerodynamic drag, D, lift, L, moment, M, interparticle force, I_p, and weight, W_t.

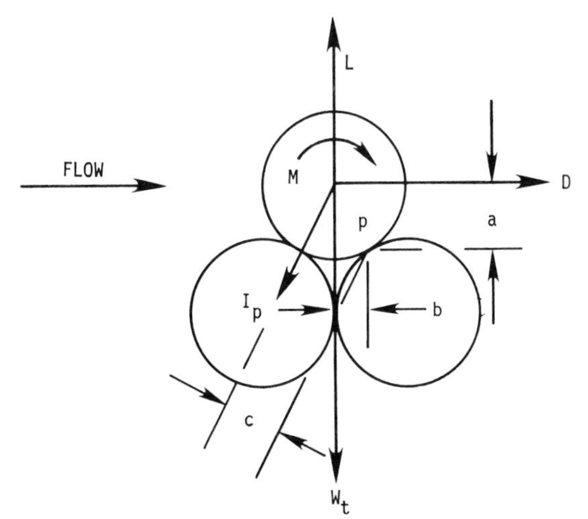

equation of equilibrium, obtained by summation of moments about the downstream point of contact, p, is

$$Da + Lb + M = W_t\, b + I_p\, c \tag{3.3}$$

Although some investigators have ignored the effect of particle lift in attempting to predict threshold friction speed, the effect of lift is apparently important. Bagnold (1956, 1973) has stated that the initial upward acceleration of a particle (not struck by another particle) from the bed is due to a lift force, but that the lift force becomes very small shortly after the particle has lifted off the bed. The lift force is due to the extremely high mean velocity gradient very near the surface. The friction Reynolds number $R_* = u_* D_p/\nu$, where D_p is the particle diameter and ν the kinematic viscosity. If R_* is less than 5, a laminar sublayer exists and the velocity profile within it is as given by Eq. (2.9). The shear, $\partial U/\partial z$, is uniform and equal to u_*^2/ν. The aerodynamic drag force and moment for a sphere resting on a plane surface and in a uniform shear have been derived by Goldman *et al.* (1967) and O'Neill (1968) for $R_* < 0.45$. A similar approximation for lift was found by Saffman (1965, 1968), also for very low friction Reynolds numbers.

Experimental values of lift on hemispheres lying on a surface and immersed in a fluid were obtained by Chepil (1958) and Einstein & El-Samni (1949). In both cases the lift and drag forces were found to be of the same order of magnitude. However, the measurements of lift and drag in both cases were made at rather high Reynolds numbers in fully turbulent flow. Measurements of drag force on a sphere resting on a bed of similar spheres were conducted through quite a large range of Reynolds numbers by Coleman (1972) and Coleman & Ellis (1976*a*, *b*). The preceding results for experimental and theoretical lift and drag forces can be compared by defining the forces in terms of dimensionless coefficients, as, for example,

$$\begin{aligned} D &= K_D\, \rho\, u_*^2\, D_p^2 \\ L &= K_L\, \rho\, u_*^2\, D_p^2 \\ M &= K_M\, \rho\, u_*^2\, D_p^3 \end{aligned} \tag{3.4}$$

The values of the lift, drag and moment coefficients, K_D, K_L, and K_M, are summarized in Table 3.1.

The discrepancy for larger Reynolds numbers between Coleman's drag coefficient and Chepil's – which is much less – is probably due to the fact that Coleman measured total force with a force transducer while Chepil

Table 3.1. *Experimental and theoretical values of small particle shear flow force coefficients*

Investigator	Drag coefficient (K_D)	Lift coefficient (K_L)	Moment coefficient (K_M)	Reynolds number (R_*)	Fluid medium
Goldman *et al.* (1967) and O'Neill (1968)	8.01		0.74	<0.45	Theoretical
Saffman (1965, 1968)		$0.808\ R_*$		<0.45	Theoretical
Coleman & Ellis (1976*b*)	5.44			0.95	Hydroxyethyl-cellulose solution
Coleman (1972) and Coleman & Ellis (1976*a*)	15.42			130–13 200	Water
Einstein & El-Samni (1949)		2.42		3600	Water
Chepil (1958)	3–4.7	2.2–5		1000–1400	Air

measured only pressure drag. The values of the coefficients in Table 3.1 are compared with those deduced from threshold experiments in Section 3.3.2.

Interparticle forces may be due to moisture, electrostatic forces, and other forces of cohesion which are known to be relatively more important for small particles and relatively independent of particle density (Iversen *et al.*, 1976*a*). Particles of any solid, if small enough, cohere on contact, even when thoroughly dry, and particularly well in a vacuum. Electrostatic forces appear to be particularly important for grains in motion on both Earth and Mars (Greeley & Leach, 1978).

The angle of repose for a group of particles is one indicator of the effect of cohesive forces. For ordinary dune sand, the angle of repose is about 34° from the horizontal. For small particles, it can be much higher and can even approach the vertical. Snow particles are sufficiently light so that negative slopes can be formed, due to electrostatic effects and to their sintering capability (i.e., the ability of the particles to form solid bonds by freezing).

3.3.2 *Theoretical expressions for threshold*

The moment arm lengths in Fig. 3.3 are, of course, proportional to particle size, i.e., $a = a_1 D_p$, $b = b_1 D_p$ and $c = c_1 D_p$. Thus, if the equations of Eq. (3.4) are substituted into Eq. (3.3), the threshold coefficient, A, becomes

$$A^2 = \frac{u_{*_t}{}^2 \rho}{\rho_p g D_p} = \frac{(\pi b_1/6)[1 + (6c_1/\pi b_1) I_p / \rho_p g D_p{}^3]}{K_D\, a_1 + K_M + b_1\, K_L} \tag{3.5}$$

Coleman (1967) derived expressions for a_1 and b_1 (from which c_1 can be derived) for a sphere resting on a closely packed bed of like spheres for geometry resulting in minimum threshold. Substituting his expressions into Eq. (3.5) and letting the cohesive force, I_p, be zero results in an expression for a cohesionless threshold coefficient

$$A_1{}^2 = (\pi/24) \,/\, [K_D/(2)^{1/2} + (3)^{1/2}\, K_M + K_L/4] \tag{3.6}$$

Experimental threshold values were obtained for a large range of values of atmospheric density (Greeley *et al.*, 1976, 1980*a*). Cohesionless threshold values were extrapolated from these results (Iversen & White, 1982) and for values of $R_{*_t} = u_{*_t} D_p/\nu$ between 0.03 and 0.3, Eq. (3.6) was curve fit to the data by linear regression. The result is illustrated in Fig. 3.4. The coefficients found by curve fitting are

$$\begin{aligned} K_D + (6)^{1/2}\, K_M &= 4.65 \\ K_L &= 32.8\, R_{*_t} \end{aligned} \tag{3.7}$$

The drag parameter value, 4.65, compares favorably with Coleman's direct measurement, 5.44, and the theoretical value of 9.82. The lift parameter value, however, is 40 times Saffman's theoretical result, and the deduced value of lift force is thus of the same order of magnitude as the drag force, perhaps because of shielding of the lower half of the particle, not accounted for in the theory. If the lift force were truly negligible, as Saffman's result indicates, the curve in Fig. 3.4 would be horizontal, which it appears to be approaching only at the lower limit of friction Reynolds number, R_{*_t}. At large R_{*_t} (>70), when Coleman's and Chepil's drag and lift values of 15.42 and 5 are substituted into Eq. (3.6), the threshold coefficient, A^2, is 0.0108. This value is within experimental results for large particle threshold.

Chepil's analysis (1959) is similar to the above except that the lift and drag forces are reinforced by so-called turbulence and packing factors and that the cohesive force is not included. His analysis is thus strictly appropriate only for large particles. Chepil used his experiments involving large Reynolds numbers to determine the ratio of lift-to-drag force (0.85) on the particle. Again equating moments about the contact point, p,

$$D = 0.445\left(\frac{\pi}{6}\rho_p\, g\, D_p^3 - L\right) \tag{3.8}$$

where the number 0.445 is derived from the empirically determined vertical location of the drag force (equivalent to determining the ratio of moment-to-drag factors, K_M/K_D). Next, the lift is assumed to be 85% of the drag, and both forces are reduced by the ratio of packing, η_p, to turbulence,

Fig. 3.4. Cohesionless friction speed parameter (A_1^2 of Eqs. (3.6) and (3.11)) versus Reynolds number. Data adapted from MARSWIT (Greeley *et al.*, 1980*a*; Iversen & White, 1982). This diagram shows how threshold data would appear if cohesion forces did not exist.

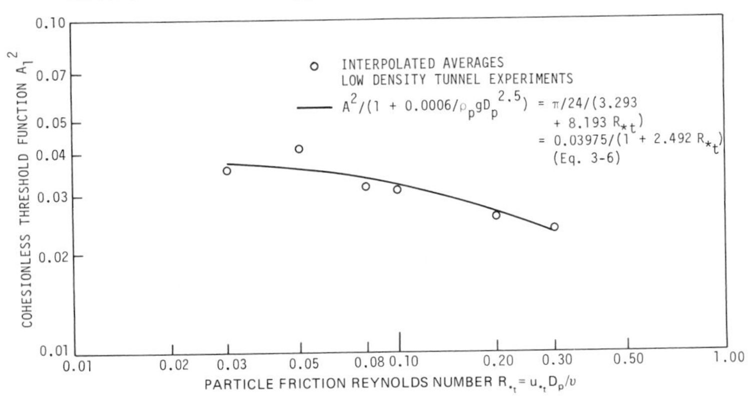

T_t, factors. The packing factor, η_p, is the ratio of aerodynamic force taken by the top grains to that for the whole bed, and the turbulence factor, T_t, accounts for the fact that the maximum instantaneous shear stress is greater than the average value. The drag is assumed equal to the surface shear stress, ρu_*^2, times projected area, $\pi D_p^2/4$ (factor of 20 less than Coleman's measured value, see Table 3.1). Thus

$$u_{*t}^2 = \frac{[0.445(2/3)\eta_p]\ \rho_p g\ D_p}{[1+0.85\ (0.445)]T_t\rho} \tag{3.9}$$

With the packing factor equal to Chepil's value of 0.21 and turbulence factor equal to his 2.5, the Bagnold coefficient, A, becomes 0.134, somewhat higher than the usually accepted value for large particles in air.

In introducing his turbulence factor, T_t, Chepil assumes that initial movement of the particle is facilitated by the maximum impulse of lift and drag forces caused by turbulence. The analysis of Dyunin (1954*a*) is also based on the concept of initial particle motion because of the presence of peak turbulent fluctuations. His analysis, however, is completely different from the foregoing. The turbulence impulse is represented by a solution to the viscous equations of motion, called the elliptical vortex, assumed to represent a turbulent eddy. The pressure difference, which is assumed to cause the particle motion, is the sum of two terms, one proportional to wind speed squared and a viscous term independent of wind speed. The result, in terms of wind speed at 1 m height, rather than friction speed at threshold, is

$$U_t = \{3.5\ g\ (\rho_p-\rho)\ D_p/\rho + 550\nu\ [g(\rho_p-\rho)/\rho_p D_p]^{1/2}\}^{1/2} \tag{3.10}$$

The interesting facet of this equation is that it predicts a minimum threshold speed at an optimum diameter, as does an equation of the form of Eq. (3.5), arrived at in a quite different manner. Agreement of Eq. (3.10) with experimental values of threshold, however, is not good, even with altered values of the numerical coefficients, which Dyunin obtained empirically. It can thus be concluded that the cohesive forces are not primarily due to viscosity, as Dyunin's analysis assumes.

3.3.3 *Wind tunnel experiments*

Several wind tunnels have been constructed to investigate the effects of wind on sand and soil movement. The primary facility for early investigation of the threshold of motion for sand particles was built by R. A. Bagnold (1941). He derived an empirical curve for the dimensionless threshold friction speed, $A = u_{*t}\ (\rho/\rho_p g\ D_p)^{1/2}$, as a function of friction

Reynolds number, $R_{*_t} = u_{*_t} D_p/\nu$, by determining the friction speed at threshold for sand particles of various average diameters.

Another facility used mostly for studying the wind erosion of soil is described by Zingg & Chepil (1950) (see also Chepil & Woodruff, 1963; and Lyles & Krauss, 1971).

Experimental threshold results obtained for materials of different densities in a one-atmosphere environment are illustrated in Fig. 3.5. The existence of the optimum diameter for minimum threshold is clearly illustrated, as is the fact that for large particles the threshold parameter, A, is essentially constant (if it were exactly constant, however, all data points would lie on a straight line through the origin). Fig. 3.6 shows the data plotted in dimensionless form and compared with other investigators' results; agreement is satisfactory for friction Reynolds number, $R_{*_t} \geqslant 3$. The scatter for smaller Reynolds numbers is due to forces of cohesion (causing A not to be a function only of R_{*_t}), due to differences in particle-size distribution, and due to the difficulty in measuring threshold naturally. The data in Fig. 3.5 were obtained in the wind tunnel shown in Fig. 3.7(a).

Fig. 3.5. Threshold friction speed at one atmosphere. (From Iversen *et al*., 1976*a*.)

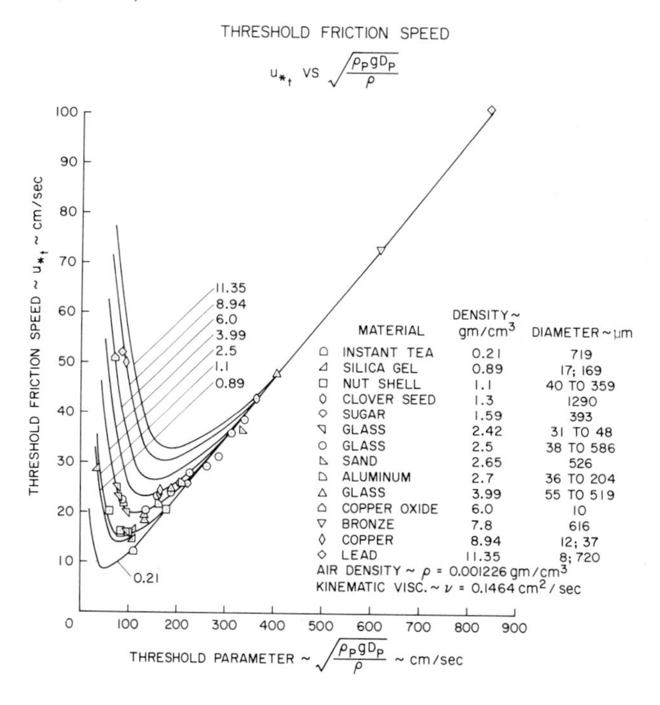

Fig. 3.6. Threshold friction speed parameter versus friction Reynolds number. Data points are from Iversen *et al.* (1976*a*), compared to the data curves of Bagnold (1941), Chepil (1945, 1959), and Zingg (1953). Chepil's two curves are for different particle-size distributions.

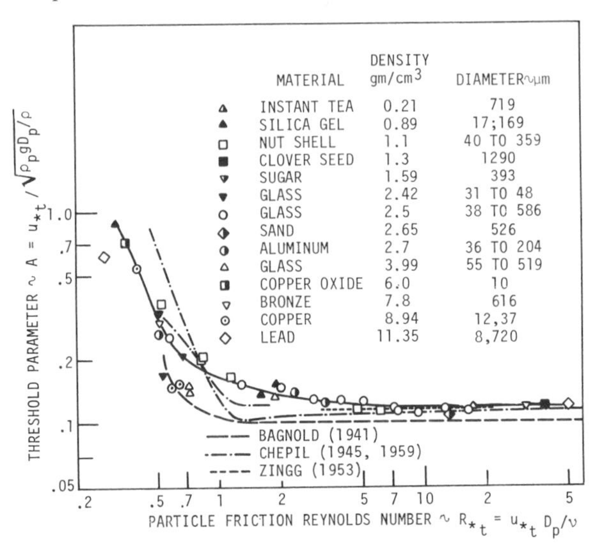

Fig. 3.7. (*a*) The Iowa State University boundary-layer wind tunnel located at Ames, Iowa. The test-section is 1.2 m by 1.2 m in cross-section. The wind tunnel is 19.5 m long and is powered by a 75 kw

(a)

The MARSWIT (Mars Surface Wind Tunnel), illustrated in Fig. 3.7(*b*) (Greeley *et al.*, 1981), was built for the purpose of studying particle motion at fluid densities appropriate for Mars. A typical set of data is shown in Fig. 3.8 where the threshold parameter, A, is plotted as a function of particle friction Reynolds number, since they are plotted for a single value of

Caption for Fig. 3.7 (cont.)

electric motor with variable-pitch fan. A baffled storage area is located downwind of the test-section to remove particles from the stream. Maximum speed is about 23 m/sec. (*b*) The MARSWIT (Martian Surface Wind Tunnel) is located at the NASA Ames Research Center, Moffett Field, California. The wind tunnel test section cross-section area is 1.1 m^2 and the tunnel length is 14 m. The tunnel is located in an environmental chamber which is capable of attaining atmospheric pressures from one atmosphere (1013 mb) down to 3.8 mb. The tunnel is powered by a high-pressure nozzle-ejector system and has a maximum speed capability of 13 m/sec at one atmosphere, increasing to 180 m/sec at 5 mb pressure.

(*b*)

particle diameter (154 μm). The experiments were run by measuring threshold speed for stepwise increases in ambient pressure. As the pressure is increased, the threshold friction Reynolds number, R_{*_t}, increased, causing corresponding changes in A. The increase in R_{*_t} is due to a greater reduction in viscosity, v, than in threshold friction speed, u_{*_t}, as ambient pressure increases.

The threshold speed data were analyzed in the following manner (Iversen & White, 1982): curves were plotted as in Fig. 3.8 for each of 21 test runs (eight particle diameters, 37–673 μm; two fluids, air and carbon dioxide; and two particle densities, 1100 and 2650 kg/m³). It was first assumed that the form of the threshold equation is

$$A = A_1 f(R_{*_t}) (1 + K_1/\rho_p g D_p{}^n)^{1/2} \tag{3.11}$$

which is a generalization of Eq. (3.5) in terms of friction Reynolds number and which assumes that the interparticle force, I_p, is proportional to particle diameter to the exponent, $3-n$. Values of A were interpolated from the experimental curves for constant values of friction Reynolds number, R_{*_t}, for which the product [A_1] times [$f(R_{*_t})$] would be constant. Fig. 3.9 illustrates these cross-plots for five values of R_{*_t}. If the threshold parameter, A, were a function only of R_{*_t}, as some investigators have

Fig. 3.8. Threshold parameter, A, as a function of particle friction Reynolds number, R_{*_t}, from Iversen & White (1982). The data (MARSWIT, Greeley *et al.*, 1980*a*) are for a constant value of particle diameter and density (154 μm sand), but with differing values of fluid density. The curve is Eq. (3.12).

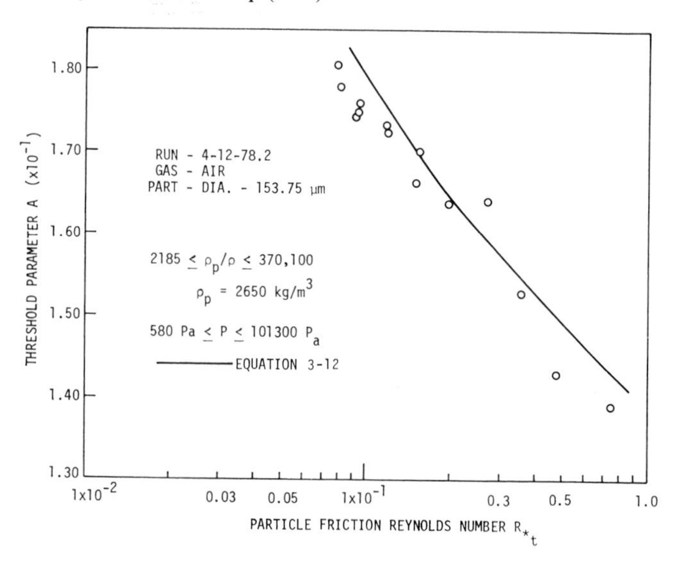

assumed, then the curves in Fig. 3.9 would have to be horizontal. The rapid increase in A^2 as diameter decreases below 80 μm demonstrates the powerful effect of cohesive forces for small particles. If the exponent, n, in Eq. (3.11) is positive, then for constant R_{*_t}, A should approach an asymptotic limit as particle diameter becomes large. This appears to be the case in Fig. 3.9. Linear regression analysis was used to determine best fits for the coefficient, K_1, and exponent, n. The results are $K_1 = 0.006\ g/(\mathrm{cm}^{0.5}\ \mathrm{sec}^{-2})$ and $n = 2.5$. The semiempirical equations obtained from the entire set of data are

$$A = 0.2\ (1+0.006/\rho_p g\ D_p^{2.5})^{1/2}/(1+2.5\ R_{*_t})^{1/2}$$

for $0.03 \leqslant R_{*_t} \leqslant 0.3$

$$A = 0.129\ (1+0.006/\rho_p g\ D_p^{2.5})^{1/2}/(1.928\ R_{*_t}^{0.092}-1)^{1/2} \tag{3.12}$$

for $0.3 \leqslant R_{*_t} \leqslant 10$

$$A = 0.120\ (1+0.006/\rho_p g\ D_p^{2.5})^{1/2}\{1-0.0858\ \exp[-0.0617(R_{*_t}-10)]\}$$

for $10 \leqslant R_{*_t}$

Fig. 3.9. Threshold parameter, A^2, as a function of particle diameter, D_p, from Iversen & White (1982). The data (MARSWIT, Greeley *et al.*, 1980*a*) show clearly that A^2 is not a function of the Reynolds number alone. The curves are Eq. (3.12).

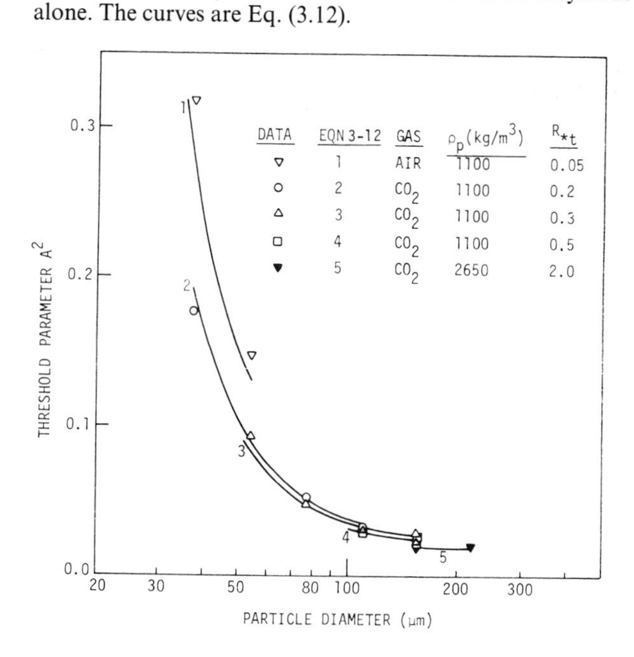

For particles of size between 50 and 600 μm, Eq. (3.12) gives essentially the same result as those of Iversen *et al.* (1976*b*).

3.3.4 *Roughness effects*

The effects of non-erodible roughness on threshold speed can be classified in two ways. For the first, let us consider sand and dust mixed with pebbles or other large particles which are too big to be moved (i.e., non-erodible) by a given wind. In soils or sands containing both erodible and non-erodible fractions, the amount of material removed by the wind is limited by the size and number of non-erodible particles left exposed on the surface.

When first exposed to a strong wind, removal of material will continue, while the height of non-erodible projections above the surface and their number per unit area will increase until all the erodible grains are sheltered. The rate of removal of the grains decreases with time until removal ceases, as shown by the curves in Fig. 3.10. If there are enough non-erodible pebbles, the surface eventually consists only of pebbles which protect the smaller sizes lying underneath. Fig. 3.11 illustrates the apportionment of surface shear stress between the sand grains and the non-erodible pebbles.

In the desert, this type of pebble-covered surface is termed desert pavement or desert floor. If the non-erodible pebbles are of different color than the sand grains, fascinating surface patterns can occur. Fig. 3.12 illustrates an approximation to the two-dimensional cavity discussed in

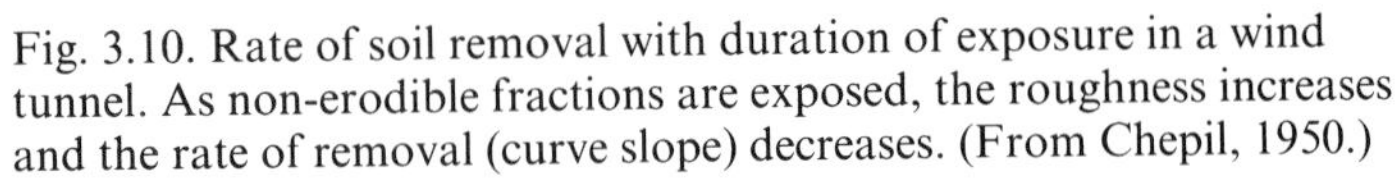

Fig. 3.10. Rate of soil removal with duration of exposure in a wind tunnel. As non-erodible fractions are exposed, the roughness increases and the rate of removal (curve slope) decreases. (From Chepil, 1950.)

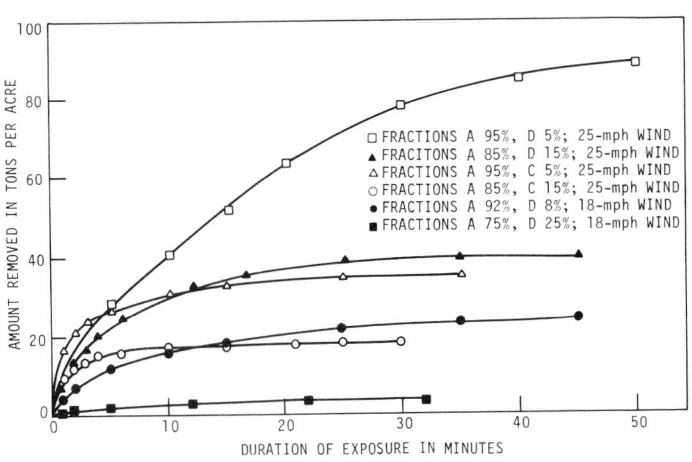

Chapter 6. The wind direction in Fig. 3.12(*b*) is from left to right. A roller vortex is formed within the separated flow region within the trough, which scours the right side of the trough, desposits material on the left side, and moves sand in the direction opposite to that of the prevailing wind. Of special interest is the sharp distinction between the dark and light portions of the trough. The dark strip, which is the high surface stress region, exists because of the basalt pebbles lying on top of the lighter colored sand.

The second kind of non-erodible roughness element is one which is 'permanently' in place, as, for example, a geometrical array of large rocks among which is distributed erodible particulate material (such as at the Viking Lander sites). The most comprehensive set of wind tunnel experiments to determine the effect of this kind of roughness element was conducted by Lyles *et al.* (1974), although there are additional data available in Iversen *et al.* (1973). Lyles and his colleagues reported data for

Fig. 3.11. Surface shear stress as a function of ratio of height, H_r, to spacing distance, L_x, of non-erodible roughness elements. Roughness increases with H_r/L_x. As roughness increases, not only does total surface stress increase, but a greater percentage is taken up by the non-erodible elements, leaving less stress to move the erodible material. (From Lyles, 1977.)

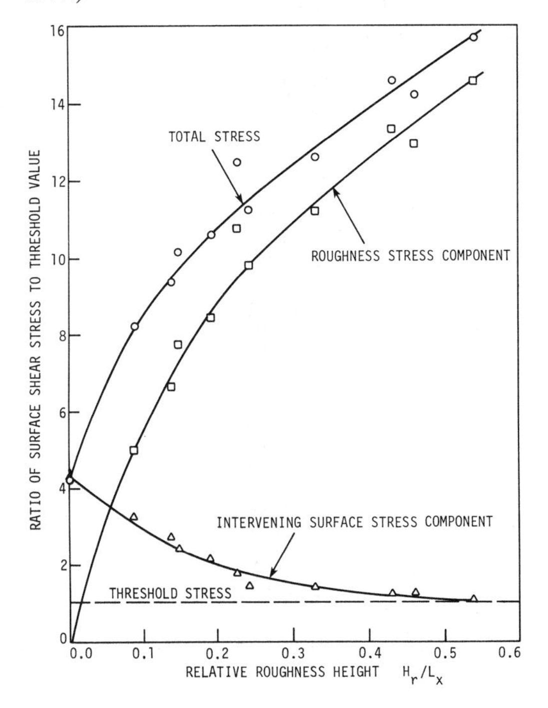

Fig. 3.12. (*a*) Desert pavement in the Amboy lava flow area, Mojave Desert. The light-colored sand is covered by larger, darker lava pebbles in the regions of high surface shear stress. (*b*) Lava channel in the Amboy lava flow, Mojave Desert. Prevailing wind is left to right. The very dark channel levees are solid basalt. Note that the right side of the channel floor is dark, indicating high surface shear stress where the light-colored sand has been removed, leaving the lava pebbles. A separated flow region is formed within the channel, with upwind flow near the channel floor, depositing material on the left side of the floor. (From Iversen & Greeley, 1978.)

(a)

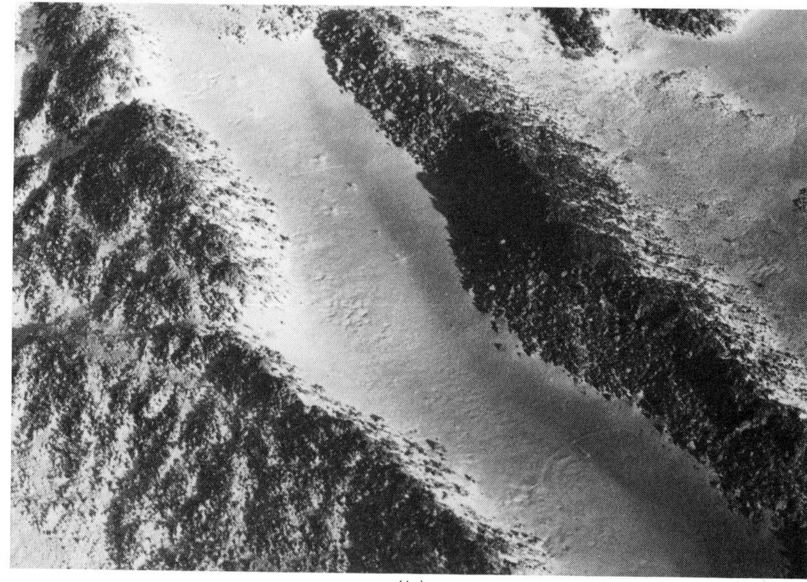

(b)

threshold friction speed as a function of the ratio of roughness height to roughness element spacing (geometrical arrays of cylinders of spheres) and of ratio of area covered by the roughness elements to total surface area. They present equations for these functional relationships, but another possibility is to relate the increase in threshold speed to the ratio of particle diameter to equivalent roughness height, D_p/z_o. An approximate fit to their data can be written

$$u_{*_{tR}}/u_{*_t} = 2(D_p/z_o)^{-1/5} \tag{3.13}$$

where the subscript, R, refers to rough surface threshold. This equation should not be construed as having universal validity as it has not been tested for other kinds of non-erodible roughness elements such as desert pavements. The equation would be valid only for $D_p/z_o \leqslant 30$, at which limit the friction speed ratio is equal to one corresponding to the erodible surface roughness only ($z_o = D_p/30$), i.e., without the presence of non-erodible elements.

The ideal of deployment of arrays of non-erodible roughness elements is of practical value for preventing soil erosion (Lyles, 1977) and for trapping drifting snow or sand (Tabler & Jairell, 1980).

3.3.5 *Effects of moisture and other cohesive forces*

It is more difficult for the wind to pick up wet sand than dry sand, but the moisture effects on soils are even more noticeable as sand can be dried more rapidly by the wind. The effect of moisture on the threshold speed of sand has been investigated both in nature and in the wind tunnel, as shown in Table 3.2.

There are additional bonding agents besides moisture. For example, Nickling & Ecclestone (1981) show the effects of bonding by precipitation of dissolved salts in drying sands. Schmidt (1980) analyzed the increase in threshold friction speed for snow particles due to their sintering, and Gillette *et al.* (1982) have measured the effect of crusting on the threshold friction speeds for some desert soils.

3.3.6 *Vortex threshold*

Examination of the threshold curve (Fig. 1.2) shows that particles smaller than about 80 μm require progressively stronger wind speeds to move, as the particle diameter becomes smaller. Yet dust storms on Earth and Mars involve small particles. On Mars, wind speeds in excess of the speed of sound (250 m/sec) would be required for particles in the size range estimated for martian storms. Since such strong winds are inconceivable

Table 3.2. *Effect of moisture on threshold friction speed*

Moisture content (% volume water)	Threshold speed u_{*_t} (cm/sec)	
	Wind tunnel, Belly (1964) (D_p = 400 μm)	Natural sand beach, Svasek & Terwindt (1974) (D_p = 250 μm)
Dry	23	19
0.1	35	20
0.3	42	42
0.6	47	55

(nor have any been measured by the meteorology instruments aboard the Viking Landers), we can ask if there might be other mechanisms by which very small particles are lifted from the surface.

One such mechanism is saltation impact (Fig. 1.12), in which large grains dislodge smaller grains or create them by abrasion, spraying them into the boundary layer where turbulence can transport them upward. Another mechanism is the dust devil. The dust devil, while it does not account for a large portion of the dust injected into the atmosphere on Earth, has been suggested by Sagan and colleagues (1971) to be a possible mechanism for entrainment of dust and sand particles on Mars and, therefore, to be a possible cause of dust storms on the 'Red Planet'.

Experiments were performed at Iowa State University (reported in Greeley *et al.*, 1981) to determine the strength of such vortices necessary to lift particles from the surface. A vortex generator (Hsu & Fattahi, 1976) designed to simulate the dust devil, or tornado vortex, was used to perform the experiments. The vortex, which is formed so that its axis is perpendicular to the ground, is produced by rotating a honeycomb disk situated within a stationary nozzle which is located some distance above the surface. The vortex formed is an approximate small-scale model of atmospheric vortices such as the tornado or dust devil.

The swirling motion due to the atmospheric vortex (dust devil) causes a radial pressure gradient at the surface:

$$\frac{\mathrm{d}p}{\mathrm{d}r} = \frac{\rho u_\theta^{\,2}}{r} \propto \rho r \omega_0^{\,2}$$

or

$$\Delta p = \int \frac{\rho u_\theta^{\,2}}{r}\,\mathrm{d}r \propto \rho r_0^{\,2} \omega_0^{\,2}$$

or

$$r_0 = (\Delta p/\rho\omega_0^2)^{1/2} \tag{3.14}$$

where p is pressure, r is radial distance from the vortex center, ρ is air density, u_θ is tangential (swirl) speed, ω_0 is the angular speed of the generator honeycomb, and the characteristic radius, r_0, is as defined in Eq. (3.14) where Δp is the maximum surface pressure difference from the center of the vortex to that at large radius.

The maximum pressure difference, Δp, was measured on the surface of the vortex generator with a differential pressure transducer for the range of angular speeds, ω_0, of the generator. The values of characteristic radius, r_0, were calculated from the measured values of Δp. The rotor diameter of the generator used was 46 cm, and the calculated values of characteristic radius, r_0, varied from 6.9 to 9.5 cm.

Let us assume that the top layer of particles of thickness equal to particle diameter, D_p, is lifted by the vortex at threshold. The forces acting on this layer per unit area are then $a_p\Delta p$ and τ upward and σ_I and $\rho_p g D_p$ downward, where a_p is the effective fraction of the pressure difference, Δp, σ_I is the stress due to interparticle force, τ is an upward normal stress due to the effects of viscosity and $\rho_p g D_p$ is just equal to the weight of the particle layer per unit area. If the surface stress, τ, is defined as $\tau = C_\tau \rho r_0^2 \omega_0^2$, then the value of $r_0\omega_0$ at threshold, where forces are in equilibrium, can be written as

$$r_0\omega_0 = [1/(a_p + C_\tau)^{1/2}]\,(\rho_p g D_p/\rho)^{1/2}\,(1 + \sigma_I/\rho_p g D_p)^{1/2} \tag{3.15}$$

The characteristic threshold speed, $r_0\omega_0$, is not a true speed but is proportional to the square root of the pressure difference, Δp, i.e., $r_0\omega_0 = (\Delta p/\rho)^{1/2}$. However, the value of $r_0\omega_0$ should be representative of the maximum wind speed within the vortex very near the surface. Unfortunately, neither the function, $\sigma_I\,(D_p)$, nor the function, $C_\tau(r_0^2\omega_0/v)$, nor the ratio, a_p, are known and sufficient data to determine these functions empirically are not available. The values of $r_0\omega_0$ at threshold were measured for a range of particle densities and diameters. A photograph of one of the experiments is shown in Fig. 3.13 and the quantitative results of the measurements are shown in Fig. 3.14. If the sum of the coefficients, $a_p + C_\tau$, is set equal to 1, an equation can be fit to the data

$$r_0\omega_0 = \left(1 + \frac{150}{\rho_p g D_p}\right)^{1/2} \left(\frac{\rho_p g D_p}{\rho}\right)^{1/2} \tag{3.16}$$

where the coefficient, 150, has units of dynes/cm^2 which is the value of tensile stress, σ_I, apparently at least approximately independent of particle diameter. For small particles the tensile stress appears to be much greater than the weight per unit area of the surface layer (according to Eq. (3.16)), and the threshold speed is thus nearly independent of the particle diameter, in marked contrast to the rapid increase in the boundary-layer threshold speed, with reduction in particle diameter below 80 μm. Only for very large and very heavy particles does the threshold speed, $r_0\omega_0$, become approximately proportional to the square root of the product of particle density and diameter, as for the boundary-layer threshold. The entrainment mechanism for raising surface particles with a vertical vortex (dust devil) is obviously quite different than for the boundary-layer case, as the variation with particle diameter is much less pronounced for vortex threshold than it is for boundary-layer threshold. The primary lifting mechanism appears to be the sudden pressure difference between the bottom and top of the surface

Fig. 3.13. Laboratory-scale dust devil. The dust cloud is $\approx$20 cm in diameter. The vortex generator is visible at the top of the photograph. (Courtesy of Cheng-Ting Hsu.)

rather than the lift and drag produced by viscous shear as in the boundary-layer case. (The vortex generator produces a vortex which moves around over the surface as a function of time in similar fashion to a naturally produced atmospheric vortex.) The tensile stress due to cohesion, which bonds smaller particles to the surface, also seems to be a different function of particle diameter than that for the boundary-layer case.

3.3.7 *Threshold predictions for other planets*

Telescopic observations of Mars from Earth led to the conjecture of dust storms and to predictions of threshold friction speeds on that planet (Ryan, 1964; Sagan & Pollack, 1969). Additional predictions were made after the return of the Mariner 9 images. All of these early predictions, including the authors' (Iversen *et al.*, 1973; Greeley *et al.*, 1974*a*), assumed that the threshold parameter, A, of Eq. (3.5) is a function only of friction Reynolds number, R_{*_t}, although the MARSWIT data later showed that is not the case (see Fig. 3.8).

Sagan (1975) and Sagan & Bagnold (1975) realized that the existence of an optimum diameter (minimum threshold friction speed) might be due to the variation with diameter of particle cohesion forces instead of, or in addition to, viscous forces (Reynolds number), and made threshold predictions for Mars and Venus. These predictions were made, first, by

Fig. 3.14. Vortex threshold data. Threshold parameter versus particle density–diameter–gravitational acceleration product. The threshold speed (proportional to $r_0\omega_0$) is nearly independent of particle diameter. The primary mechanism in vortex threshold appears to be the pressure difference due to the high rotational speed within the vortex, rather than to direct viscous effects.

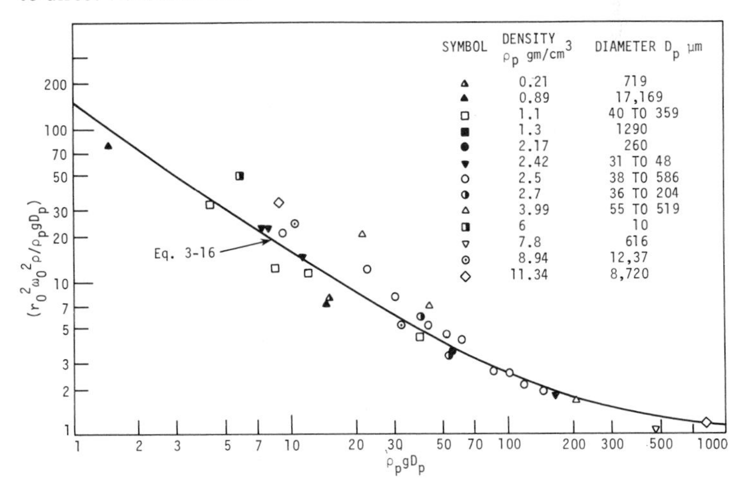

extrapolating a cohesionless threshold estimate for air on Earth from White's (1970) data on particle entrainment in water; then they assumed that $A = A(R_{*_t}, \rho_p/\rho)$ and made their predictions for Mars and Venus by extrapolating to the martian and venusian values of density ratio, ρ_p/ρ. Unfortunately, as shown in Iversen *et al.* (1976*a*), because the forces of cohesion are not a direct function of fluid density (being primarily a function of particle diameter), the threshold coefficient, A, cannot be written as an explicit function of density ratio, ρ_p/ρ. A comparison of various predictions for Mars is shown in Fig. 3.15. The predictions for particles greater than 500 μm in size agree fairly well.

Low atmospheric density threshold data obtained experimentally clearly show the existence of a threshold minimum, and it would thus be expected that cohesion forces and the resulting minimum also exist on Mars. Eq. (3.12) was used to predict threshold speeds for given conditions on the three planets – Earth, Mars, and Venus (Figs. 3.16, 3.17, and 3.18). Fig. 3.16 illustrates the effect of typical ranges of ambient temperature and pressure on Mars. The highest pressure and coldest temperature (the combination giving the largest value of atmospheric density) results in the lowest

Fig. 3.15. Threshold friction speed predictions for Mars. The prediction by Hess (1973) assumes that the threshold parameter, A, is a function only of friction Reynolds number. Sagan & Bagnold's (1975) extrapolated estimate does not result in an optimum diameter. The solid curve (Iversen *et al.*, 1976*b*) is plotted for both standard and cohesionless particles.

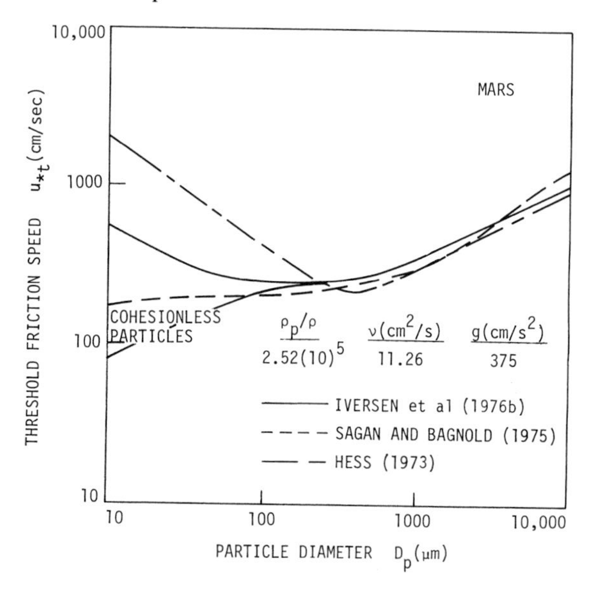

threshold speed (about 1 m/sec for monodisperse particles of 115 μm average diameter). The particle density is assumed to be 2650 kg/m^3, appropriate for quartz grains.

Comparisons of threshold speeds for the three planets, as well as Saturn's moon, Titan, are shown in Fig. 3.17. Threshold speeds on Mars are about an order of magnitude higher than those on Earth, which in turn are an order of magnitude higher than those on Venus. Near optimum diameter, the threshold prediction of Eq. (3.12) is a stronger function of interparticle force than of friction Reynolds number, so that an approximate determination of optimum diameter can be made by differentiating with respect to diameter, holding R_{*_t} constant. The result shows that the value of optimum diameter is proportional to $(\rho_p g)^{-2.5}$. Thus, the smaller gravitational attraction on Titan results in a larger optimum diameter, compared to the other planets, as shown.

Fig. 3.18 presents the same calculations for the three planets in dimensionless form. The dimensionless threshold speed and particle diameter are those defined by Eq. (3.17)

$$\begin{aligned} u_{*_t}(\rho/\rho_p g v)^{1/3} &= (A^2 R_{*_t})^{1/3} \\ D_p(\rho_p g/\rho v^2)^{1/3} &= (R_{*_t}/A)^{2/3} \end{aligned} \tag{3.17}$$

The early estimates of martian threshold speeds assumed that the A versus

Fig. 3.16. Threshold friction speed predictions for Mars using Eq. (3.12). (From Iversen & White, 1982.)

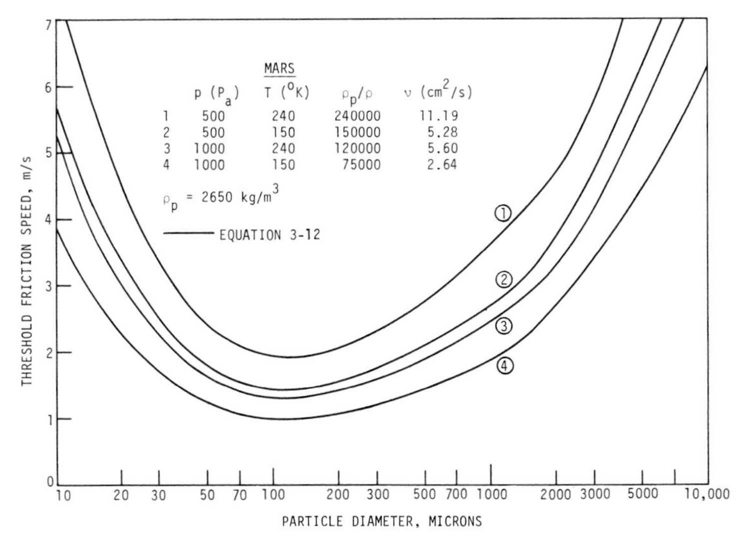

R_{*_t} curve is universal, but such is clearly not the case. As shown in Fig. 3.18, if threshold variation due to the forces of cohesion is improperly interpreted as due to a friction Reynolds number variation only, then martian estimates of threshold are overpredicted and venusian estimates are underpredicted.

3.4 Particle trajectories

The trajectory of an individual grain once it is injected into the atmosphere depends not only upon the wind speed and atmospheric density, but can vary widely depending on the particle's size and density. Tiny particles are suspended by turbulence and can be carried to great heights and for distances of hundreds of thousands of kilometers. Sand-size

Fig. 3.17. Threshold friction speed predictions for Venus, Titan, Earth, and Mars using Eq. (3.12). Gravitational and atmospheric data for Titan from Stone & Miner (1981) and Tyler *et al.* (1981). Particle densities for Venus, Earth, and Mars were 2.65 g/cm^2, appropriate for silicate minerals; particle density for Titan was 1.9 g/cm^3, the average density for the satellite.

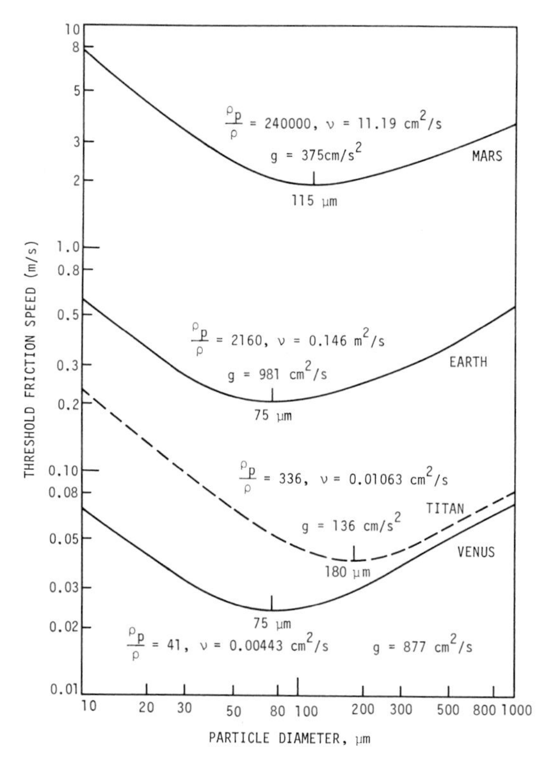

particles, on the other hand, moving in saltation, reach maximum heights and distances measured in the order of meters or less.

3.4.1 *Particle forces*

Forces on the particle during its journey through the air include the major forces of particle weight and aerodynamic lift and drag, and perhaps lesser forces due to pressure gradient. Additional forces include the so-called Basset and apparent mass forces. The Basset force is an aerodynamic force caused by particle acceleration and the apparent mass force is due to motion of the fluid as the particle moves through it. These latter two forces and the pressure gradient forces are usually negligible for particles much denser than the fluid medium through which they are moving, and so are negligible for all but the very tiniest particles moving within a planetary atmosphere.

For static threshold conditions, there must exist an appreciable lift force, as discussed previously (see Section 3.3.1). This force gives rise to a vertical acceleration of the particle but must decrease rapidly after take-off

Fig. 3.18. Dimensionless threshold friction speed versus dimensionless particle diameter. If threshold parameter, A, were a unique function of friction Reynolds number, R_{*_t}, there would be only one curve. The effect of interparticle cohesion forces is dominant for particles below ≈ 100 μm and causes the large differences in the curves for large differences in atmospheric density. (From Iversen and White, 1982.) Predictions were made using Eq. (3.12).

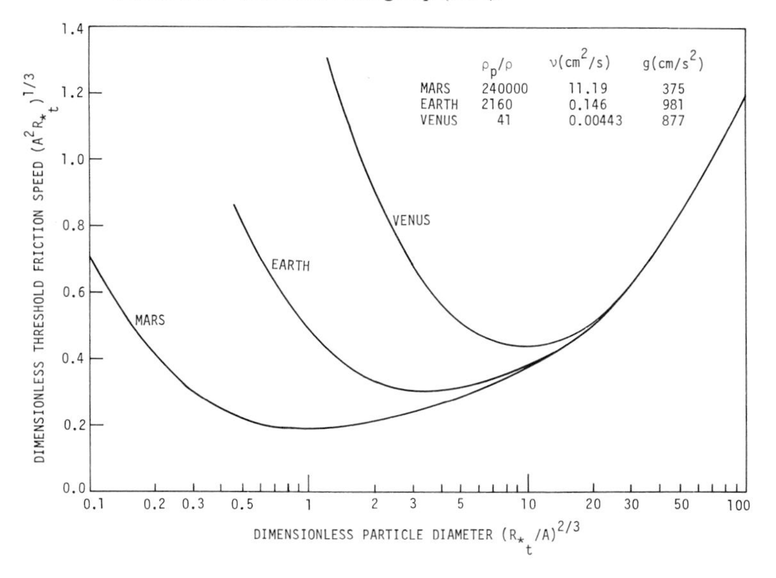

as the particle leaves the high-shear region close to the surface, and as the relative velocity between the particle and fluid diminishes.

Maeno *et al.* (1979) measured trajectories of snow particles in a cold wind tunnel. Their data showed that the horizontal velocities of ascending particles are less than the mean wind speed but that descending particles generally had horizontal velocities greater than the mean wind speed. The vertical decelerations of ascending particles and acceleration of descending particles were both found to be greater in magnitude than gravitational acceleration. The effects of lift due to shear and drag can explain the vertical decelerations, but the downward acceleration magnitudes are more difficult to explain.

There can be significant lift forces during portions of the trajectory, however, if the particle is spinning. This was shown to be true by fitting calculated particle trajectories to those obtained with a high-speed motion picture camera in the Iowa State University environmental wind tunnel (White & Schulz, 1977). Lift forces caused by a spinning body moving through the air are called *Magnus forces*. When terms accounting for the Magnus effect were included by White & Schulz in the equations of motion, theoretical trajectories in much better agreement with the observations were obtained.

3.4.2 *Saltation trajectories*

A typical particle trajectory in saltation is shown in Fig. 3.19. White & Schulz measured the characteristics of 100 particle trajectories for which the average particle diameter was 586 μm. For a friction speed of about 40 cm/sec, the average lift-off angle measured was about 50°, the

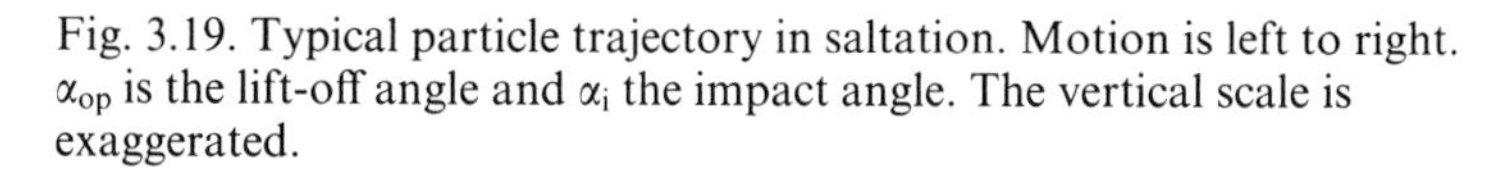

Fig. 3.19. Typical particle trajectory in saltation. Motion is left to right. α_{op} is the lift-off angle and α_i the impact angle. The vertical scale is exaggerated.

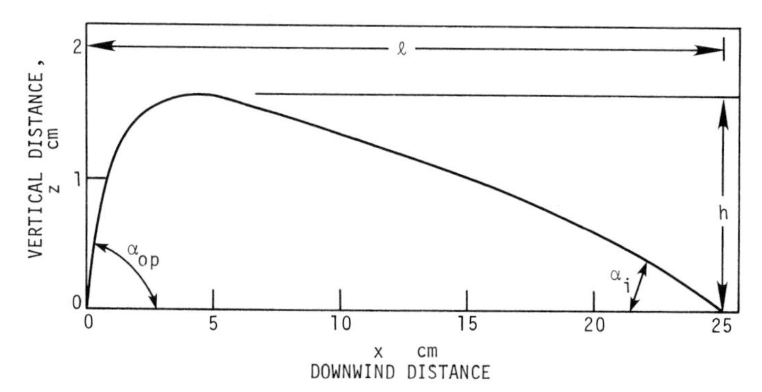

lift-off speed was 69 cm/sec, the average impact angle was 13.9°, and the average impact velocity was 161 cm/sec. There were wide variations from average values in each of these measurements (see Table 3.3).

Bagnold (1941) estimated the initial vertical speed to be of the order of the friction speed, u_*. (This is seen to be the case for White & Schulz's data.) Owen (1964) thus indicates that the maximum height of particle trajectory should be somewhat less than $u_*^2/2g$, which is the height attainable with an initial vertical speed equal to u_*, and with no drag force on the particle. The data of White & Schulz show that the average trajectory would attain a height of the order predicted by Owen but that a few particles, because of greater lift-off speeds, could attain heights considerably greater than $u_*^2/2g$ without the benefit of assistance by turbulence. The greater lift-off speeds are probably caused by impact from airborne particles striking the surface. A particle as large as 586 μm is relatively unaffected by turbulent fluctuations at these speeds. White & Schulz also show that the Magnus effect can cause the trajectory height to be greater than $w_o^2/2g$, where w_o is the measured effective initial vertical speed.

Bagnold (1973) states that a solid particle should become liable to suspension at a stage, u_*/u_{*t}, approximately equal to 0.8 U_F/u_{*t}, or $U_F/u_* = 1.25$ (see Eq. (3.2)). Thus, on Earth, according to Fig. 3.2, particles of diameter 60 μm or less, upon becoming airborne, are susceptible to suspension, since the vertical speed is greater than or equal to particle terminal speed. If turbulence characteristics on Mars and Venus scale to those on Earth, the corresponding diameter on Mars is 240 μm and on Venus is only 33 μm. As Bagnold (1960) states, in a desert sand storm where few particles less than 50 μm in diameter are present, all particles move in a low cloud whose uppermost boundary rarely reaches shoulder height, so that people's heads may be visible while the rest of their bodies are not. He indicates that particles of diameter 100 μm rise only some centimeters, whereas grains of diameter 1000 μm may rise to heights of 1.5–2 m. Tabler (1975) states that for drifting snow occurring in a strong wind, as soon as particles become small enough to be suspended by the wind, they begin to sublime and will disappear, so that most airborne particles are travelling in the saltation mode. Radok (1977), however, reports having seen snow particles in the Antarctic being carried by 'trombes' (vertical axis eddies) to heights of 100 meters and remaining airborne for many tens of seconds. Such particles would have to be considered to be in suspension.

For those particles which are light or small enough to be relatively subject to turbulent fluctuations of the wind, it is impossible to describe a

Table 3.3. *Particle trajectory measurements (from White & Schulz, 1977)*

	Number of trajectories measured	Mean value	Minimum value	Maximum value	Standard deviation
Lift-off angle, α_{0p} (°)	57	49.9	20	100	19.6
Lift-off speed (cm/sec)	57	69.3	20	200	32.5
Impact angle α_i (°)	43	13.9	4	28	3.31
Impact speed (cm/sec)	43	161.2	40	320	45.9

particle trajectory because the turbulent fluctuations themselves are random. A dust particle can thus be carried vertically upwards and downwards many times and can reach great heights (of the order of kilometers) and distances (of the order of hundreds of kilometers) before deposition on the surface, as described by Gillette (1977) for dust storms on the Great Plains of North America.

3.4.3 *Predictions of saltation trajectories*

Predictions of saltation trajectories must be considered as typical or average since the particles leave the surface with a variety of initial speeds and lift-off angles. Owen (1980) describes a stylized calculated trajectory height of 0.81 u_*^2/g and a length of 10.3 u_*^2/g. These numbers are within the range of those measured by White & Schulz.

Owen's trajectory height prediction is based on an assumed vertical speed at lift-off. Rather than assuming an initial vertical speed, White *et al.* (1976) calculate particle trajectories by letting a lift force act on the particle due to the shear very near the surface. Their results give maximum trajectory heights less than Owen's for Mars and greater than Owen's for Venus. The calculated predictions are compared in Table 3.4. A significant discrepancy exists between the two sets of calculations for the results for trajectory height, and sufficient uncertainties exist in both methods of calculation to preclude determining which result is close to reality. It is probably certain that particles fly faster, higher and farther on Mars than on Earth, and faster, higher and farther on Earth than on Venus.

Measurements of particle velocity by a photoelectric cell method at pressures corresponding to those on Earth and Mars have been made in the MARSWIT wind tunnel. The results for a fairly large particle are shown in Fig. 3.20, which show that the particle speeds are only 10–20% of the wind tunnel free stream speed (wind speed above wind tunnel boundary layer).

Table 3.4. *Saltation trajectory path lengths and heights for Earth, Mars, and Venus*

Planet	Particle diameter (D_p)	Friction speed u_* (twice threshold) (cm/sec)	Saltation height (cm)		Saltation length (cm)	
			Owen (1980)	White *et al.* (1976); White (1979, 1981*b*)	Owen (1980)	White (1981*b*)
Mars	100	390	328[a]	1–5	4170[a]	—
Earth	100	42	1.5	0.8–2.5	18.5	—
Venus	100	4.9	0.022[a]	0.2	0.28[a]	0.4

[a] Based on measured trajectories on Earth which may not be valid for other planets.

Similar experiments at one Earth atmosphere pressure result in particle speeds of 50–60% of free stream speed whereas particles on Venus should achieve nearly 100% of free stream speed (Greeley *et al.*, 1983). However, because the wind speeds at martian pressures are an order of magnitude higher than on Earth, the particle speeds, and thus the erosive capabilities of saltating particles, are much greater on Mars than on Earth (Greeley *et al.*, 1982), which, in turn, should be much greater than on Venus.

3.5 The saltation layer

Of extreme importance in the saltation phenomenon is the vertical distribution of particles, as well as total flux, as functions of the wind speed. The formations of all scales of bed formations, from centimeter-size ripples to kilometer-size dunes, are all due to the saltation process. In this section we discuss particle flux and concentration, as well as the effect of saltation on the wind speed profile near the surface.

Fig. 3.20. Particle speed distributions for saltating particles at martian pressure conditions in the MARSWIT wind tunnel. Speeds are shown as both the percentage of free-stream wind speed (above the boundary layer) and actual speed. The pressure was 6.6 mb (0.0065 atmosphere), the average particle diameter was 715 μm, and free-stream speed was 115 m/sec (Greeley *et al.*, 1980*b*.)

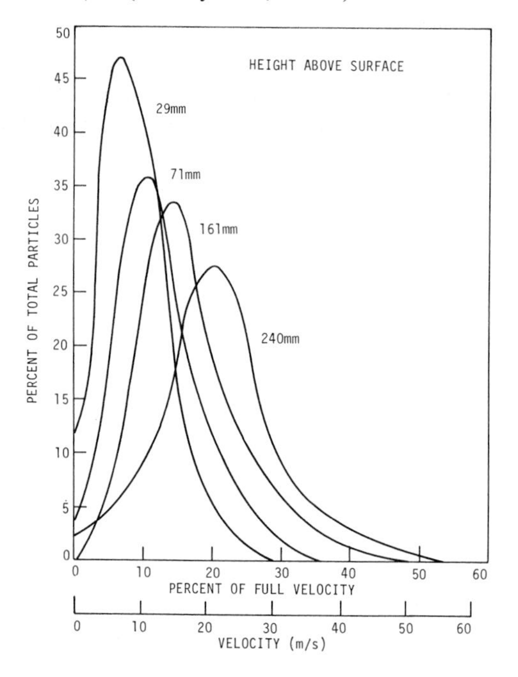

3.5.1 *Particle flux*

Bagnold (1941) derived the expression for momentum loss of the air due to sand in saltation as

$$q\,(u_2 - u_1)/L_p \approx q\,u_2/L_p = \tau = \rho\,u_*^2 \tag{3.18}$$

This quantity represents the momentum loss per unit time per unit length of travel per unit lateral dimension, i.e., momentum loss per unit area per unit time, which is equal to force per unit area or surface stress, τ. Mass of sand per unit lateral dimension per unit time is q, u_2 is final horizontal particle speed on impact, u_1 is initial horizontal speed (assumed small), L_p is distance traveled per grain, and τ is the surface stress. Bagnold assumes that $u_2/L_p \approx g/w_o$ where w_o is initial vertical velocity and, further, that w_o is of the order of friction speed u_*, thus

$$q \propto \rho\,u_*^2\,w_o/g \propto \rho u_*^3/g \tag{3.19}$$

Bagnold found that he could fit his experimental data for different particle diameters by

$$q\,g/\rho\,u_*^3 = C\,(D_p/D_{p_o})^{1/2} \tag{3.20}$$

where C is a function of particle size distribution (with values ranging from 1.5 to 2.8) and D_{p_o} is 250 μm. He indicates that the transport rate is greater (as much as twice) for mixed particles than for particles of uniform size. In addition, the dimensionless transport rate of Eq. (3.20) is probably a function of several parameters, including size distribution, the ratio of friction speed to threshold, and possibly density ratio, terminal speed/threshold friction speed ratio, and surface roughness. Unfortunately, accurate transport rate data are very difficult to obtain, and many different formulae have thus been derived to fit experimental data. The difficulty of establishing an appropriate formula to fit all cases is emphasized by the number and variety of derived equations, listed in Table 3.5.

Several of the equations in Table 3.5 (Eqs. (3.21), (3.23–31), (3.33)) account for the fact that mass transport rate should be zero at threshold ($r_t = 1$). The value of u_{*_t} in these equations should probably be the impact threshold, which is lower than the static threshold value.

Experimental data for mass transport rate taken in the laboratory and in the atmosphere are shown in Fig. 3.21. The data taken in the atmosphere naturally have more scatter but in both cases the data seem to fit Eq. (3.27) fairly well.

In many situations in nature (the Viking Lander sites on Mars are examples), there are large non-erodible elements scattered through an area

Table 3.5. *Mass transport rate expressions*

Source	Expression	Equation number
Bagnold (1941)[a]	$qg/\rho u_*^3 = C(D_P/D_{P_o})^{1/2}$	3.20
Kawamura (1951)[b]	$qg/\rho u_*^3 = C(1+r_t^2)(1-r_t)$	3.21
Zingg (1953)[a]	$qg/\rho u_*^3 = C(D_P/D_{P_o})^{3/4}$	3.22
Dyunin (1954*b*)[c]	$qg/\rho U^3 = C(1-r_u)$	3.23
Kuhlman (1958)[c]	$qg/\rho U^3 = C(1-r_u^3)$	3.24
Owen (1964)[b, d]	$qg/\rho u_*^3 = (0.25+0.33\ r_t p_t)(1-r_t^2)$	3.25
Dyunin (1959)[c]	$qg/\rho U^3 = C(1-r_u)^3$	3.26
Lettau & Lettau (1978)[b]	$qg/\rho u_*^3 = C(1-r_t)$	3.27
Kind (1976)[b]	$qg/\rho u_*^3 = C(1-r_t^2)$	3.28
Iversen *et al.* (1976*c*)[b,d]		
Schmidt (1982)	$qg/\rho u_*^3 = Cp_t(1-r_t)$	3.29
Maegley (1976)[a, b]	$qg/\rho u_*^3 = C(D_P/D_{P_o})^{3/4}(1-r_t^2)$	3.30
Maegley (1976)[a, b]	$qg/\rho u_*^3 = C(D_P/D_{P_o})^{3/4}(1-r_t^{13.72})$	3.31
Radok (1977)[c, f]	$qg/\rho U^3 = (g/\rho U^3)\exp(C_1+C_2 U)$	3.32
Lyles *et al.* (1979)[b, e]	$qg/\rho u_*^3 = (C/A^2)(1-r_t)/r_t$	3.33
Takeuchi (1980)	7 equations for snow transport	

[a] $D_{P_o} = 250\ \mu m$
[b] $r_t = u_{*_t}/u_{*_t}$
[c] $r_u = U_t/U$, U = wind speed at a given reference height, U_t = wind speed at threshold
[d] $p_t = U_F/u_{*_t}$
[e] A = Bagnold's threshold coefficient
[f] C_1 and C_2 are not dimensionless

which is otherwise covered partially or wholly with sand or erodible soil. The effect of non-erodible elements, which contribute to aerodynamic roughness, has been investigated by Chepil (1950), Kuhlman (1958), and Gillette (1977). They found: (1) that the deflation rate decreases with time if there are non-erodible elements initially buried in the sand or soil, and (2) that the deflation rate is smaller for rougher surfaces than for smoother surfaces at equal values of friction speed, u_*.

The vertical flux of an eroding soil has been calculated, from measurements of horizontal flux, by Gillette (1977). For the horizontal flux, he correlated the data using Eq. (3.27). The ultimate vertical flux is assumed to be only fine particles capable of being transported long distances by the wind.

3.5.2 *Concentration distribution*

For very fine particles, the variation of particle concentration with height above the surface would be dependent on turbulent diffusion. If there are no changes in concentration or wind speed with either horizontal direction, the differential equation for mass transport of suspendible material is (Gillette & Goodwin, 1974)

$$\frac{\partial}{\partial z}\left(K\frac{\partial \tilde{C}}{\partial z}\right)+U_{\mathrm{F}}\frac{\partial \tilde{C}}{\partial z}=0 \tag{3.34}$$

Fig. 3.21. (*a*) Mass transport rate as a function of friction speed. Wind tunnel data from Williams (1964). (*b*) Mass transport rate as a function of friction speed. Atmospheric data of sand transport on a natural beach, from Svasek & Terwindt (1974), and on a river delta, from Nickling (1978). The average particle diameter on the beach was 250 μm, significantly greater than the river delta sand. The beach sand thus exhibits greater transport rate on the average.

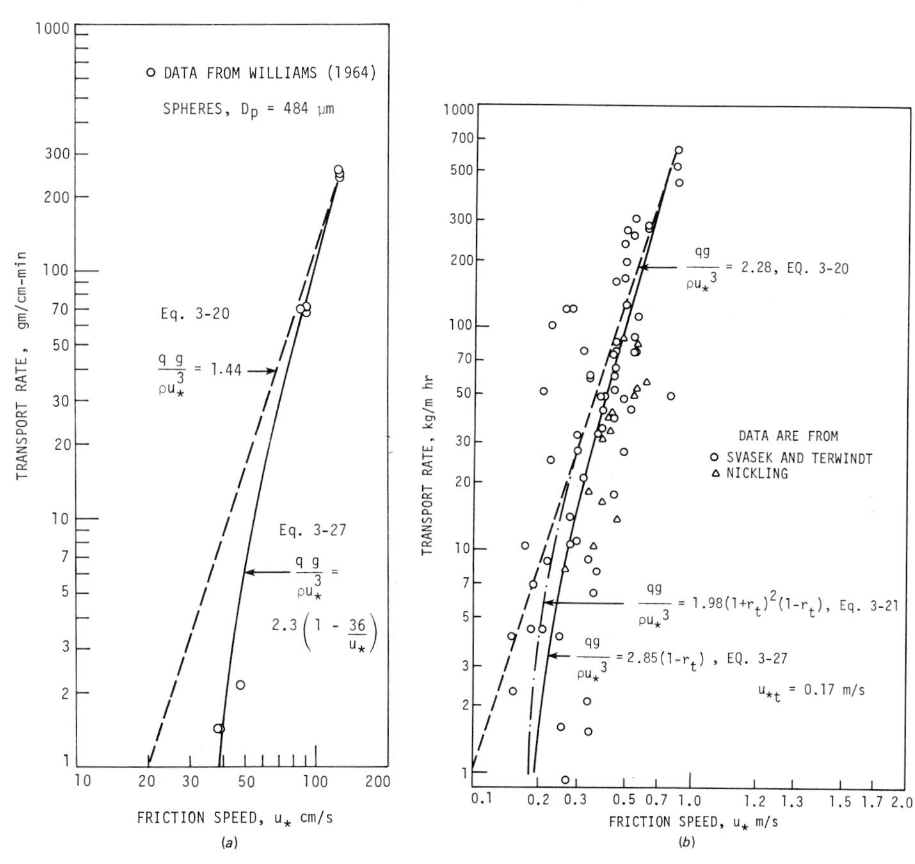

In this equation, U_F is the terminal speed of the particle, $\tilde{C}$, the particle concentration (mass/volume), and K, the turbulent diffusivity. In the logarithmic wind speed profile (Eq. (2.11)), the diffusivity, K, is equal to $0.4\, u_* z$ and Eq. (3.34) can be integrated to obtain the concentration distribution

$$\tilde{C} = \tilde{C}_1 \, (z/z_1)^{-U_F/0.4\, u_*} \tag{3.35}$$

Data for snow in saltation are shown in Fig. 3.22. The data approximately follow the form of Eq. (3.35) but the exponent is different from that in the equation. The reasons for the discrepancy are: (1) that the terminal speed varies with height because the particle size varies with height, and (2) the particles are large enough so that their trajectories are only partially affected by turbulence. The empirical exponent is closest to the value of Eq. (3.35) for the largest wind speed because the ratio of terminal speed, U_F, to friction speed, u_* (or vertical turbulent fluctuation speed), is then closer to

Fig. 3.22. Particle concentration curves for blowing snow as a function of wind speed (U_{10} is wind speed at 10 m above surface) and height above the surface. Note that the curves appear to coalesce at a point very close to the surface at a value of particle concentration approximately equal to the fluid density (as predicted by Owen, 1964). Note that the exponent, b, is smaller in magnitude than that predicted from turbulent diffusion (Eq. (3.35)). (Data adapted from Mellor, 1965.)

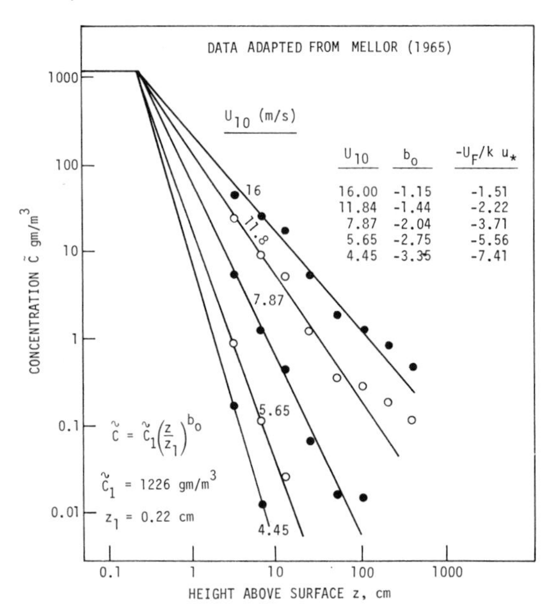

the suspension–saltation boundary, and the particles are affected to a greater extent by turbulence.

Multiplying the concentration at a given height, z, by the wind speed at that height results in the mass flux (mass per unit time per unit area), providing the particles are moving at the same speed as the air. That is approximately true if the wind speed is near or above the saltation–suspension boundary. For large particles moving in saltation, the average particle speed can be significantly less than the wind speed. If the wind speed is high enough or particles small enough so that particle trajectories are significantly affected by turbulence, the total mass transport is the integral of the concentration–wind speed product,

$$q = \int \tilde{C}\, U \,\mathrm{d}z$$
$$= \int \frac{\tilde{C}_1 u_*}{0.4}\left(\frac{z}{z_1}\right)^{-b_o} \ln\left(\frac{z}{z_0}\right)\mathrm{d}z \tag{3.36a}$$

Where Eq. (2.11) is used to define the variation of wind speed with height and the exponential factor in Eq. (3.35) is replaced by the number b_o. Eq. (3.36a) can be integrated to obtain

$$q\Bigg|_{z_2}^{z_3} = \frac{u_*\tilde{C}_1}{0.4}\left(\frac{z_1}{1-b_o}\right)\left(\frac{z}{z_1}\right)^{1-b_o}\left\{\ln\frac{z}{z_o}+\frac{1}{b_o-1}\right\}\Bigg|_{z_2}^{z_3} \tag{3.36b}$$

Eq. (3.36b) gives the mass transport per unit width per unit time between heights z_2 and z_3. If we assume that the concentration is constant $= \tilde{C}_1$ from $z = z_o$ to z_1, and that the exponent, b_o, is a constant greater than 1, from z_1 to infinity, the total mass transport from Eq. (3.36b) becomes

$$q = \frac{u_*\tilde{C}_1 z_o}{0.4}\left\{\left(\frac{b_o}{(b_o-1)}\right)\frac{z_1}{z_o}\ln\frac{z_1}{z_o}-\frac{z_1}{z_o}+1+\frac{1}{(b_o-1)^2}\frac{z_1}{z_o}\right\} \tag{3.36c}$$

Gillette & Goodwin indicate that Eq. (3.36b) is valid for saltating particles over a certain height above the surface, even though the equation is derived only for particles suspended by turbulence. Although Eq. (3.36c) is, strictly speaking, not valid for saltation, it is interesting to note that if roughness, z_o, is proportional to u_*^2/g (for saltation, see (Eq. 3.39) below), and $\tilde{C}_1$ is proportional to fluid density, then Eq. (3.36c) is of the same form as Eq. (3.20), the mass transport equation for saltation.

Visibility in snow or dust storms, of course, is a function of particle concentration. Particle sizes near the surface in a snow storm are sufficiently large so that attenuation of light is a direct function of particle projected area and number of particles along the sight path (Schmidt,

1982). For dust storms, concentrations of 100–400 mg/m^3 are needed to reduce visibility to 200 m (Hagen & Skidmore, 1977).

Average particle diameter as a function of height above the surface has been measured in snow (Mellor, 1965), for blowing sand (Williams, 1964), and in dust storms (Chepil & Woodruff, 1957). Williams shows particle diameter decreasing with height to about 10 cm and then increasing again, but it must eventually decrease with height as the saltation layer height is exceeded. Data for snow show that the average diameter varies from about 165 μm at 3 cm above the surface, down to 87 μm at 2 m above the surface.

3.5.3 *Mass transport prediction for Mars and Venus*

One series of wind tunnel tests has been performed in the MARSWIT tunnel to determine mass transport values at low atmospheric density. The results are shown in Fig. 3.23. The data are correlated with mass transport rate at one earth atmosphere using Eq. (3.21). Using the equations of Table 3.5, one can predict that, if the ratio of threshold to friction speed r_t is the same on two planets, the ratio of transport rates on the two planets would be

$$\frac{q_1}{q_2} = \left(\frac{A_1}{A_2}\right)^3 \left(\frac{\rho_2 g_1}{\rho_1 g_2}\right)^{1/2} \left(\frac{\rho_{p_1} D_{p_1}}{\rho_{p_2} D_{p_2}}\right)^{3/2} \tag{3.37}$$

For the conditions of Fig. 3.17, and for large particles of the same diameter and density (so that $A_1 = A_2$, $\rho_{p_1} = \rho_{p_2}$, $D_{p_1} = D_{p_2}$), the ratio of mass transport rate on Mars to that on Earth would be about 6.5, and the ratio of that on Earth to that on Venus would be about 6.9.

3.5.4 *Effect of saltation on wind speeds near the surface*

The effect of the saltation layer is to greatly increase the effective aerodynamic roughness height over that of a quiescent surface, provided there are no non-erodible roughness elements present. Bagnold (1941) noted in his experiments that there appeared to be a focus, i.e., a given height above the surface at which the wind speed does not change with increase in wind speed above that point. Bagnold fit his data to

$$(U/u_*) = (1/0.4)\ln(z/k') + (u_t/u_*) \tag{3.38}$$

His focus height, k', was set at 3 mm, and u_t = 250 cm/sec.

Owen (1964) derived a different function for the wind speed, by assuming that the effective roughness height should be proportional to saltation-layer height, which should be of the order of $u_*^2/2g$, since the initial vertical velocity of the particle is of the order of u_*, thus

$$\frac{U}{u_*} = \frac{1}{0.4} \ln\left(\frac{2gz}{u_*^2}\right) + D' \tag{3.39}$$

The quantity, D', which was established empirically at 9.7, is equivalent to setting the proportionality constant between roughness height in saltation, z_o', and $u_*^2/2g$ at 0.020 65 (i.e., $z_o' = 0.020\,65\, u_*^2/2g$). Tabler (1980*a*) made many careful measurements in blowing snow over a snow covered frozen lake and found the data to fit Eq. (3.39) if D' is set equal to 9.1 ($z_o' = 0.026\,50\, u_*^2/2g$). With that value, Eqs. (3.38) and (3.39) are almost indistinguishable for normal particle diameters on Earth.

Fig. 3.23. (*a*) Particle mass transport rate as a function of surface shear stress ($\tau = \rho u_*^2$). Data from MARSWIT (White, 1979). The triangles are at one atmosphere pressure (10^5 Pa) and the circles are at greatly reduced pressure (2300 Pa). (*b*) Same data as Fig. 3.23 (*a*). Eq. (3.21) is used to correlate the data. (White, 1979.)

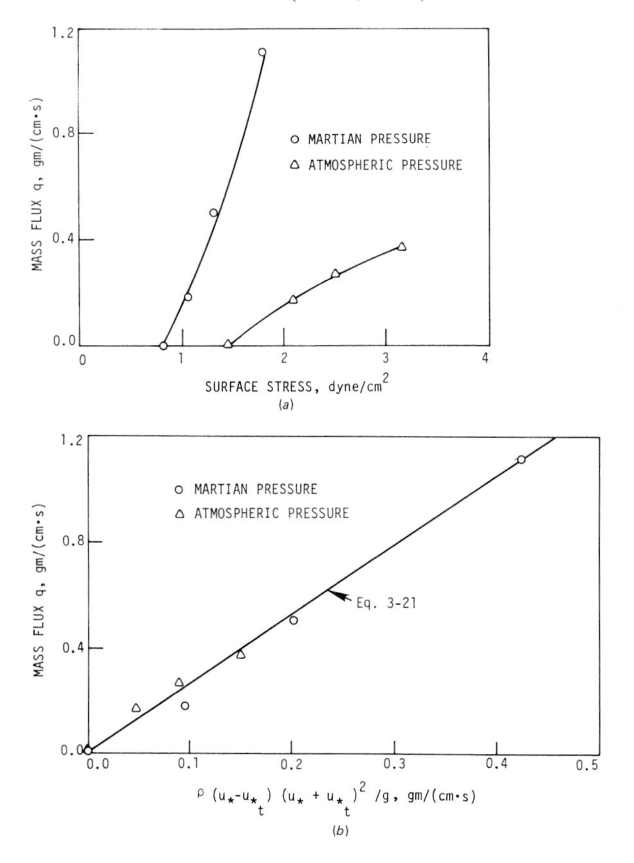

Other equations which have been derived include those of Bagnold (1973)

$$U/u_* = (1/0.4)\ \ln\{z/[D_p(0.03+(u_* - u_{*_t})/u_{*_t})]\} \tag{3.40}$$

and White & Schulz (1977)

$$U/u_* = (1/0.4)\ \ln(z/D_p) - 2.29 + 10.79\ (u_{*_t}/u_*) \tag{3.41}$$

A comparison of these equations is illustrated in Fig. 3.24. The wind speeds near the surface are obviously much lower during saltation because of the increased roughness due to saltation. Since Eqs. (3.38) and (3.39) fit Bagnold's (1941) and Zingg's (1953) data well, they should be more reliable than Eqs. (3.40) and (3.41).

If martian conditions are substituted into Eqs. (3.38) and (3.39), the

Fig. 3.24. Effect of the saltation layer on wind speed profile near the surface. The curves with steepest slope are all for $u_*/u_{*_t} = 1.5$ and the rest are all for $u_*/u_{*_t} = 4$. Probably the most accurate equations are those of Bagnold (1941), Owen (1964), and Tabler (1980*a*).

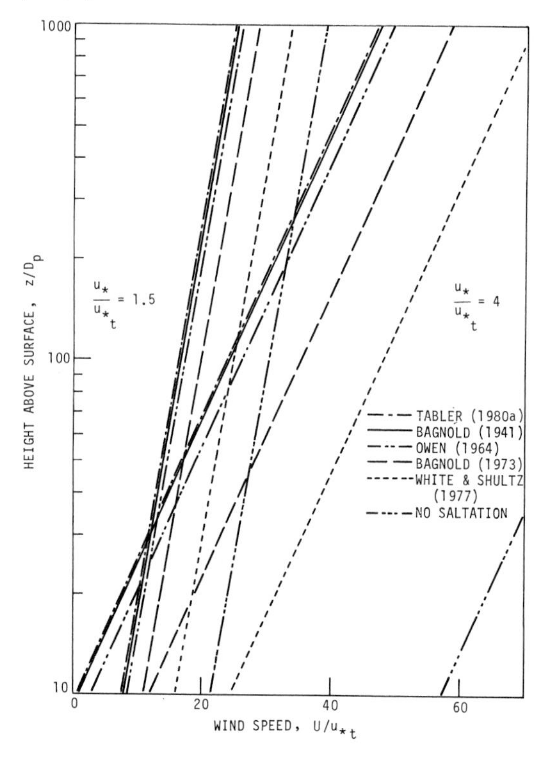

results are quite different, as shown in Fig. 3.25. Eq. (3.39) gives more-likely results, as that equation should be more universal in application. If the saltation-layer thickness is also a function of density ratio, ρ_p/ρ, as well as $u_*^2/2g$, however, then the prediction of Eq. (3.39) may be somewhat in error.

Fig. 3.25. Prediction of wind speed near the martian surface with Eqs. (3.38) and (3.39). On Earth these equations give practically the same result but they give quite different predictions for Mars. It is most likely that Eq. (3.39) is a better prediction for the conditions on Mars ($u_* = 290$ cm/sec).

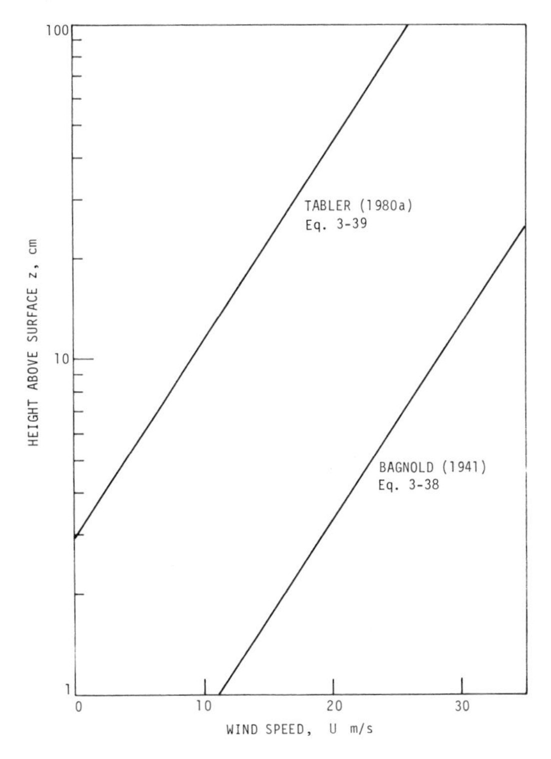

4

Aeolian abrasion and erosion

4.1 Introduction

Anyone who has experienced a sand storm is acutely aware of the effectiveness of windblown sand as an agent of abrasion. Every year countless cars and trucks are caught in sand storms where, in exposures sometimes as short as a few minutes, windows can be frosted and paint can be stripped to bare metal. Multiply this high rate of abrasion over geological time and some appreciation can be gained of the effectiveness of wind erosion (Fig. 1.5).

Wind erosion takes place either through *abrasion*, the wearing away of relatively solid rock or indurated sediment, or through *deflation*, in which loose particles, such as sand, are blown away. Wind erosion commonly occurs through windblown particles that act as *abradants* – the agents of wind erosion.

In this chapter we first discuss the mechanics of wind abrasion on the microscopic scale and review the formation of ventifacts. We then present results from various laboratory simulations and estimate rates of wind abrasion on Earth and Mars. In the last section we discuss large wind-eroded landforms.

4.2 Aeolian abrasion of rocks and minerals

Studies of wind abrasion have generally been carried out either by materials scientists and engineers who are concerned with problems such as the abrasion of turbine blades by airborne particles, or by geologists who are concerned with the erosion of rock materials. Engineering studies typically are well documented and quantitative, but do not usually involve materials appropriate for geological applications. And, although geological studies involve analyses of rock abrasion, most of these studies are qualitative or, at best, only partly quantitative for terrestrial environments; thus the results are not appropriate for extrapolations to other planets.

Beginning in the mid-1970s, however, several investigations were carried out to assess aeolian abrasion on Mars. From necessity, these studies considered the mechanics of abrasion from a quantitative approach and shed light on aeolian abrasion in general.

4.2.1 *Mechanics of abrasion*

Wind abrasion results from the impact of windblown grains on a target and basically involves the transfer of kinetic energy. The resulting collision may damage either the grain or the target, or both. In principle, the greater the kinetic energy – resulting from a more massive grain, or a higher grain speed – the greater the amount of abrasion. We shall see, however, that the problem is not quite this simple.

For now, let us consider abrasion of a simple homogeneous, brittle material, resulting from the impact of a typical, rounded windblown grain. As discussed by Marshall (1979), upon impact the grain tends to flatten slightly and to indent the target surface. The contact area increases slightly with particle size and if the loading exceeds a certain critical limit, which is a function of the target, then a circular crack develops at the indentation site. This crack – called a *Hertzian fracture* (Fig. 4.1) – has a radius 10–30% greater than the contact radius (Johnson *et al.*, 1973). Results from experiments show that there is a tendency for the diameter of the fracture to increase with an increase in either the grain diameter or the impact velocity.

Experiments carried out by Marshall (1979) involving quartz plates abraded by rounded grains show a sequence of stages that are probably appropriate for most brittle materials (Fig. 4.2):

Embryonic stage: 'Softening' of the surface for abrasion; isolated pits of various shapes (irregular, V-shaped, polygonal and crescentic) develop around the periphery of Hertzian fractures where material is easily chipped away because of an underlying fracture plane; where Hertzian fractures are closely spaced, the removal of material is facilitated by the intersection of the conical frustra at depth.

Youthful stage: Pits increase in number and coalesce due to lateral growth; the cone slopes become exposed and parts of the original surface can survive on the cone tops. Paradoxically, the first areas to have been impacted (i.e., the sites of the cones) are often the last areas abraded.

Advanced stage: Exposed conical frustra progressively degrade until they are reduced to about the elevation of the surrounding areas. The

Fig. 4.1. Examples of Hertzian fractures produced in various targets. (*a*) diagram of Hertzian fractures produced by the impact of rounded grains (typical of windblown particles); the fractures are roughly conical, expanding with depth into the impacted surface. (*b*) upper right, transmitted light photomicrograph of vitreous silica impacted by a grain ≈800 μm in diameter; the white central zone is the contact area of the grain on the surface; the concentric patterns indicate fracturing at depth; (*c*) middle left, SEM image showing Hertzian fracture on the surface of the target; (*d*) middle right, SEM enlargement of concentric fractures on the target surface; (*e*) lower left, multiple Hertzian fractures produced on a quartz plate by the impact of quartz grains ≈2 mm in diameter (coarse sand); (*f*) lower right, SEM image of multiple Hertzian fractures showing removal of target material from between the concentric fractures. (Courtesy of J. R. Marshall.)

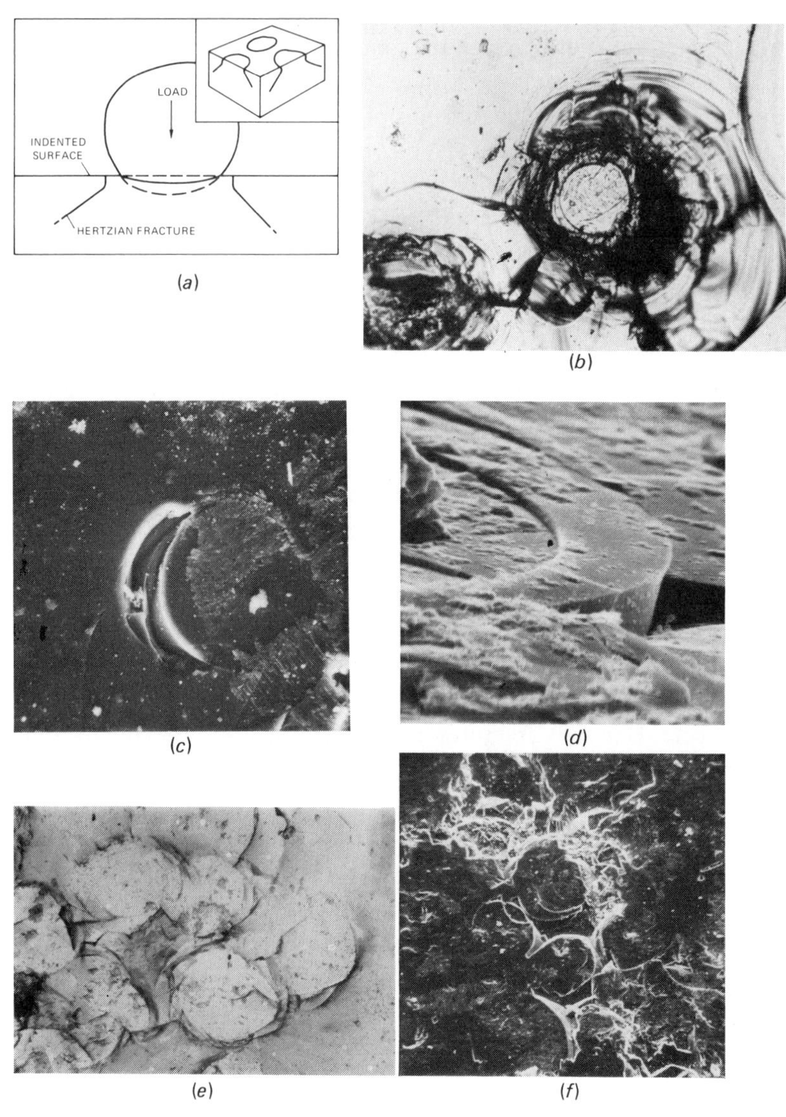

cones provide resistant projections that tend to protect surrounding areas by preventing impacting particles from making direct contact with the floor of the pitted region.

Mature stage: Complete removal of the original cone structures. Hertzian cracks are not capable of redeveloping on a surface that is blocky. Erosionally and texturally, this is a steady-state surface and most erosion takes place by the loosening of small blocks and plates.

Fractures produced by angular particles are different from those resulting by impact of rounded grains. Lawn & Wilshaw (1975) showed that a vertical crack will form beneath the impact of the pointed or angular edge of a grain for stresses above some initial threshold value. After the

Fig. 4.2. Sequence showing the abrasion of quartz plate by quartz particles ≈ 780 μm in diameter at speeds of 20–26 m/sec: (*a*) embryonic stage, (*b*) youthful stage, (*c*) advanced stage, and (*d*) mature stage. (Courtesy of J.R. Marshall.)

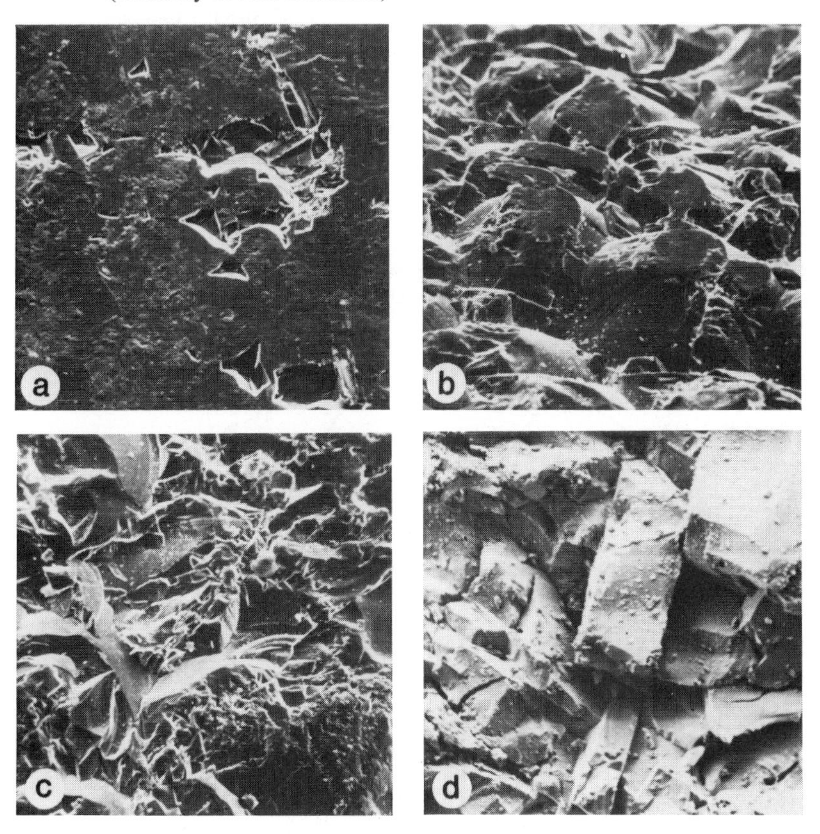

grain bounces away, the crack may close and lateral fractures may be generated. Chipping can result if the cracks reach the surface of the target and can produce star-shaped fractures (Fig. 4.3).

Whether caused by the impact of rounded or angular grains, the fractures generated have a strong influence on the rate of aeolian abrasion. Where Hertzian fractures develop on a fresh surface, abrasion is inhibited. At some critical spatial density of fractures, material becomes easily removed, but abrasion diminishes thereafter because isolated conical frustra are difficult to remove. Impact on top of the remaining cones only deepens the structure, whereas cone spacing prevents contact of rounded particles with surface between cones.

In contrast to abrasion by rounded grains, abrasion involving primarily angular particles results in the chipping of the surface by intersection of vertical fractures and proceeds at a fairly uniform rate, as shown in Fig. 4.4.

4.2.2 *Susceptibility to abrasion (S_a)*

The susceptibility to abrasion, S_a, is a measure of the erodibility of different materials and is expressed as the ratio of the material mass eroded to either the mass of impacting particles or the number of impacting particles. S_a is not a coefficient of abrasion, as it varies with parameters such as impacting particle velocity, U_p, and angle of impact.

Fig. 4.3. Fracturing and chipping produced by an angular grain, showing radiating patterns produced by the intersection of lateral and median vents.

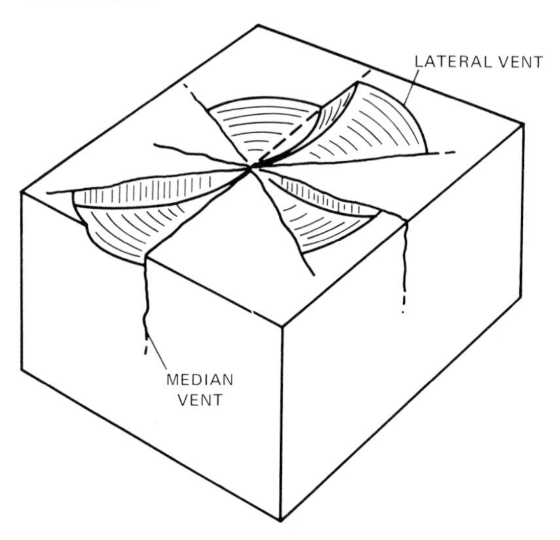

In order to determine the susceptibility to abrasion for various rocks under a wide range of conditions, an apparatus (Fig. 4.5) was fabricated by Greeley and colleagues (1982), which allows impact velocity, impact angle, impacting particle size and type, sample type, and atmospheric density and composition to be controlled. Using quartz grains, basalt sands, and sand-size aggregates of fine grains as the abrading particles, susceptibilities to abrasion were assessed in terms of four parameters: (1) particle size, (2) particle velocity, (3) angle of particle impact, and (4) atmospheric density.

Figure 4.6 shows S_a as a function of diameter of impacting quartz particles for basalt, obsidian, and hydrocal (gypsum cement). The effect of particle size in the range analyzed is fairly well defined by a slope of approximately 3 on a log–log plot, giving the relationship

$$S_a \propto D_p^3 \tag{4.1}$$

Fig. 4.4. Diagram showing the sequence of abrasion by impact of angular particles. (Courtesy of J. Marshall.)

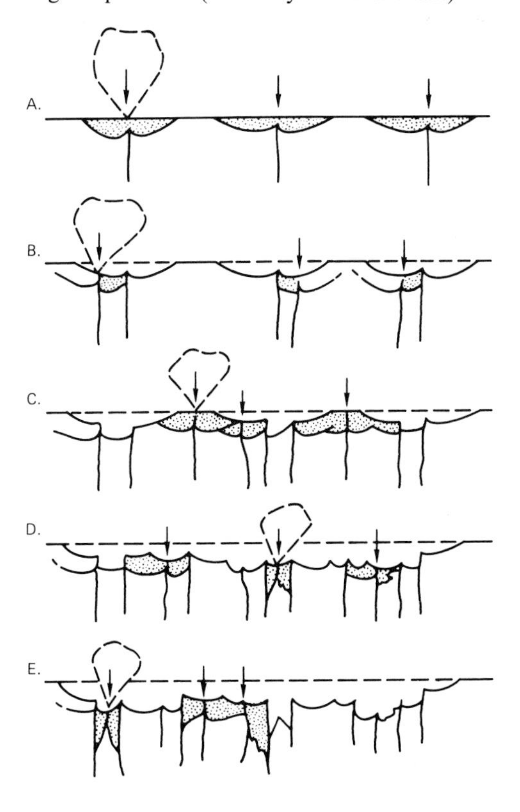

Fig. 4.5. Apparatus at Arizona State University, used for determining susceptibilities to abrasion, S_a, of various rocks, minerals, and other materials as function of particle speed, size, impact angle and composition, and atmospheric environment; samples are placed in holders (A), particles drop from the hopper (B) to the rotating area (C) where they are slung against the targets at known velocities and impact angles. Atmospheric composition and density can be controlled. Chamber (D) is contained by lid (E) which swings into place. S_a is determined as a ratio of the target mass lost (determined by weighing samples before and after the experiment) to the particle mass impacted. The apparatus is based on a design by Argonne National Laboratories for materials testing.

Fig. 4.7 shows the effect of impact velocity on S_a for a variety of rocks and materials. Although the effect of particle velocity, U_p, is poorly defined for volcanic tuff (both welded and non-welded) and rhyolite, the graphs for most materials show a slope of 2 on a log–log plot, giving the relationship

$$S_a \propto U_p^2 \quad (4.2)$$

Combining these relationships gives

$$S_a \propto D_p^3 U_p^2 \quad (4.3)$$

which shows that mass lost (m) per impact is directly proportional to the kinetic energy (KE) of the impacting particle:

$$KE = \tfrac{1}{2} m U_p^2 \propto \frac{\pi \rho_p D_p^3}{3} U_p^2 \propto D_p^3 U_p^2 \quad (4.4)$$

This result is in agreement with experiments by Dietrich (1977*a*,*b*) involving the abrasion of minerals by windblown dust. He concluded that the fundamental parameters which control aeolian abrasion are the effective kinetic energy of the impacting grain and the bond strength of the

Fig. 4.6. Susceptibility to abrasion of basalt, obsidian, and Hydrocal (gypsum cement) expressed as weight of target lost (*g*) per impacting sand grain (*i*) as a function of impact particle diameter. (From Greeley *et al.*, 1982.)

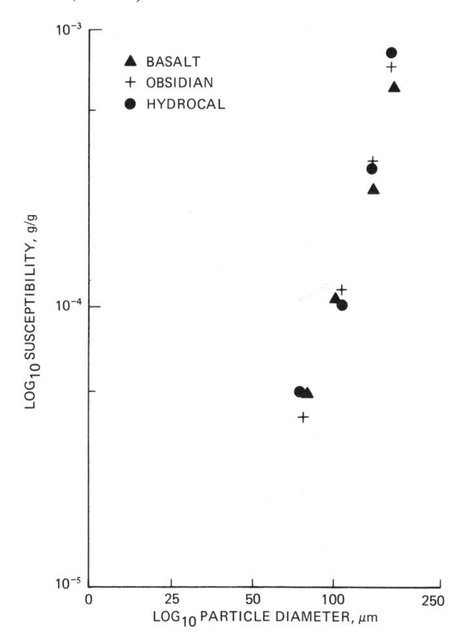

Fig. 4.7. Susceptibility to abrasion of various materials as a function of impact velocity. (From Greeley *et al.*, 1982.)

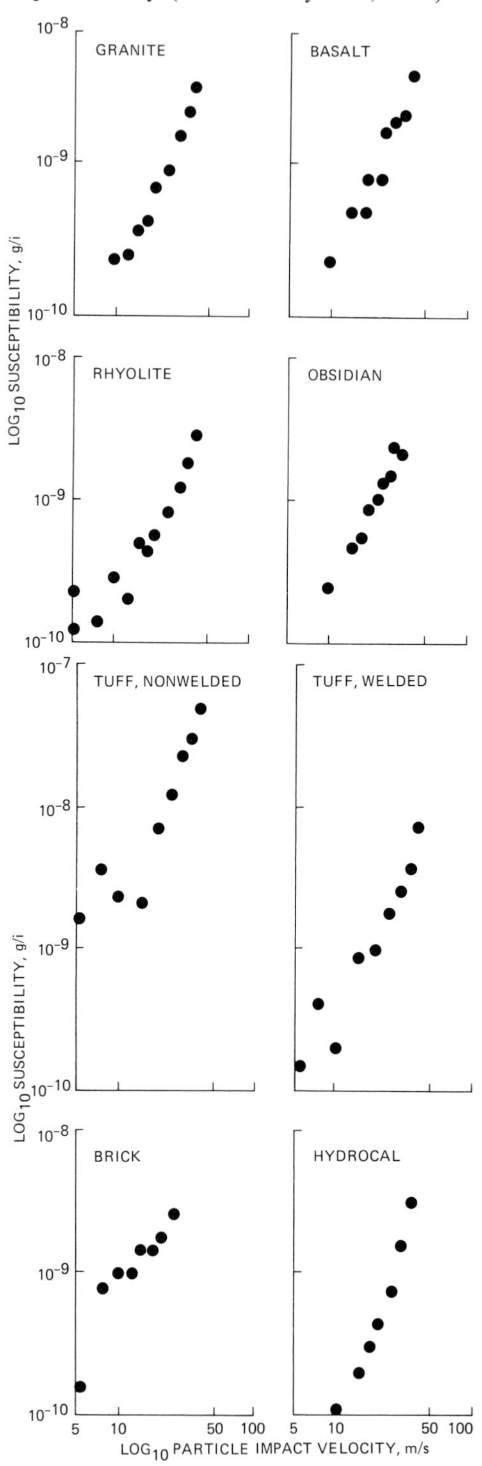

target, and that Moh's Hardness Scale (the standard scale used by geologists for the relative hardnesses of minerals) is not a good indicator of S_a for various materials. It was on this basis that Dietrich concluded that 'hard' materials could be abraded by dust and even snow, given enough time.

Increasing the rate of abrasion, however, is not as simple as increasing the flux of impacting grains, even though this would increase the net kinetic energy. Wood & Espenschade (1965) conducted experiments to assess the effects of dust concentration on S_a. As shown in Fig. 4.8, they found that, above a certain concentration, S_a declined. They attributed this result to the accumulation of dust particles into masses that were ineffective in causing erosion.

Similar results were obtained from experiments by Suzuki & Takahashi (1981) for sand. They found that the rate of abrasion increased with the flux of sand up to a flux of about 1.0 g/cm^2 sec, at which value the abrasion rate began to decrease. They attributed the result to buffering effects of rebounding grains off the target face.

The angle of the abrading grains with respect to the target must also be taken into account in assessing rates of aeolian abrasion. Fig. 4.9 shows the effect of particle impact angle on S_a. Most of the rocks and materials tested appear to mimic a combination of typical brittle and typical ductile substances, as described by Oh *et al.* (1972), except for obsidian which behaves as a brittle material. High values of S_a at low impact angles suggest a cutting or gouging mechanism, but it is unlikely that erosion occurs by plastic deformation as is the case for true ductile materials (Oh *et al.*, 1972). For abrasion granite specimens, it is also unlikely that individual mineral grains are separated and 'gouged out' by impactors because granite has crystals many times larger than the impacting particles. It may be possible,

Fig. 4.8. Susceptibility to abrasion of steel and aluminum as a function of dust concentration, showing that above a certain value abrasion decreases with increasing mass of the impacting particles. (After Wood & Espenschade, 1965.)

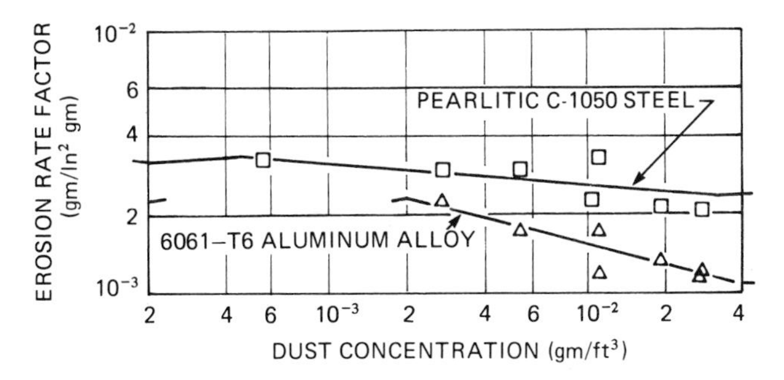

however, that the mineral grains in the rocks are fragmented into microblocks and plates which are sufficiently loosened to be gouged out by later impacts.

In summary, S_a is highly dependent on the particle velocity, size, and impact angle. In addition, glassy materials will erode very quickly for surfaces perpendicular to the wind; crystalline materials such as granite and basalt erode more quickly when surfaces are at an acute angle to the wind.

4.3 Ventifacts

Sand blasting has long been recognized as an effective means of erosion, both in nature and as a means for cleaning buildings and stripping surfaces in preparation for painting. The first recorded observations in the geological context are by Blake (1855) who described wind-eroded features in the Coachella Valley of southern California, an area that continues to be of interest for the study of aeolian processes. Blake later described various grooves and flutes formed in crystalline rocks and attributed them to sand blasting.

Fig. 4.9. Susceptibility to abrasion of materials, as a function of particle impact angle. (From Greeley *et al.*, 1982.)

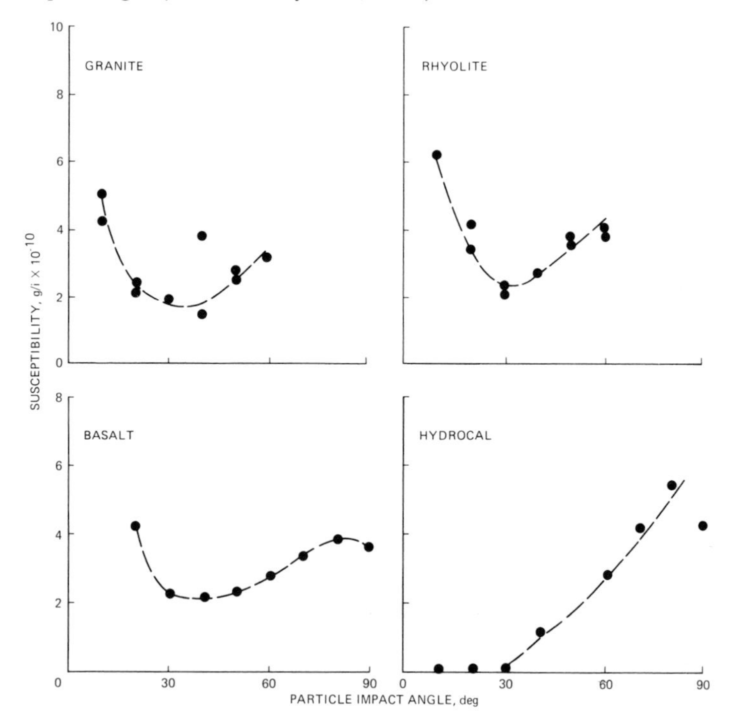

The term ventifact – literally meaning *wind-made* – was coined by Evans (1911, cited in Bates & Jackson, 1980), a British geologist who carried out extensive field work in arid regions of Asia and Africa. Although some definitions restrict the term to 'sand' blasted rocks, we prefer the broader context suggested by Whitney & Dietrich (1973) to include any wind-modified object, without restriction to erosion by windblown sand.

Ventifacts can be recognized by one or more features, including wind cut faces – called *facets* – polished, lustrous, or etched surfaces, and various features such as pits, flutes, and grooves. Ventifacts (Fig. 4.10) range in size from faceted grains a few millimeters across to rocks exceeding 3 m long. Typically, they occur wherever aeolian processes are, or have been, active,

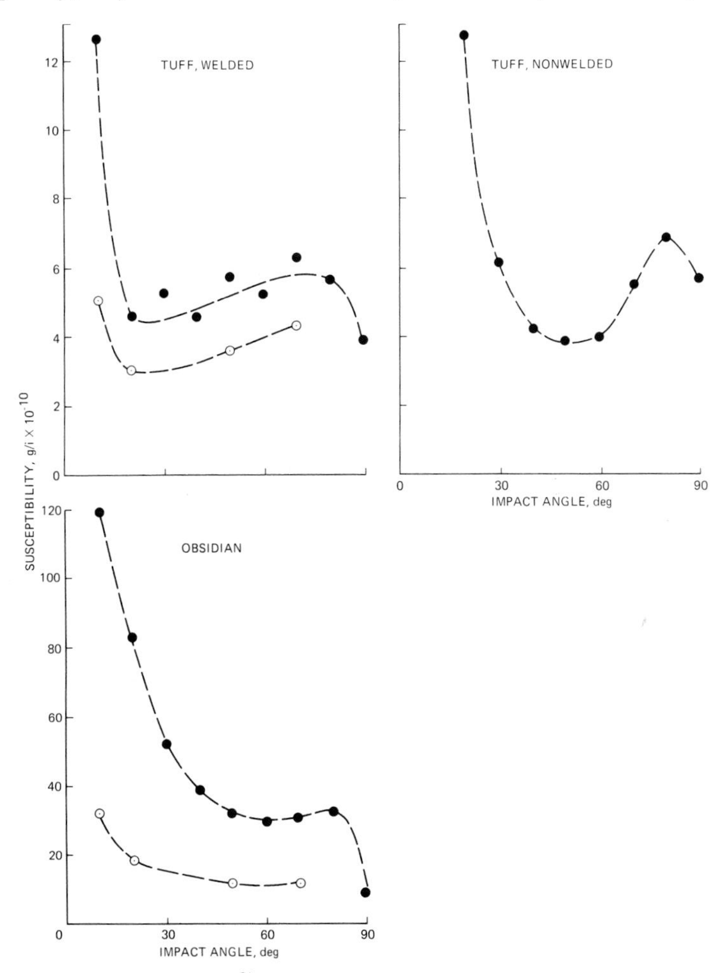

Fig. 4.9 (continued)

such as in arid regions, glacial outwash plains, and along coastlines. Conditions favorable for their formation include: (1) a ready supply of sand and silt particles as the agents of abrasion (but not a too-abundant supply, as discussed in Section 4.2), (2) winds of sufficient strength to move particles, and (3) an absence or scarcity of vegetation, which would otherwise impede the windblown particles before impacting the potential ventifacts. Glacial outwash plains meet these criteria especially well, as evidenced by abundant ventifacts in parts of Iceland and Antarctica (Lindsay, 1973; Whitney & Splettstoesser, 1982). Given these conditions on Mars, we might expect abundant ventifacts, but they may not be easily recognized, as we will discuss later.

Ventifacts are reported throughout the geological column, including the Precambrian (Sugden, 1968), but most descriptions are for Pleistocene and

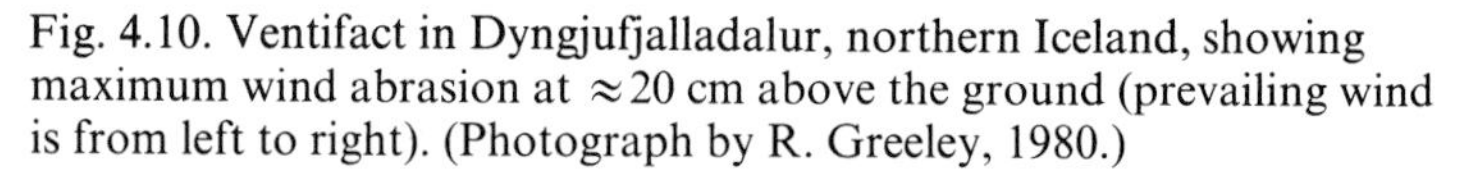
Fig. 4.10. Ventifact in Dyngjufjalladalur, northern Iceland, showing maximum wind abrasion at $\approx$20 cm above the ground (prevailing wind is from left to right). (Photograph by R. Greeley, 1980.)

Recent forms. Since Bryan's (1931) review, there has been no general survey of ventifact localities, mode of formation, etc., although many aspects of ventifacts have been discussed in detail. Sharp (1949) gives a good description of ventifact morphology and discusses the environments favorable for their formation, along with the presentation of ventifacts in a site in Wyoming, which was the primary topic of his paper.

The only semiquantitative field investigation of ventifact formation is that of Sharp (1964, 1980). This is probably the single most important study of ventifacts done thus far. He established a small experimental plot in the floodplain of the Whitewater River of Coachella Valley, near Palm Springs, an area well known for its ventifacts and frequent wind storms. The plot was established in 1952 and monitored frequently until 1969 when it was destroyed by flooding. The plot contained various natural rocks, plus man-made materials including common bricks, blocks of hydrocal (gypsum cement) and lucite rods. These items were photographed and measured to determine the amount of erosion and changes in shape during the 16-year period of observation. During the initial 11-year period (Sharp, 1964), four time intervals ranging from 30 to 223 days involved collection of windblown particles as a function of height near the plot. These data provided estimates of flux for correlation with rates of abrasion determined from the samples. Unfortunately, wind data concurrent with the observations of the experimental plot are not available at the site.

Several laboratory experiments have been carried out, which shed light on ventifact formation. Kuenen (1928) and Schoewe (1932) abraded rock fragments under controlled conditions to determine – among other factors – the role of the original rock shape on the development of ventifact morphology. More-recent experiments by Dietrich (1977*a*,*b*) and Whitney & Dietrich (1973) were conducted to study the effectiveness of abrasion by materials other than sand-size quartz. By using small, primitive wind tunnels, Whitney (1978, 1979) examined air flow around ventifacts and suggested that small vortices are important in developing features such as pits and grooves.

4.3.1 *Ventifact morphology*

Ventifacts are found in a wide variety of shapes, including forms that are prolate, oblate, triaxial ellipsoidal, pyramidal, or irregular (Fig. 4.11). Wind cut surfaces may be curved or flat ('facets'), and edges may be angular or rounded. Multiple facets are common, and a hierarchy of terms describing the facet arrangements has been proposed: einkanter, zweikanter, driekanter, etc. (one, two, or three corners, or edges). Usage of these terms has not been consistent, and because ventifacts can have as many as 20 facets, the usefulness of this hierarchy is questionable.

Perhaps more important than the terms applied to them are the reasons why multiple facets develop. Various studies have demonstrated that they can result from many circumstances, including multiple wind directions,

Fig. 4.11. (*a*) Small ventifact of the 'Brazil-nut' variety, from the Wind River District near Shoshoni, Wyoming. This ventifact, carved from dolerite, is ≈18 cm long. (Photograph courtesy of M. Whitney.) (*b*) Siliceous ventifact from central Egypt showing small-scale fluting presumed to result from wind abrasion. The ventifact is ≈10 cm long. (Photograph courtesy of M. Whitney.)

(a)

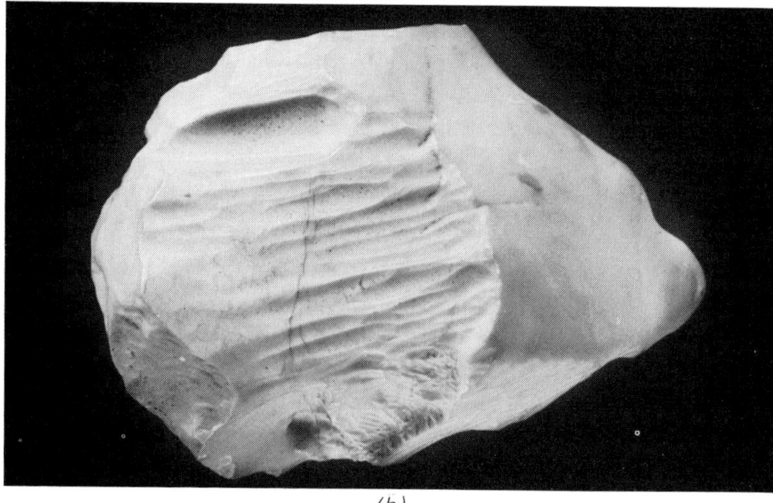

(b)

shifting and overturning of the rock during its evolution to present new surfaces to abrasion, or from complicated patterns of air flow and abrasion by suspended particles.

Pits, flutes, and grooves are carved in some ventifact surfaces, as shown in Fig. 4.12. Pits may be cylindrical, hemispherical, conical, or irregular. Although they are common on surfaces which are inclined $\approx 55–90^\circ$ to the wind, they can occur on all ventifact surfaces. The formation of most pits is probably initiated at some inhomogeneity or disruption in the rock such as a small crack or an inclusion of a soft mineral grain. Once initiated (as may form in volcanic vesicles, Fig. 4.13), pits may be enlarged and sculpted by further wind abrasion, perhaps involving small vortices carrying abrading particles in suspension (Whitney, 1978).

Grooves are U-shaped, open-ended depressions having fairly parallel sides which occur almost exclusively on the top surfaces of ventifacts. They range in size from less than 1 mm to several centimeters wide and can be up to a half-meter long. Some coincide with fractures, veins, or laminations in the rock, but most cut directly across mineral grains (Powers, 1936).

Flutes commonly occur on rounded edges between the windward facet and the top surface of ventifacts. In this respect they appear to be transitional between pits and grooves. Many flutes appear to originate in pits and have diverging sides downwind as they wrap over the edge of the ventifact, becoming shallower on the top surface. Those described by Sharp (1949) in Wyoming are broadly U-shaped and can be as long as 15 cm. They are commonly found on surfaces inclined $\approx 40^\circ$ to the wind and are apparently formed independently of mineral hardness or rock structure.

Rills are small channel-like furrows that range in size from submillimeter to 6 mm across. They appear to result from both aeolian abrasion and from chemical weathering processes such as solution, but there is little agreement on which process is predominant, or even on the criteria for identifying rills formed by one process or the other. It has been observed, however, that rills seldom occur on windward faces (Whitney & Dietrich, 1973), and thus it is unlikely that they could form by the impact of saltating grains.

Ventifact morphology has been used to map modern wind patterns from currently active ventifacts and to map paleowinds from 'fossil' ventifacts. The predominant facet generally faces the prevailing wind and, as described above, pits commonly occur on windward surfaces. As we have seen, however, pits can occur on any surface, and ventifacts can shift their orientation; thus, the use of a single ventifact, or even several ventifacts, to determine wind directions can be risky. Mapping a large population of

Fig. 4.12. Ventifacts at Garnet Hill, southern California, showing (*a*) pits on the windward face of the gneissic boulder, grading into grooves and flutes on those faces parallel to the wind, (*b*) grooves cutting across mineral grains and gneissic texture. (Photographs by S. H. Williams, 1980.)

(a)

(b)

ventifacts, however, does appear to produce a wind pattern which coincides with measurements of the active winds, as shown in Fig. 4.14.

4.3.2 *Factors involved in ventifact formation*

In Fig. 4.15 we show a stylized diagram of the parameters that appear to be important in the formation of ventifacts. Major parameters involve wind characteristics, particle supply and characteristics, target rock (the potential ventifact) characteristics, terrain characteristics, and other (non-wind) weathering influences. Although we will discuss these somewhat independently for simplicity, it must be kept in mind that they all tend to be interrelated.

The size, shape, and composition of windblown grains all affect the abrasion of rock, as discussed earlier. Thus, the availability of particles that

Fig. 4.13. Basalt surface at Amboy, California, showing vesicles that have been modified by wind abrasion. (Photograph by R. Greeley, 1976.)

can be transported by the wind is obviously an important factor in ventifact formation. Sharp (1964, 1980) demonstrated a direct correlation between the rate of aeolian abrasion (at his experimental plot) with rainfall in the area and attributed the result to the influx by fluvial processes of sand which could be picked up later by the wind.

Most cutting of ventifact facets is accomplished by sand-size grains traveling in saltation. An analysis of the speed of grains in saltation (Fig. 3.20; Table 3.3) shows that the grain continues to accelerate continuously along its path. From an impact energy consideration, the maximum erosion should occur on the parts of ventifacts closest to the ground. Sharp (1964), however, found in his test site that maximum abrasion occurs at a height of about 10–15 cm above the ground (see also Fig. 4.10), and he attributed the result to the combination of grain sizes, fluxes, and particle velocities, all of which vary with height and wind speed (Greeley *et al.*, 1982). Suzuki & Takahashi (1981) obtained the same result as Sharp, using a theoretical approach.

Dust traveling in suspension is also capable of abrasion to form facets and features such as pits, flutes, and grooves. The relative importance of saltating sand versus suspended dust, as agents of abrasion, is contentious. Higgins (1956) probably surmised the situation best when he concluded that the relative importance is simply a function of what is available to do the abrading. If sand is available, it will dominate in the cutting of major features, with suspended dust cutting smaller features and abrading

Fig. 4.14. Map of Garnet Hill, showing the orientation of ventifact surface features which, taken together, show the wind flow over the hill. (From Hunter, 1979.)

surfaces not accessible to saltating grains. If sand is not available, then dust alone in suspension can be responsible for ventifact formation, but it would probably proceed at a much slower rate than if sand were involved. It must also be kept in mind that simply increasing the flux of particles is not sufficient to increase the rate of abrasion, particularly if dust-size particles are involved (Fig. 4.8).

Wind-related factors in ventifact formation include the strengths and directions of the winds, and the nature of the surface over which the wind blows. These factors determine the sizes of particles to be moved, whether by saltation or suspension (grains in traction are not likely to cause much abrasion), and the flux and velocity of windblown grains. In general, stronger winds move more material, at higher velocities and at greater heights, than gentler winds. For example, during a major windstorm in 1977 in California, steel power poles were frosted to heights exceeding 10 m, and the height of maximum abrasion was raised from that noted by Sharp (1964), of 10–15 cm, to an average of 24 cm (Sakamoto-Arnold, 1981).

Surfaces containing rocks ('roughness elements') too big to be moved by the wind affect not only the wind velocity profile (see Section 2.4.2), but also the saltation trajectory. Grains tend to bounce off fixed, hard elements such as boulders, reaching greater heights than they would if they were traveling over a sandy surface. Because some grains reach greater heights, they are also accelerated to greater speeds, both because they travel in a stronger

Fig. 4.15. Diagram showing the three principal factors required for calculating rates of wind abrasion: (1) wind frequency (strengths, durations), (2) various particle characteristics (velocity, flux, etc., for various winds), and (3) susceptibilities to abrasion for various rocks.

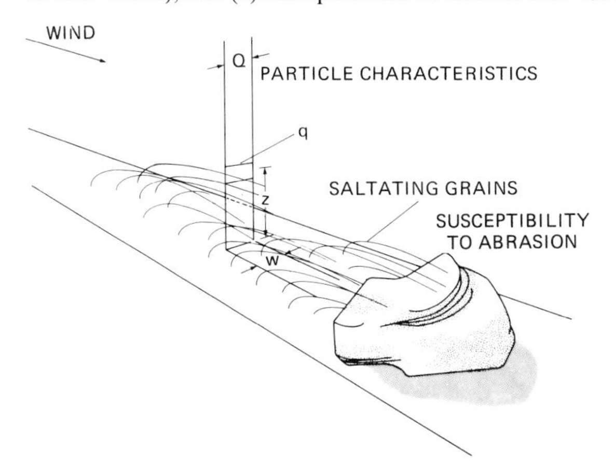

wind regime and because they are aloft for a longer period. Thus, there is opportunity for more-vigorous abrasion in a boulder strewn setting. The same result may occur on exposed bedrock surfaces, where grain saltation is also enhanced.

Various characteristics of the potential ventifact itself are important in its evolution. On small rocks (less than 15–25 cm), facets seem to be cut without regard to the original shape of the rock fragment, but this is not the case on larger rocks in which original shape is important. Regardless of the processes by which the rock is originally produced – fracturing, jointing, impact cratering, or other erosional processes – the faces so produced, and their orientation, affect not only rates of abrasion and many of the features generated by the wind, such as pits, but also the air flow field and the subsequent abrasion by dust. Maximum cutting occurs on faces oriented slightly off-axis normal to the wind, as observed in the field (Sharp, 1980), experimentally (Whitney, 1979), and in controlled abrasion experiments (Greeley *et al.*, 1982). Rock faces pointing squarely into the wind abrade slower, probably because rebounding grains hit incoming grains, slow them, and result in lower impact energies. Thus, there is initial rapid cutting of surfaces inclined slightly to the wind; with progressive cutting, the facet becomes more inclined and abrasion is retarded (King, 1936).

The composition and texture of the abraded rock fragment influences the abrasion textures produced on ventifact surfaces. Higgins (1956) noted that dense, homogeneous rocks lead to smooth faces and sharp, well-defined edges between the faces. With an increase in grain size, there is a progressive increase in pitting and grading into fluting. Rock hardness seems to govern only the degree of polish and, to some degree, the rate of abrasion.

Inhomogeneities in the ventifact are important in the development of various surface features such as etches, pits, and 'bosses.' Mineral grains that are harder than their matrix can produce raised areas, or 'bosses' (Whitney, 1978) and can lead to the formation of finger-like projections facing into the wind (Fig. 4.16). Similar projections may develop as a result of variations of induration or cement in the host rock (McCauley *et al.*, 1977*a*).

During their evolution, ventifacts may shift their position, roll, or completely overturn, presenting new or altered surfaces to the wind abrasion regime. These movements can result from a number of processes (Sharp, 1964), including undercutting of the ventifact, deflation of the material on which it rests, impact by some large windblown object such as a tree limb, or from other processes including animal activity, frost heaving, and flooding.

4.3.3 Ventifacts on Mars and Venus

Few rocks are so intensively studied as those found at landing sites on the Moon, Mars, and Venus. Samples returned from the Moon via the Apollo and Luna missions have been subjected to nearly every conceivable analysis; unmanned lunar missions have almost always involved imaging systems to obtain views of the landing site, which permit analyses of the rocks, fragments, and particles within view. The same has been true of the unmanned Viking 1 and 2 Landers on Mars and the Venera 9, 10, 13 and 14 Landers on Venus. Yet, despite the many hours spent analyzing the rocks seen on the surfaces of Mars and Venus, a great many questions remain unanswered regarding the effects of winds on the evolution of the rocks. There are, however, some general observations that can be made. Rocks at both martian landing sites range in size from 1 cm to more than 1 m across. In contrast to the Viking Lander 1 site where there is a wide range of rock shapes, colors, and textures, those at Viking Lander 2 are similar to each other and occur with about one-half more areal density than that found at the Lander 1 site.

Although rocks at both martian sites are pitted, the occurrence of pits is more frequent at Viking Lander 2 (Fig. 4.17); Mutch *et al.* (1977) noted that at least 90% of the rocks close to Lander 2 were pitted. Most of the pits range in size from a few millimeters to a few centimeters across and are

Fig. 4.16. Finger-like projection ≈10 cm long in loosely cemented sediments, which result from shielding from wind abrasion by small inclusion in the sediments. (Photograph courtesy of S. H. Williams.)

equidimensional, although some tend to be elliptical. Mutch and his colleagues considered the possible mechanisms for the formation of pits and concluded that most of the pits are vesicles and that the rocks are volcanic. On the other hand, comparison of martian rocks with ventifacts in the Western Desert of Egypt led McCauley and his colleagues (1979) to consider most of the pits seen on the Viking Lander images to be the result of wind activity. They attribute much of the abrasion to windblown dust rather than to the result of saltating sand.

Nearly all investigators who have studied the Viking Lander images have reported various surface features on the rocks, which they attribute to wind abrasion. Features include facets and grooves (Fig. 4.18), some of which are aligned with the prevailing wind directions.

Rocks revealed by the Venera cameras on Venus are even more difficult to analyze than the martian rocks. Garvin *et al.* (1981) studied the morphology of the venusian rocks imaged by Venera 9 and 10 and compared the results with similar analyses for Mars. They found that the rocks on Venus are much more rounded and that there are very few features that can be attributed directly to wind abrasion, although the resolution of the images is significantly lower than that of the Viking Lander images, and

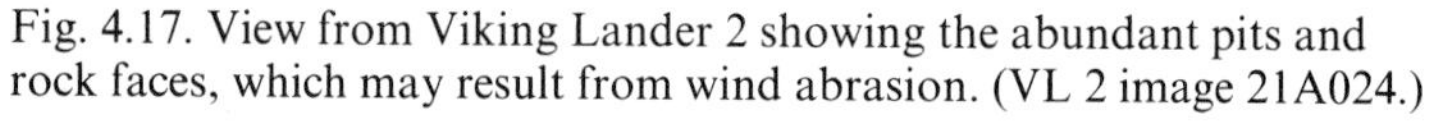

Fig. 4.17. View from Viking Lander 2 showing the abundant pits and rock faces, which may result from wind abrasion. (VL 2 image 21A024.)

it may not be possible to discern features indicative of ventifacts. Abrasion by particles would not be expected to be as extensive on Venus because the particle speeds would be lower.

4.4 Rates of aeolian abrasion: Earth and Mars

Various attempts have been made to determine the rate of formation of ventifacts. Hickox (1959) noted ventifacts in Nova Scotia, formed in material which could be dated approximately. He concluded that the facets were cut over a period less than ten years. Attempts to correlate results from field studies with laboratory experiments generally have been frustrated because of the lack of adequate wind frequency data for the field sites. However, Sharp (1964, 1980) demonstrated that during a 16-year period in the Coachella Valley of California – one of the most vigorous

Fig. 4.18. View of possible ventifacts on Mars at the Viking Lander 2 site, showing sediment-filled pits and shallow flutes (rock in center right); note also the rock in the center having an abrupt change from abundant pits to few pits on the lower part, which may represent burial and subsequent exhumation, possibly by deflation. (From McCauley *et al.*, 1979.) (VL 2 image 21B021.)

aeolian areas known – abrasion of crystalline rocks was barely recordable. He showed that the rate of ventifact formation is strongly influenced by the availability of sand and the occurrence of storm-strength winds.

In order to estimate rates of wind abrasion, whether it be for a ventifact or outcrop exposure, knowledge is required for three primary variables (Fig. 4.15): (1) wind frequency, (2) particle flux, and (3) target susceptibilities to abrasion. Wind frequency refers to the distribution of wind strengths and durations per unit time; these data, typically, are available from meteorology recording stations on Earth. On Mars the Viking landers measured winds for two locations for more than three years, although only once, or possibly twice, have wind speeds exceeded threshold. No wind frequency data are available for Venus, although some 'spot' measurements of winds have been obtained, which might be combined with numerical models of atmospheric circulation to derive wind frequencies.

Parameters relating to particle flux include the availability, velocity distributions of particles of different sizes as functions of free-stream wind speed, surface roughness, and height above the ground, as defined in Chapter 3. Susceptibility to abrasion, S_a, is a measure of the resistances of different rocks and minerals. As shown in Section 4.2.2, the S_a values are not constant but depend upon the combination of target conditions and conditions of impacting particles. Thus, S_a values must have the various conditions stated to be meaningful.

Using data from a wide range of sources, including experimental, field, and computational results, Greeley and colleagues (1982) estimated possible rates of wind abrasion on Mars. They found that the rate of abrasion should be substantially higher on Mars than on Earth, primarily because of the higher wind velocities required for particle entrainment. These high winds result not only in greater particle velocities, but in increased fluxes as well (see Chapter 3). Calculations of the present rates on Mars yield values ranging up to 2.1×10^{-2} cm/yr, based on abrasion by quartz, basalt, or basaltic ash grains in the size range most easily transported by wind and assuming an abundant supply of particles. However, estimates of the ages of the martian surface, based on preservation of impact craters (Arvidson *et al.*, 1979), suggest hundreds of millions of years with relatively little modification of any sort, including that of aeolian erosion.

This conflict between the estimated present rate of wind erosion and the antiquity of the surface implies one or more of the following possibilities (Greeley *et al.*, 1982): (1) the material composing the surface is orders of magnitude more resistant to erosion than any known rocks (unlikely), or (2) that the climate for most of the geological history of Mars was less favorable for aeolian activity than the present environment (possible, but unlikely), or (3) that the ancient surfaces have been protected from aeolian abrasion for long periods of time (evidence suggests this to be a possibility for some regions of Mars, but it is unlikely to account for all areas of the planet), or (4) that the primary agent of aeolian abrasion is in very short supply or is very inefficient, as would be the case for sand-size aggregates of fine particles. Although all the above factors may contribute to the low rate of abrasion, the last possibility is considered to be the primary factor in accounting for the apparently low rate of wind abrasion on Mars.

Although little can be said concerning the rate of wind abrasion on Venus, because of poor knowledge of the factors required to make calculations, it seems likely that the rate would be relatively low because of the low velocities of the winds (and the low kinetic energies of the particles), as well as the 'cushioning' effects of the dense venusian atmosphere.

4.5 Yardangs

Thus far, we have been concerned primarily with aeolian erosion on the scale of individual rocks. Let us now consider wind-eroded features on the scale of landforms. Around the turn of the nineteenth century, a Swedish explorer, Sven Hedin, carried out a series of geographical expeditions through central Asia. During one trip through the Taklimakan Desert of China (Hedin, 1903), his party encountered a rugged terrain of subparallel ridges separated by troughs cut out of ancient lake sediments (Fig. 4.19). Hedin described one such area as a labyrinth of clay terraces and noted that the local inhabitants referred to the ridges as *yardangs*. The name is derived from the Turkistani word *yar*, which means ridge or bank from which material is being removed. This usage struck Hedin as altogether appropriate in the geological context and applied the term to all such aeolian landforms.

The yardangs Hedin described range up to 6 m high and are hundreds of meters long, all oriented approximately parallel to the wind. Ancient trade routes wound through the yardangs, generally along the troughs parallel to the wind. Hedin found travel in any other direction through the yardang field to be exceedingly difficult.

Hedin described the yardangs as being eroded from compact, yellow clay beds that were deposited in an ancient lake named Lop-nor. He considered the ridges to have been carved first by running water, then to have been shaped later by the wind. Thus started a debate that continues today concerning yardang formation – which is more important, water or wind erosion?

4.5.1 *General characteristics*

Excellent reviews of yardangs are provided by Peel (1970) and McCauley *et al.* (1977*a*,*b*). Jack McCauley and his colleagues of the US Geological Survey became interested in yardangs after the discovery of ridges, resembling yardangs, on Mars during the Mariner 9 mission. Prior to these reviews, most references to yardangs were incorporated in regional or local studies. McCauley *et al.* surveyed the available literature and documented the known terrestrial yardangs. Applying the term *yardang* to all streamlined, aerodynamically shaped hills, regardless of size and type of bedrock in which they form (Fig. 4.20), McCauley and colleagues noted that yardangs occur in desert regions on all continents except Australia and seem to be restricted to the most arid parts of sand-poor deserts where vegetation and soil development are minimal.

Fig. 4.19. Yardangs of the Lop Nur region, Taklimakan Desert, China, derived from sketches by Hedin (1905) who first described these wind erosion features and applied the term *yardangs*. Most of the yardangs in this area are flat topped. (From McCauley *et al.*, 1977*a*.)

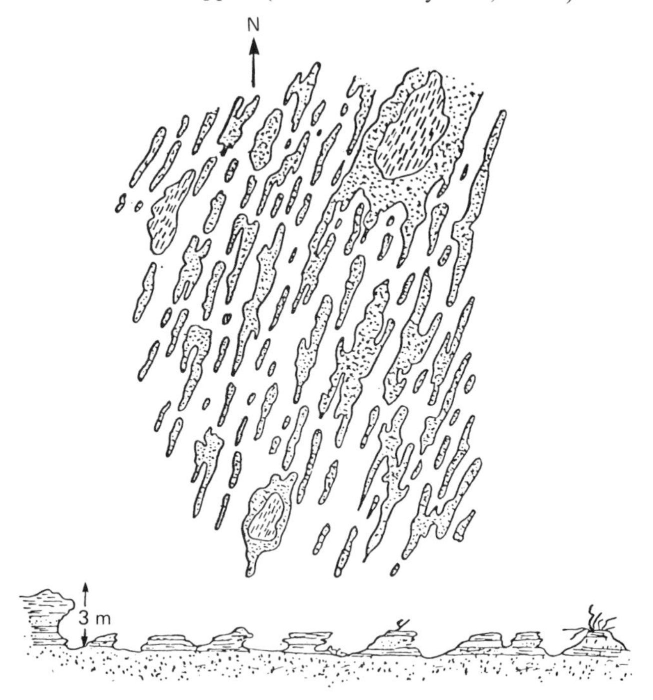

Typical yardangs have the form of an inverted boat hull (Fig. 4.21), with the highest and widest part of the ridge about one-third of the distance from the 'bow' to the 'stern'. The downwind ('stern') end is, typically, a gently tapering bedrock surface, or it may have an elongate sand 'tail'. Yardangs range in size from less than a few meters long by 1 m high to those reported by Mainguet (1968) in Chad, which are nearly 200 m high and several kilometers long. Yardangs typically occur in clusters and are oriented parallel to the prevailing winds which formed them. Many yardangs are undercut at the upwind end (Fig. 4.22), which may be a reflection of the zone of maximum aeolian abrasion, as discussed in Section 4.3.2.

Although yardangs can develop in any type of rock, most are found in slightly compacted, fine-grained sediments. It would appear that deposits which are relatively soft and easily sculpted, yet cohesive enough to retain steep slopes, are ideal for yardang formation. Thus, lakebed silts and clays, and some volcanic ash deposits are common compositions found in yardangs (Ward, 1979).

4.5.2 *Yardang localities*

Fig. 4.23 shows the localities of yardangs on Earth, compiled by

Fig. 4.20. View of a small yardang in Rogers Lake, California, showing the 'keel' or ridge crest. The meter stick resting on the ridge indicates the scale. (Photograph courtesy of W. Ward, US Geological Survey.)

McCauley *et al.* (1977*b*). In this section, we describe some of the better-known of these localities.

The Lut Desert of Iran contains some of the largest and best-known yardangs on Earth (Fig. 4.24). First described by Gabriel (1938), the ridges are up to 60 m high and extend for hundreds of meters in length – all trending north–northwest parallel to the prevailing wind. The yardangs are separated by 'boulevards' ≈ 100 m wide, which, like those in China, have controlled travel since ancient times. Gabriel noted that the Iranians of the area referred to the terrain as *Shahr Lut*, meaning desert cities, because of the fancied resemblances of the eroded terrain to buildings.

All the major deserts of Africa contain yardangs. It has been suggested that the Sphinx of Egypt was carved from an already streamlined hillock,

Fig. 4.21. Yardangs of the Ica Valley, Peru, eroded from marine sediments; the longest yardang is ≈ 500 m long; granule ripples trace wind flow. The arrow indicates prevailing wind directions. (From Ward & Greeley, 1983.)

Fig. 4.22. Yardangs at Rogers Lake, California, first described by Blackwelder (1934); (*a*) photograph taken in 1932 (copyright, Geological Society of America), (*b*) photograph of the same area in 1975, showing undercut zone in the upwind end of a yardang (arrow). (From Ward & Greeley, 1984.)

(a)

(b)

Fig. 4.23. Major yardang localities on Earth: (1) Taklimakan Desert, China (type locality), (2) Lut Desert, Iran, (3) Khash Desert, Afghanistan, (4) Sinai Peninsula, (5) Saudi Arabia, (6) Egypt, (7) Sudan, (8, 9) Libyan Desert, (10, 11) Tibesti region, Chad and Niger, (12) Algeria, (13) Namib Desert, Namib, (14) Mojave Desert, California, (15) Talara region, Peru, and (16) Paracas-Ica region, Peru. Confirmed yardangs have not been reported in Australia. (From McCauley *et al.*, 1977*b*.)

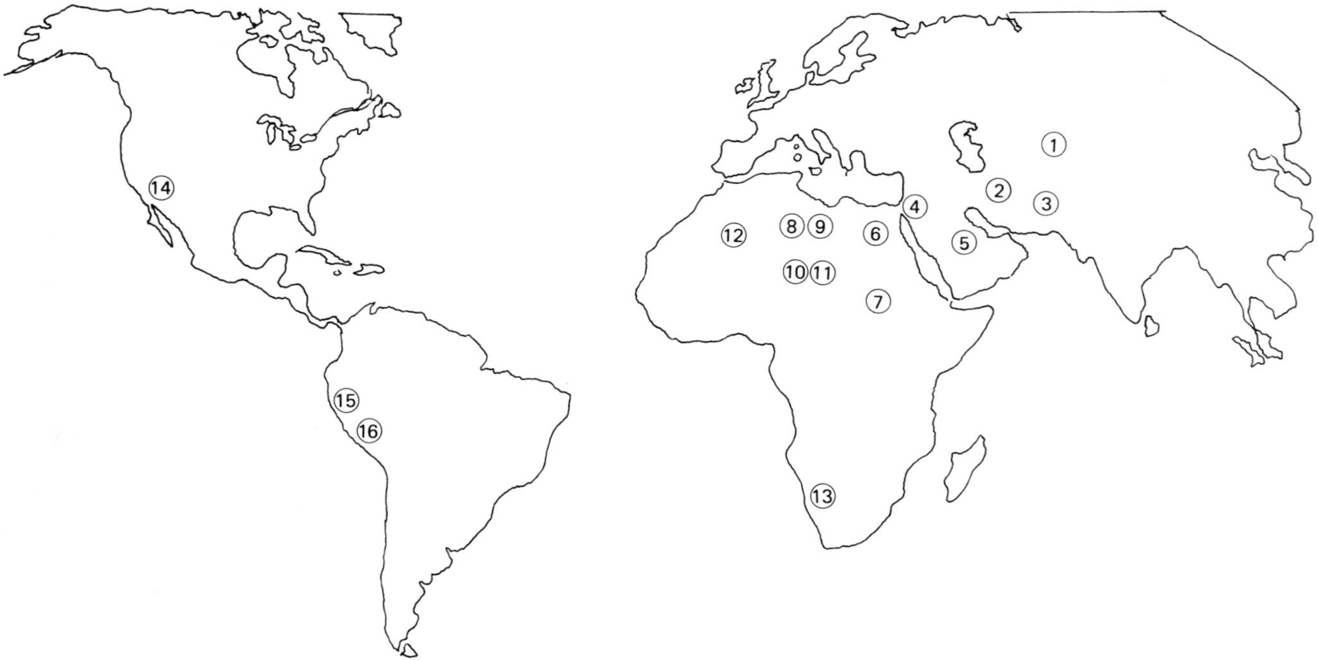

similar to numerous other natural sphinx-like forms found in the area. In fact, early British explorers referred to the features as 'mud-lions.' Southwest of Egypt in the Borkou region of Chad, Monique Mainguet (1968) noted yardangs exceeding 20 km in length by 1 km wide, separated by troughs 0.5–2 km wide (Fig. 6.14). After several years of careful study and detailed mapping, Mainguet has concluded that the orientations of both the yardangs and the sand dunes in the area are controlled by winds which are diverted around the topographical mass of the Tibesti highlands.

Some of the most impressive yardangs anywhere to be found occur in the coastal deserts of Peru. From his studies in the region, Bosworth (1922) was first to describe yardangs as having the shape of an inverted boat hull. Later

Fig. 4.24. Landsat image showing the yardangs of the Lut Desert, Iran.

studies in Peru, by Grolier *et al.* (1974) and McCauley *et al.* (1977*b*), provide detailed descriptions and superb photographs of these classic yardangs. The best locality is found on the Paracas Peninsula, west and south of Pozo Santo. Here, hundreds of yardangs – some up to 1 km long – have developed in bedded siltstones and clays. Some parts of the field show numerous sinuous channels that represent relict water courses which have been taken over by the winds to scour troughs between the ridges. McCauley and his colleagues found that the length-to-width ratio of the yardangs range from 3:1 in the north part of the field to 10:1 in the south.

Some of the Peruvian yardangs are locally interspersed with small, meter-size knobs called *quesos* (Spanish for cheese). McCauley *et al.* (1977*b*) suggested that these may be large concretions which are more resistant to wind erosion than the silts and clays in which they occur.

4.5.3 *Yardang formation*

Bosworth's (1922) classic analysis of desert processes in Peru includes a consideration of the factors involved in wind erosion. He noted that landforms evolve to a shape which offers minimum resistance to the fluid in which it is immersed. It was this consideration that led him to liken the shape of yardangs to a ship's hull. Bosworth also noted that streamlining of topography was brought about by both erosion and deposition.

Following the reasoning suggested by Bosworth, Ward & Greeley (1984) theoretically modeled the evolution of yardang shape through considerations of fluid mechanics. Drawing upon principles outlined by Fox & McDonald (1973), they noted that the drag force exerted by a fluid passing around an object reaches a minimum when the width-to-length ratio is about 1:4 in plan-view (Fig. 4.25). Furthermore, flow over a hill viewed in cross-section is seen to have a distribution of velocities and pressure gradients, as shown in Fig. 4.26. Thus, Ward & Greeley concluded that, in flow over a hill, erosion will occur in distinctive zones until the 'ideal' shape is attained, i.e., an elongate hill of 1:4 width-to-length ratio, asymmetric in profile, with the highest part in the upwind one-third of the hill.

To test the idea of an equilibrium shape for yardangs, a series of qualitative experiments was carried out (McCauley *et al.*, 1977*c*; Ward, 1978; Ward & Greeley, 1984). Using easily eroded materials, such as wet sand and sawdust, various forms (mounds, cylinders, cubes, etc.) were molded and subjected to uniform winds under controlled conditions. Ward

& Greeley found that most models evolve through a series which include: (1) erosion of windward corners, (2) erosion of the front slope and upper surface, (3) erosion of the rear flanks, and (4) erosion of the downwind upper surface, eventually producing a classic yardang shape (Fig. 4.27). Erosion was at a maximum early in the experiment and diminished

Fig. 4.25. Drag coefficient for flow past a streamlined object, as a function of thickness ratio, showing that total drag is minimal at a ratio of ≈0.25. (After Fox & McDonald, 1973.)

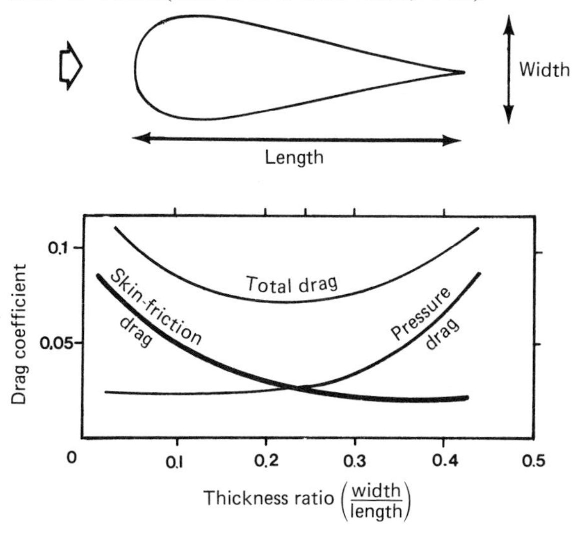

Fig. 4.26. Pressure variations along a surface of variable height. (From Ward & Greeley, 1984, after Fox & McDonald, 1973.)

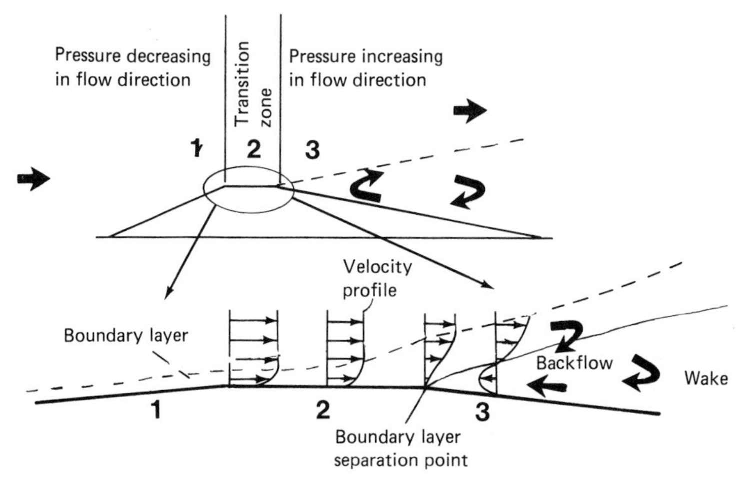

asymptotically with time. In general, the experiments appear to bear out the theoretical models regarding flow over and around an obstruction.

Nature, however, is seldom as simple or ideal as theory and laboratory modeling would suspect. As numerous investigators have noted, the initial shape of yardangs may owe more to stream erosion than to wind. Differences in lithology, both horiziontal and vertical, in the material to be sculpted may also control degrees of erosion and the evolution of yardang shapes. The role of abrasion by windblown grains (which is controlled by a host of factors, as we have discussed) versus deflation of grains weathered *in situ* will contribute to the shape of the evolving yardang.

Thus, the shape of yardangs is the result of a complex interplay of internal factors, such as lithology and structure, and external factors, including flow field, surrounding topography, and the supply of agents of abrasion.

4.5.4 *Yardangs on Mars*

Mariner 9 images of Mars first showed linear ridges which suggested the presence of yardangs (McCauley, 1973; Sharp, 1973). The improved quality, higher resolution, and broader coverage of Mars provided by the Viking mission has greatly expanded the available data on martian yardangs and their geological settings (Ward, 1979). Detailed

Fig. 4.27. Side-view of a wind tunnel model composed of slightly compacted particles. Initial shape was an elongate block; after being subjected to 130 hours of wind, the block evolved to this typical yardang shape (wind is from the left; the model is ≈ 30 cm long).

mapping of aeolian features on Mars shows that yardangs occur in several areas.

Fig. 4.28 shows yardangs in the western Tharsis area (Fig. 1.14), a region characterized by extrusive volcanism. The morphology of the linear ridges and flat-topped mesas closely resemble that of the type yardangs described by Hedin (1903) in China, but they are considerably larger – individual ridges are tens of kilometers long, separated by valleys up to 200 m wide. Although initially interpreted as longitudinal dunes (Breed & Ward, 1979), Ward later (1979) concluded that the ridges are yardangs. Analyses of the geological setting and inferences, based on thermal inertia data from the Viking Orbiters (which provide clues to particle grain sizes), led Ward to conclude that the lithologies for the yardangs might be ignimbrites (a type of volcanic ash), highly porous lava flows, or mudflows.

Fig. 4.29 shows yardangs in the Aeolis region of Mars at about 4°S, 206°W. The orientation of the largest ridges suggests that they were eroded by southwest winds; however, a second, smaller set of ridges is oriented

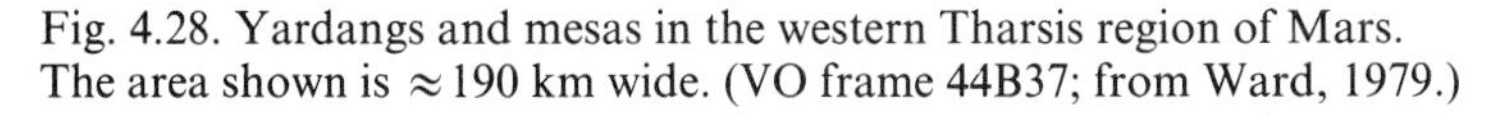

Fig. 4.28. Yardangs and mesas in the western Tharsis region of Mars. The area shown is ≈ 190 km wide. (VO frame 44B37; from Ward, 1979.)

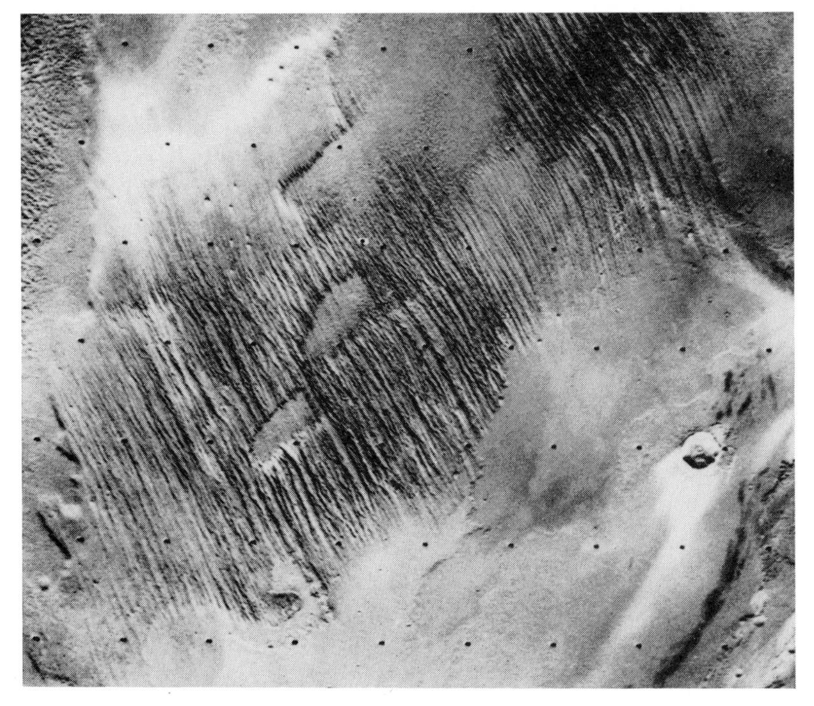

perpendicular to the larger ridges, suggesting either another wind regime or some sort of control, such as differences in rock fabric.

From his study and mapping of ridges on Mars, Ward (1978, 1979) concluded that features shown in Figs 4.28 and 4.29, and observed elsewhere, could not have been formed by any other processes than wind, and must, therefore, be yardangs. The martian yardangs occur primarily in equatorial plains (Fig. 7.16), regions interpreted to be relatively young in geological history.

Fig. 4.29. Yardangs in the Aeolis region of Mars. The area shown is about 14 km wide. (VO frame 724A26; from Ward, 1979.)

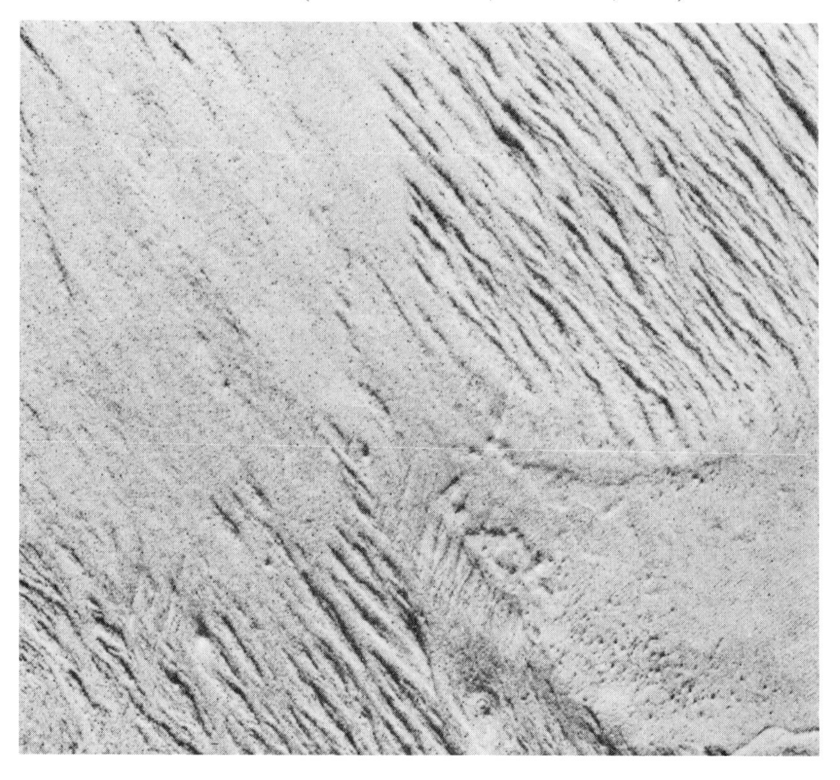

5

Aeolian sand deposits and bedforms

5.1 Introduction

Sand dunes and deserts are intimately linked in the minds of most people and epitomize aeolian processes. The most impressive sand dunes occur in vast sand seas, often called *ergs* – an arabic word for wind-laid sands. Almost all ergs occur in desert basins, generally in areas downwind from terrain experiencing high rates of deflation. Sand then tends to collect where net transportation is low, as in topographical basins or traps.

Wilson (1971, 1973) surveyed all ergs on Earth that are larger than 12 000 km^2 and identified 58 such areas, mostly within regions receiving less than 15 cm precipitation per year. Despite the link between sand dunes and deserts, the two need not occur together. Many deserts lack dunes entirely. And dunes may occur in non-desert areas, such as along coastlines and in river flood plains. Although dunes are aesthetically pleasing, they represent only one form of aeolian deposit.

Bagnold (1941) first described the various processes involved in dune formation and sand migration. His model has been modified slightly and amplified by Hunter (1977) and Kocurek & Dott (1981), who describe three basic processes of deposition by windblown sand: (1) *tractional deposition* (Bagnold's 'accretion' process), in which grains moving by saltation and impact creep come to rest in a sheltered position, (2) *grainfall deposition*, in which particles settle out of the air, usually in zones of flow separation; Bagnold distinguished (1) from (2) in that, during grainfall, the grains do not move forward when they reach the ground, whereas in tractional deposition, they may bounce along until they find a stable position among other grains, and (3) *grainflow deposition*, or avalanching, in which grains reach the brink of a dune then avalanche down the slip face (Fig. 5.1). The deposits formed on the slip face in this manner are steeply inclined downwind and make up *foreset beds*. Anyone who has climbed a dune knows the characteristics of these two types of deposits; loose, steeply dipping beds of the encroachment deposits cause constant backward sliding on the slip face. In contrast, it is relatively easy to walk up the dune along the crest on the well-packed tractional deposits. Knowledge of these and related sand deposits in the vast sand seas of North Africa enabled Brigadier Bagnold to operate effective, quick-strike desert patrols during the Second World War. Using specially outfitted desert trucks, the patrols could navigate the supposedly impassable sand seas by knowing where the tight-packed sands occurred in relation to the dune masses.

Rubin & Hunter (1982, 1983) have studied various sedimentary structures in aeolian deposits and have derived criteria to interpret the processes of sedimentation and the geometry of the bedforms involved in their formation. Their models were derived primarily from study of active sand waves and were then applied to ancient sediments.

Numerous classifications of aeolian deposits and related landforms have been proposed; Table 5.1 shows the scheme used here, modified from Bagnold (1941) and Wilson (1972*a*,*b*). In this chapter we discuss the various types of aeolian sand deposits and the forms which they may assume, including dunes and ripples and their counterparts on the planets.

5.1.1. *Sand waves*

Many students of desert regions have described the wide variety of

Fig. 5.1. Barchan dunes in the Sechura Desert of Peru; prevailing wind is from upper left to lower right; the 'horns' or 'arms' point in down-wind direction; the slipface forms at the angle of repose (≈30–34°) as dune migrates downwind; the brink defines the top of the slipface; the crest refers to the highest part of the dune. (US Geological Survey photograph, courtesy of E.C. Morris.)

Table 5.1. *Sand deposits*

Ripples
Normal ripples
Fluid drag ripples
Megaripples
Sand dunes (simple)
Simple dunes
Compound dunes
Complex dunes
Sand shadows and drifts
Sand sheets and streaks

aeolian sand wave bedforms, such as ripples and dunes, and discussed the processes involved in their formation. Wilson (1972*a*) provided a working definition of *bedform* as 'a regularly repeated pattern which forms on a solid surface because of the shearing action of a fluid'. In addition to features formed from windblown sand, Wilson noted that the definition applies equally well to windblown snow and loess, and erosional features in compact alluvium or solid rock. The term also applies to subaqueous features.

During the late-nineteenth and early-twentieth centuries, ripples were thought to develop from some sort of wave motion, perhaps analogous to water waves (reviewed by Cooke & Warren, 1973). It was suggested that ripples might result from Helmholtz waves, which are waves generated by differential motion (or shear) of two adjacent fluid layers. The idea was that the sand bed and the sand-laden wind above the bed could be treated as 'fluids'. However, this concept cannot apply because (among other reasons) the sand bed beneath the surface is not in motion and cannot be regarded as a fluid.

Another misconception of the period was the notion that ripples were simply small-scale dunes which would grow into full-scale dunes given enough time (Cornish, 1914; and others). It was also noted, however, that sand waves of different scales can occur together, such as short wavelength ripples on dune slopes of much larger wavelength, and by 1941 Bagnold had firmly established the fundamental differences in both the characteristics and mechanisms of formation between dunes and ripples.

The most extensive treatment of the heirarchy of various sand waves was carried out by Ian Wilson. By the early 1970s Wilson had completed extensive studies on the morphometry of sand waves, but unfortunately he was killed in an automobile accident in Nigeria in 1972. Although some of his work had been published, several of his key papers only came to print after the accident through the efforts of his colleagues.

Wilson (1972*c*) measured the areal frequency of different wavelengths observed in sand waves from more than 270 localities in 12 countries and found a possible three-fold clustering of forms into ripples, dunes, and draas. *Draa* is an Arab word meaning arm and is often used to describe very large sand waves. Wilson also plotted grain sizes from various Saharan sands as functions of wavelengths and again suggested a clustering that corresponds to ripples, dunes and draas. Although most investigators agree with the separation between ripples and dunes, the separation between dunes and draas is not generally accepted. We use the term draa to refer to very large sand dunes.

Wilson concluded that the absence of intermediate forms between ripples and dunes means that they are not gradational stages, but rather represent different mechanisms of formation – in support of Bagnold's statements regarding the differences between ripples and dunes – and in so doing put to rest the notion that ripples were simply small-scale dunes. Ripples result primarily from the impact of grains in saltation, whereas for dunes, the saltation pathlength is so small compared to the dune dimensions that the shape of the dune is controlled by aerodynamic effects.

In general, ripples can change their form on timescales of minutes, whereas dunes usually reflect the wind patterns of seasonal and longer time scales. Chorley & Kennedy (1971) note that bedform wavelengths of 1 km size, and larger, require formation times of the order of thousands of years; thus, draas possibly reflect long-time changes such as might accompany shifts in climate.

Wilson (1972*a*) noted six factors regarding the development of aeolian bedforms, regardless of size:

(1) *Presence of nucleus:* Some discontinuity, or disturbance is required to initiate the bedform pattern. Such a nucleus can be a pebble or a bush, the remnants of a previous bedform, or an irregularity on the surface, as might result from differential erosion.
(2) *Spontaneity of formation:* The development of bedform occurs spontaneously and does not depend on preexisting rhythms of flow.
(3) *Grain-by-grain action of the wind:* The development of the pattern occurs at the surface by erosion and deposition of

individual grains, not by mass internal deformation such as folding.

(4) *Interaction between form and flow:* There is a continuous interaction between the surface topography and the fluid flow patterns, which converges toward a system in dynamic equilibrium.

(5) *Bedform migration:* So long as winds are sufficient to move particles, there will be a downwind migration of the bedform and its pattern. This is in response to general erosion of the windward (*stoss*) sides of bedforms, and deposition on the leeward sides.

(6) *Bedform equilibrium:* As long as wind conditions, topography, and supply of particles are constant, an equilibrium in the bedform and its pattern is eventually reached and maintained.

Given these considerations for sand waves in general, let us now consider ripples and dunes in more detail.

5.2 **Ripples**

Within the heirarchy of various sand waves, ripples constitute the smallest bedform. They can develop almost wherever sand and wind occur together, including sites of typical aeolian activity such as deserts, along sandy river banks, and even in children's playground sand boxes. Wind tunnel studies by Seppala & Linde (1978) show that the speed of ripple formation on an initially smooth surface increases directly with wind velocity until an equilibrium condition is reached. Where removal of sand exceeds supply, the ripple crests become sharper, but eventually all of the sand is swept away.

Ripples generally develop approximately transverse to the direction of the wind and have crests that are straight or slightly sinuous. Except for very large ripples, crests can seldom be traced for more than a score of meters before they split into two crests, or simply end. Ripples have been used as local wind vanes to map air flow patterns on both large and small scales. Although this practice is probably valid on relatively flat surfaces, Howard (1977) has shown that the orientation of ripple crests on dune slopes is a function of both the surface wind direction and the slope, and thus may not always reflect the true wind direction, even locally.

The formation of sand ripples has been held in fascination for a long time and, despite numerous field studies, laboratory experiments, and mathematical analyses, many of the fundamental questions regarding their formation remain unanswered. In the next sections we will examine the general size and shape of ripples, consider the ideas that have been proposed for their formation, and discuss the possibility of different classes of ripples.

5.2.1 *Ripple morphology*

Ripples that are formed by winds blowing from a uniform direction are asymmetric in cross-section (Figs. 5.2 and 5.3). From studies of ripples at the Kelso dunes in California, Sharp (1963) found windward slopes as steep as 8–10° and leeward slopes from 20–30°. Leeward slopes tend to have two parts, a steeper part directly beneath the ripple crest and a

Fig. 5.2. Ripple formation related to the trajectory of in-coming grains in saltation; the surface would be preferentially eroded in zone *A–B* by saltation impact; zone *B–C* would be relatively unaffected. (From Bagnold, 1941.)

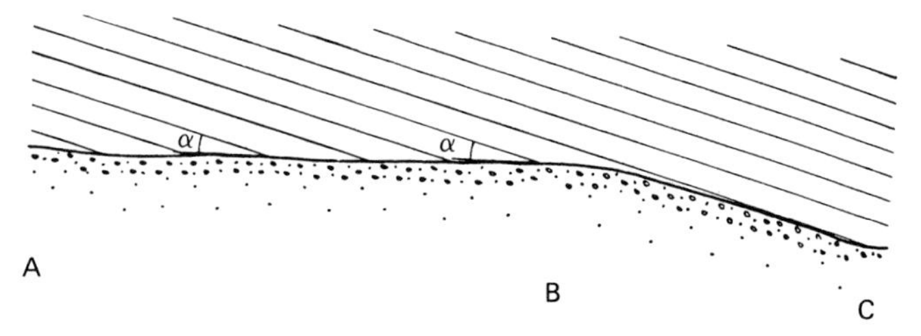

Fig. 5.3. Cross-sections of ripples at the Kelso dunes, California. (From Sharp, 1963.)

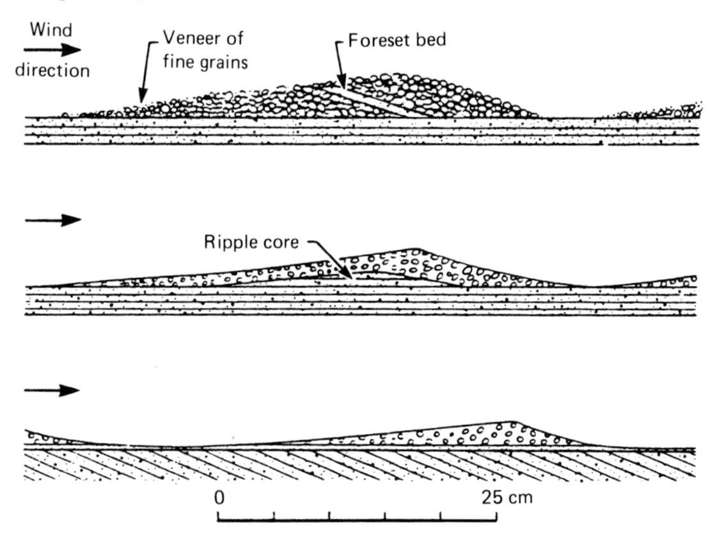

gentler, concave part that is transitional to the next ripple. This asymmetry can be reversed in only a minute or two by strong winds blowing in a direction opposite to that which formed the ripple initially.

Observations of active ripples show that grains are driven up the windward face by saltation impact until they reach the crest, at which point they tumble down the leeward slope to form a type of sand grain talus (Sharp, 1963).

Ripple wavelengths range from 0.5 cm to 25 m, with larger forms occurring in *megaripples*, which we will discuss in more detail later. However, for normal ripples, wavelength appears to be a function of wind speed, average sand size, and sand-size distribution. Stone & Summers (1972, cited in Mabbutt, 1977) give the following empirical expression

$$L_a = 63.8\, D_p^{0.75} \tag{5.1}$$

in which L_a is the equilibrium wavelength in centimeters and D_p is the particle diameter for grains on the ripple crest; however, wavelength also appears to increase with increasing wind speed, which is not taken into account in this expression. Although ripple heights, or amplitudes, may be as great as 25 cm, Sharp (1963) found typical wavelengths of 7–14 cm and heights of 0.5–1.0 cm for ripples at Kelso.

The *ripple index* is the ratio of wavelength to height and is a useful parameter for describing ripples. Cornish (1914) gave an index of 18 as typical for most sand ripples, although departures from this mean can be large. At least qualitatively, the ripple index varies inversely with grain size and directly with wind velocity. However, Sharp (1963) pointed out that in some cases the index may be misleading. For example, ripples formed on wet or frozen substrates may be separated by a flat trough, which could give a greater wavelength than for loose sands of the same size and would result in a higher ripple index.

5.2.2. *Internal structures in ripples*

There have been numerous studies of bedding structures and other sedimentary features in dune sands, but remarkably few studies have been conducted on the internal structures of ripples. Sharp (1963) impregnated ripples at the Kelso field with a quick-drying plastic in a 5–10% solution of Glyptol. Upon drying, the mass could be removed to the laboratory, sliced into thin sections, and examined microscopically. Many authors have noted that the crests of ripples seem to be marked by coarse grains. Sharp, however, found at Kelso that most ripples were homogeneous in cross-section and consisted of relatively coarse grains resting on a base of thinly bedded, finer sand (Fig. 5.3). He considered the base to result from the settling of fine grains onto the ripple where they become trapped in spaces between the coarse grains. As the coarse grains move forward in traction, the fine grains sift downward and accumulate as a basal layer. Sometimes fine grains form a thin veneer at the foot of the windward face, giving the appearance that coarse grains are concentrated on the crest. In his analysis of various sedimentary features in windblown sands, Hunter (1977) also attributed size grading to interstitial settling of finer grains.

Fryberger & Schenk (1981) carried out a series of wind tunnel experiments to investigate aeolian deposits, including sedimentary structures within ripples. They found that the shape and attitude of ripple foreset laminae were controlled by the overall form of the ripple and that the thickness of ripple strata could change abruptly or pinchout completely in downwind distances of only a few centimeters. They determined that the strata could be produced either by the advance of each ripple with all, or part, of each ripple preserved beneath the succeeding ripple, or by wind gusts that produced ripple foreset strata in which coarse grains tended to concentrate at the crest.

5.2.3 *Ripple formation*

Bagnold (1941) observed ripple formation in the field and in the laboratory and concluded that ripples originate through the impact of saltating grains. The following sequence summarizes the process as proposed by Bagnold:

(1) The key to ripple formation is the presence of chance irregularities on a sand surface (Wilson's 'nucleation' factor, given in Section 5.3). As soon as winds increase to the speed at which saltation occurs, these irregularities become emphasized; as shown in Fig. 5.2, more grains would strike surface *AB* than *BC*, causing preferential erosion by saltation impact over *AB*. Thus, *AB* marks a zone of net erosion.

(2) Grains derived from *AB* will land downwind at a distance equal to the average saltation pathlength. Meanwhile, *BC* remains in the 'shadow' of the irregularity and is a zone of net accumulation.

(3) The saltation of grains from *AB* propagates zones of erosion progressively downwind, and the whole system moves forward. Thus, the ripple wavelength corresponds approximately to the saltation pathlength.

(4) The amplitude of the ripple increases in height to a maximum which is defined by the barrage of incoming grains. Eventually, the ripple height comes into dynamic equilibrium for the given wind speed and grain size.

Sharp's (1963) observations of ripples led him to conclude that ripple height might be a more direct control of wavelength than the saltation pathlength. He proposed that grains in ripples are moved forward primarily by creep from saltation impact; again, chance irregularities and/or differential movement of the grains in creep would cause a 'piling up' of some of the grains. Once initiated, the pile would continue to grow to an equilibrium height governed by the angle of incidence for the grains in saltation. In addition, increases in ripple height would bring the deposited grains into a zone of higher wind velocity within the boundary layer, so that they would tend to be removed by the wind, even without the impact of incoming grains. As in Bagnold's model, the zone of impact by saltating grains would be propagated downwind from the pile. As the incidence angle decreases, the effective 'barrier' (the ripple height) is lowered and the wavelength increases.

The principal difference in the two models is in the governing factor. In Bagnold's model, the saltation pathlength is the main factor; in Sharp's model it is the saltation incidence angle and ripple height. It is difficult observationally to disentangle these two factors because they are so intimately related (see Section 3.3 on saltation). The incidence angle is inversely related to velocity; hence with increases in wind speed, the ripple wavelength also increases. Similarly, the saltation length is a function of wind speed, and with increases in wind, the pathlength also increases. Thus, regardless of model, an increase in wind speed will cause an increase in ripple wavelength for sands of equal sizes.

5.2.4 *'Fluid drag' ripples, 'normal' ripples, and 'megaripples'*

Most of the discussion above applies to so-called normal ripples, or those which form primarily by the impact of sands having diameters of about 0.3–2.5 mm. Ellwood *et al.* (1975) investigated possible natural classes of ripples by analyzing the grain-size distribution and morphology for a wide range of ripples in the Algerian Sahara. They found a complete gradation of wavelengths over the range 1 cm–20 m and concluded, with the aid of a mathematical model, that all ripples could be attributed to the saltation mechanism, but that grain-size distribution can have a pronounced efffect on the characteristics of the ripple geometry, which can give rise to different ripples.

The normal ripples discussed in the previous sections have been termed by various workers as 'sand ripples', 'impact ripples', or 'ballistic ripples'. Another class of ripples has also been recognized, which are variously termed 'giant ripples', 'granule ripples', 'erosion ripples', 'sand ridges', 'megaripples', or 'pebble ridges'. We prefer the term *megaripples* for this class, which can have wavelengths up to 25 m. They form typically from collections of particles having a bimodal-size distribution and under winds which are generally too light to move the large grains, but which are strong enough to set the finer grains into saltation. This results in a lag concentration of coarse grains on the surface and a concentration of the coarsest grains on the ripple crest. Megaripples have been ascribed to deflation because of this lag concentration. Saltation is also important, however, with the large grains being moved in creep by the impact of smaller, saltating grains. In comparison to 'normal' ripples, megaripples are commonly more symmetric, which probably reflects shifting wind directions.

An interesting set of megaripples was found in northern Iceland (Greeley & Peterfreund, 1981) as shown in Fig. 5.4. These ripples have wavelengths of about 7 m and amplitudes of ≈ 25 cm. They consist of two kinds of particles: light-toned, low-density pumice ≈ 1 cm in diameter, and dark, denser obsidian ≈ 0.75 cm in diameter. Unlike most megaripples, however, the smaller obsidian particles, rather than the larger ones, are concentrated along the crest. This suggests that grain density is important as well as the diameter in the development of megaripples. The occurrence of small but denser particles of magnetite on the crests of quartz sand ripples shows this can also be true for smaller wavelength ripples. The product of density and diameter may be the governing parameter since threshold speed increases with that product (see Eq. (3.5)).

A third possible category of ripple was described by Bagnold (1941) as occurring in well-sorted, fine sands that are blown by high-velocity winds. Termed 'fluid drag ripples', they apparently form when winds reach the speed at which fine grains begin to go into suspension. Evidently, the cloud of grains above the surface is a mixture of grains having very long saltation paths (a result of the small size and high wind) and grains passing into suspension. This turbulent cloud of grains generates long, flat ripples, termed by Wilson (1972*a*) 'aerodynamic ripples'. Wilson likened these to certain subaqueous ripples, described by Allen (1968) as resulting from possible secondary flow fields.

Ripples which resemble some types of subaqueous bedforms or aeolian dunes develop in wind tunnel experiments with fine (38 μm) particles (Fig. 5.5). Because the particle terminal speed for the small particles shown in Fig. 5.5 is only about one-half of the maximum vertical turbulent fluctuation speed (see Section 3.4.2), the particle trajectories are greatly affected by turbulence. The geometry of this type of ripple is thus due primarily to local variations in surface shear stress (and, thus, to variations in particle mass transport rate) rather than to the saltation trajectory as is the case for normal ripples. The configurations of dunes at larger scale are also due to local variations in surface shear stress rather than to saltation impacts; thus the resemblance to large-scale transverse dune patterns in Fig. 5.5 is perhaps not coincidental.

5.2.5 *Ripples on other planets*

Having some idea of the processes involved in the formation of ripples and the parameters governing their geometry, we can speculate on what ripples might be like in other planetary environments. In both Bagnold's (1941) and Sharp's (1963) models for ripple formation, the

Fig. 5.4. Megaripples in Iceland near Askja; dark particles are obsidian, light particles are pumice; field vehicles indicate the scale. (Photograph by R. Greeley, 1977.)

saltation trajectory is important in the control of the ripple wavelength, as are grain size and wind velocity. A mathematical analysis of the saltation trajectory for grains in a venusian environment (Greeley *et al.*, 1980*c*; White, 1981*b*) shows that the pathlength would be short and would have relatively high incidence angles (see Section 3.3). Furthermore, surface winds measured on Venus are very low. These factors should combine to produce short ripples with relatively low ripple indices.

Fig. 5.5. Dune-like ripples produced in the Iowa State University wind tunnel by blowing air over glass spheres (≈ 30 μm) at speeds just above threshold. Ripple wavelength was ≈ 8 cm; free-stream wind speed was 5.3 m/sec; the ratio of terminal speed to threshold friction speed ≈ 0.5; wind direction is from left to right.

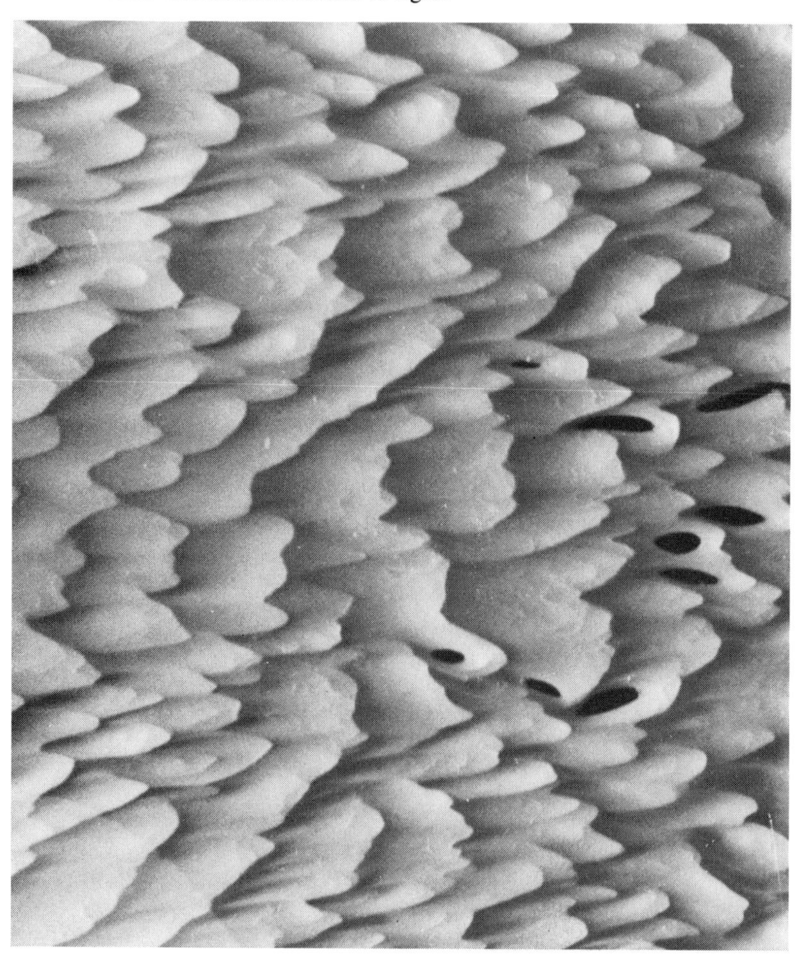

In contrast to Venus, ripples on Mars should be long, low features with high indices because even the minimum particle-moving winds are high and because the trajectories are long and have low angles of incidence. Most of the surface in the view of the Viking Landers is littered with rocks, and it is probably difficult for ripples to develop by saltation impact because most grains would hit rocks, not loose particles. One area at Viking Lander 2, however, consists of a long trough occupied by fine, granular material; patterns within the material suggest ripples (Fig. 5.6) oriented approximately transverse to the prevailing winds, as measured by the Lander meteorology experiment. Less well-defined ripples have been suggested by Sharp & Malin (personal communication, 1982) in the vicinity of Big Joe Boulder at the Viking Lander 1 site. As one would expect, these possible ripples on Mars appear to have long wavelengths and low ripple indices.

There are no images of the surfaces of Titan, nor even any direct evidence that aeolian processes occur there. We can, however, speculate on possible ripples, given what is known or estimated to be the environment on Titan's surface. It has been suggested that all three phases of methane may be present – liquid methane oceans, solid methane ice surfaces (a potential source of windblown ice particles), and gaseous methane clouds in the

Fig. 5.6. Viking Lander 2 image of the martian surface showing a trough $\approx$25 cm wide running diagonally across the field of view; arrows point to possible ripples. (VL 2 frame 21A024.)

dense (1.6 b), predominantly nitrogen, atmosphere. The density ratio of solid–gas nitrogen or solid–gas methane is about the same as snow to air on Earth; thus, aeolian ripples formed on Titan might be similar to bedforms developed in snow on Earth.

5.3 Sand dunes

Sand dunes are great fun! They are aesthetically pleasing to the eye, especially at sunrise and sunset when shadows play against the dune textures, and even the oldest of children enjoy jumping and sliding down the soft faces of dune sands. From a more practical standpoint, dunes can be both a boon and a bane to humankind. Ancient dune deposits can be important reservoirs for petroleum and water, and knowledge of the form and sedimentary characteristics of dune sands can help recover these critical commodities. The Hopi Indians and their predecessors recognized the value of dune deposits in parts of arid northern Arizona. The sands prevent run-off by allowing the scant rainfall to sink into the underlying formations where it accumulates as groundwater, and then reemerges as springs. The Indians were aware of the relationship between the dune deposits and the water supply and located their villages around the margins of the dunes. In addition, they grew their corn crops on semistabilized sand dunes and sand sheets because the sands acted as a moisture-holding mulch (Hack, 1941).

But active sand dunes can pose problems when they encroach upon towns, highways, cultivated fields, and other works of humankind. Understanding the mechanics of dune migration can help solve such problems, or at least provide insight into possible solutions.

In this section we discuss the classification of dunes in terms of their morphology, mechanisms of formation, and internal structure. We also discuss rates of dune movement and the attempts, both successful and unsuccessful, to control dune migration. We then review what is known about the various dunes and dune fields that have been recognized on Mars. Warren (1969) provides a fairly exhaustive bibliography on terrestrial sand dunes and related phenomena.

Art Bloom (1978), in his excellent text on geomorphology, wrote: 'True dunes can best be considered as deformable obstructions to air flow. They are free to move, divide, grow, and shrink, but, they are not dependent on a fixed obstruction for their maintenance'. In general, Bagnold (1941) found that in order for net accumulation of grains to build into a dune, a patch of 'unorganized' sand had to be at least 4–6 m long. The reason for this is that wind must be sufficiently retarded by the saltation cloud for deposition of sand to occur (the saltation cloud removes momentum from the wind); for distances less than about 4 m, the wind and its load of sand simply overshoot the patch. Under conditions of deposition, the patch will continue to grow as a mound of sand until a critical height is reached, depending upon the wind strength and average grain size, at which point a slip face develops on the lee side of the dune (Figs. 5.1 and 5.7). The minimum height for a slip face appears to be about 30 cm (Bagnold, 1941). With the formation of a slip face, the sand mass then can move forward through a continuous series of accumulations on the windward side of the mass, migrate by saltation and creep of grains up to the brink of the dune, then avalanche on the slip face as sands cross the *brink* of the dune. These grain flow deposits (Fig. 5.7) are tongue-shaped bodies that spread out at the base of the slip face. Generally only part of the slip face fails at any given time. Thus, the leeward dune mass consists of a series of overlapping, successive sand deposits formed primarily by grain flow deposition.

5.3.1 *Dune classification and formation*

During the mid-1930s, the Peabody Museum of Harvard University carried out a series of archeological studies in the Jeddito Valley of the Navajo county in northern Arizona. As part of these studies, J. T. Hack

Fig. 5.7. Slumps on a recently active slipface of a sand dune; these tongue-like masses generate foreset beds which consist of loosely-packed grains typical of avalanche deposits. (US Geological Survey photograph.)

was engaged to study the sand dunes of the region, the results of which form an important part of our understanding of sand dune formation. Hack (1941) defined three primary types of dunes: (1) *longitudinal dunes*, oriented parallel to the prevailing wind direction, (2) *transverse dunes*, oriented perpendicular to the prevailing wind direction, and (3) *parabolic dunes*, which are U-shaped dunes pointing downwind. He related these basic dune types to the wind strength, sand supply, and amount of vegetation, as shown diagrammatically in Fig. 5.8.

Over the past 100 years or so that dunes have been studied, countless names and classifications have been derived. Many of the names appearing in print originate from local languages and usage; some of these names, such as barchan, have come to near-universal acceptance, although the spellings may differ considerably. It seems that each time someone completes a study of dunes in a given area, the existing nomenclature and classification schemes do not quite apply, and the investigator feels obliged to devise a new, or at least modified, scheme. There is probably nothing wrong with this, but it has resulted in a rather bewildering array of terms. Carol Breed & Teresa Grow (1979) provide a very nice set of correlation tables of dune types and references which are valuable in comparing what has been written about essentially the same dune types in different areas.

McKee's landmark publication (1979*a*) provides a dune (and other sand

Fig. 5.8. Diagram showing the relationship of the three main dune forms to vegetation, sand supply, and wind. (From Hack, 1941.)

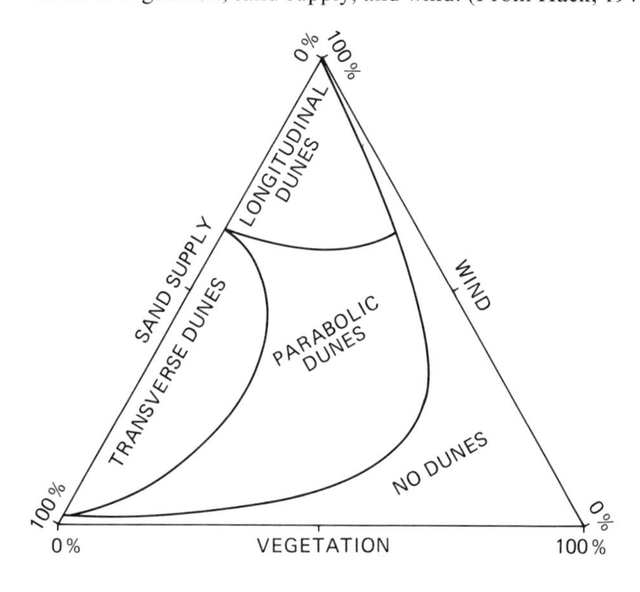

deposits) classification scheme which, although based on Hack's dune types, differs from previous studies in two important ways. First, it incorporates knowledge of dunes gained by analysis of the internal structures in dunes. McKee and his colleagues have dissected numerous dunes in order to obtain this knowledge. The second aspect of McKee's classification is the worldwide analysis of dunes using Landsat images. For the first time, Landsat allowed a uniform comparison of dunes using images of the same scale and quality. Although the study tended to be restricted to larger dune forms because of limitations of Landsat spatial resolution, it was supplemented by conventional aerial photographs and knowledge gained from field studies. These two aspects of McKee's classification make it valuable for planetary studies because it combines remote sensing of dunes from orbit with knowledge of dune formational processes. For this reason, we have followed the McKee (1979*a*) and Breed & Grow (1979) classification scheme very closely, as given in Table 5.2 and Fig. 5.9. The scheme, used to measure different parameters of dunes in planimetric view, is shown in Fig. 5.10.

Dune classification is based primarily on the general shape of the sand mass and the number and position of slip faces. The dunes in Table 5.2 are essentially *simple dunes* (single dunes and dune types); *compound dunes* are composed of two or more dunes of the same type which coalesce or overlap; *complex dunes* consist of more than one type of dune which merge or overlap. All dunes may be modified by topography, or owe their shape to obstructions, and may be further identified as *falling dunes*, *climbing dunes*, *echo dunes*, etc.

5.3.2 *Transverse dunes*

Transverse dunes are asymmetric in cross-section, tend to have a flat crest, and are oriented approximately perpendicular to the prevailing wind. This general category of dune, termed 'crescentic dunes' by Breed & Grow (1979), include three subcategories (Fig. 5.9): *barchan dunes*, *barchanoid ridges* (Fig. 5.9) and *transverse ridges*. From a global survey of Landsat images, Breed & Grow found the largest compound barchans to occur in the Taklimakan Desert of China where the average horn-to-horn width is 3.2 km. From an analysis of 45 acres, each 2500 km^2, they found that the spacing of transverse dunes varies directly with the wavelength, regardless of overall size or form, and that simple forms become compound or complex with increasing size.

Barchan dunes are crescent-shaped, asymmetric mounds of sand having slip faces bounded by two 'horns' or 'arms' oriented downwind (Fig. 5.1).

Table 5.2. *Basic dune types (modified from McKee, 1979*a*)*

Name	Form	Slipface(s)	Wind[a]
Transverse			
Barchan	Crescent in plan-view	1	Transverse
Barchanoid ridge	Rows of connected crescents in plan-view	1	Transverse
Transverse ridge	Asymmetric ridge in cross-section	1	Transverse
Longitudinal	Symmetric ridge in cross-section	2	Parallel
Parabolic	*U*-shaped in plan-view	1 or more	Parallel
Dome	Circular or elliptical mound	none, or poorly defined	—
Star	Central peak with three or more arms	3 or more	Multiple

[a] Refers to orientation of dune axis with respect to wind direction or the vector of more than one wind direction.

They tend to form in regions subjected to a single prevailing wind direction where the supply of sand is somewhat limited, and on surfaces that are relatively hard – such as desert pavement – and relatively free of vegetation. In considering the formation of barchan dunes, Bagnold (1941) reasoned that the rate of advance is inversely proportional to the height of the sand patch; because the slip face is lower at the edges than in the middle, the edges would advance faster, transforming the patch into a crescent-shaped body. Some sands leave barchan dunes as 'streamers' from the horns, forming thin, fan-shaped deposits (Lettau & Lettau, 1969).

Asymmetry of the crescent in plan-view (i.e., one arm longer than the other) has been attributed to multidirectional wind patterns, asymmetry of sand supply, or topography (Long & Sharp, 1964), perhaps aided by secondary flow patterns of the wind over the dune.

Most barchan dunes are small. Finkel (1959) reported, from his field study of dunes in Peru, an average barchan horn-to-horn width of 37 m, with the height being fairly consistently one-tenth of the width. Although some variations in the overall form of barchan dunes may occur as they move downwind, as long as the prevailing wind continues to blow and the sand supply is not altered significantly, the dune maintains its essential shape. This is well documented in several areas, including near the Salton

Sea where one of the dunes was tracked by Norris (1966) over a period of several years; even though part of the dune crossed a small ravine, it maintained its basic form. An interesting study of the variation in particle-size distribution over the surface of a small barchanoid dune in Denmark has been performed by Barndorff-Nielsen and his colleagues (Barndorff-Nielsen *et al.*, 1982). They found that the maximum average particle size occurred at the dune crest.

Barchanoid ridge dunes, or simply 'barchanoids', consist of chains of crescent-shaped dunes linked together to form a scalloped ridge. These

Fig. 5.9. Diagrams illustrating the principal dune types and the winds responsible for their formation. (From McKee, 1979*a*.)

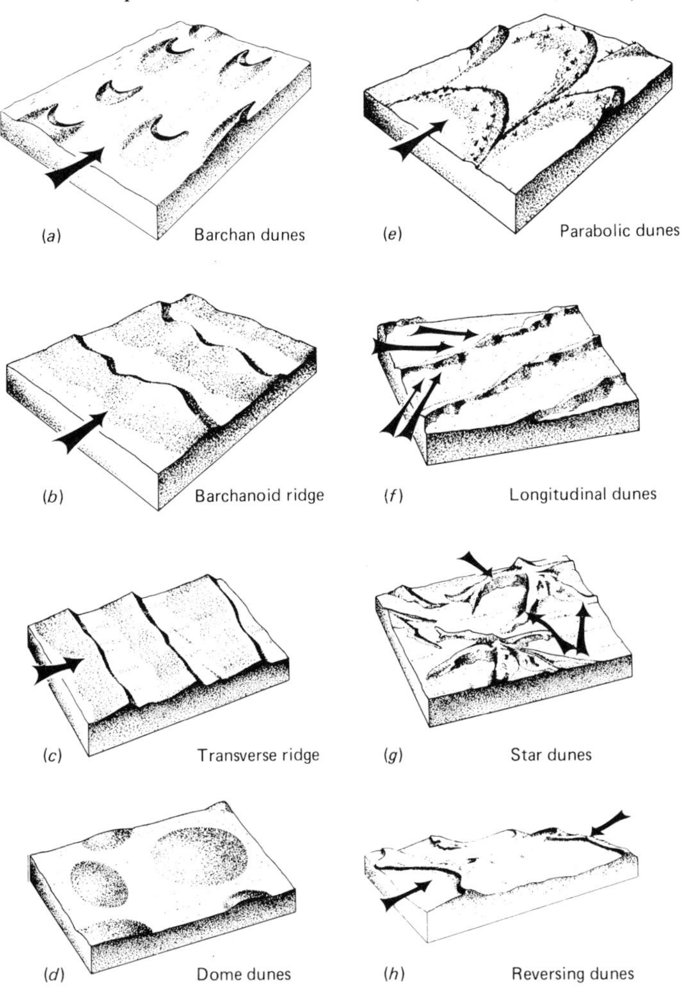

dunes appear to be gradational with *transverse ridge dunes*, in which the crescentic elements give way to a fairly straight ridge oriented perpendicular to the wind (Fig. 5.9). Mabbutt (1977) relates the three types of transverse dunes to sand supply, with single barchans forming with least supply, barchanoids with a greater supply, and transverse ridges forming in areas of maximum sand supply.

5.3.3. *Longitudinal dunes*

In contrast to transverse dunes, longitudinal dunes are symmetrical in cross-section and have two slip faces which commonly meet at a sharp crest. Longitudinal dunes are the most common type of simple dune on Earth. We use the term longitudinal dunes for those that are oriented parallel to the prevailing wind direction or to the vector of multiple wind directions. The crest often rests on a broad, convex-upward base, or plinth, having slopes of 10–20°. Some longitudinal dunes (Fig. 5.11), especially

Fig. 5.10. Diagrams illustrating common forms for transverse dunes and the conventions for measuring width, W, length, L, and wavelength, λ. (From Breed, 1977.)

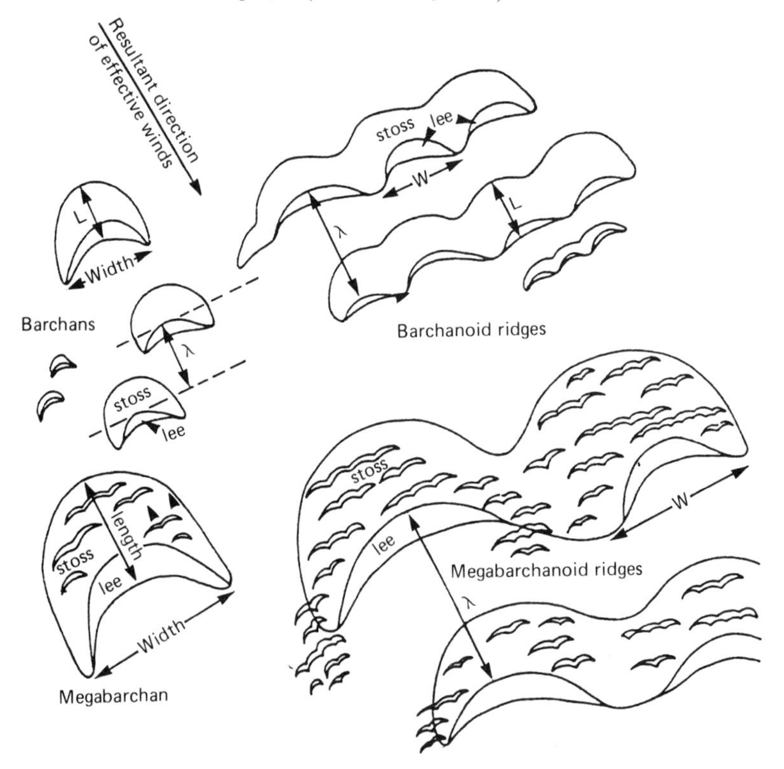

those in Australia, have 'Y' junctions which are open into the wind (Breed & Grow, 1979). Tsoar (1978) further points out that the crests of some longitudinal dunes tend to meander and that the crest line rises and falls to form peaks and saddles along the dune. Other longitudinal dune crests, however, may be more regular, and may be rounded, rather than peaked, because of vegetation cover.

From their global survey, Breed & Grow found that the spacing between longitudinal dunes tends to be twice the mean width of the dune, regardless of overall size for dunes formed in open, unrestricted areas. Where hills, playas, stream beds, or other restrictions occur, they found that the spacing of dunes increased. Some of the largest longitudinal dunes occur in Saudi Arabia where individual dunes are as long as 190 km. Mabbutt (1977) gives a height of between 5 and 30 m as common for longitudinal dunes.

Comparisons of the internal structure and general morphology suggest

Fig. 5.11. Longitudinal dunes, Simpson Desert, Australia. (US Geological Survey photograph by T. Nichols; from McKee, 1979*a*.)

to Breed & Breed (1979) that there may be at least two different types of longitudinal dunes. The first type is the *seif* (Arabic for sword) *dune*, described by Bagnold as slightly curved in plan-view; the second type is straight and narrow, characterized by the 'sand ridges', described by Madigan (1946) and others in Australia.

Numerous hypotheses have been advanced regarding longitudinal dunes and considerable controversy has grown around the mechanisms of longitudinal dune formation and growth, as reviewed by Cooke & Warren (1973), Mabbutt (1977), and Tsoar (1978, 1982). Aufrere (1930) suggested that such dunes are residual features resulting from erosion. This basic idea was adapted by King (1960) who suggested that the Australian longitudinal dunes were carved by the wind from fluvial and lake deposits laid down during Pleistocene times. Although this mechanism may apply to some longitudinal dunes, most dunes that have been trenched do not show sedimentary cores of non-aeolian origin, as would be expected if they were eroded from fluvial or lacustrine deposits.

A variation of the erosional hypothesis was advanced by Verstappen (1968) who suggested that longitudinal dunes were the remnants of the trailing arms from parabolic dunes (see Section 5.3.4). Ridge junctions in this and similar models would be the result of advancing 'blowouts' (King, 1960; Folk, 1971*a*).

Some longitudinal dunes are clearly associated either with local sources of sand or with topographical obstacles, as discussed by Twidale (1972) for central Australia. Hills (1940) demonstrated the relationship between dunes and playas, in which the dunes 'spawned' downwind from lunettes (clay dunes formed on the downwind side of playas). Fig. 5.12 shows a dune extending from the lee of a mountain, which evidently serves as a focus for the flow of windblown sand. Some investigators, however, do not consider dunes formed in association with obstacles to be longitudinal dunes but prefer the term 'lee dunes' for these forms, as we will discuss in Chapter 6. On a larger scale, some of the longitudinal dunes in northern Arizona, which occur on a broad plateau, appear to originate from erosional ravines along the margin of the plateau (Fig. 5.13). These ravines tend to funnel sands blown from river deposits up to the plateau where they are 'fed' into the longitudinal dunes.

Almost all current ideas about longitudinal dunes involve complex wind flow patterns (Fig. 5.14) generated either from unidirectional winds or from multiple winds. Bagnold (1953) suggested that helicoidal flow (longitudinal vortices) was generated parallel to unidirectional winds by differential heating of sand surfaces. Such a flow would cause preferential erosion and deposition parallel to the wind direction. Variations on this theme have been used by Hanna (1969), based on meteorological models used to explain longitudinal dunes; Glennie (1970), Folk (1971*a,b*), Mabbutt *et al.* (1969), and Wilson (1972*a*) considered helicoidal flow resulting from a unidirectional wind interacting with linear topographical forms to further modify and shape sand accumulations.

It is difficult to imagine, however, that the birth of a sea of longitudinal dunes could be originated by the formation of longitudinal vortices. The vortex size is generally determined by the relatively constant depth of the boundary layer. Thus, at the start of formation, one would have to hypothesize very small longitudinal dunes starting out at regular distances of ≈ 1 km apart. The vortices would have to be positioned at exactly the

Fig. 5.12. Lee dune on the northern edge of Pampa Coscomba, Peru; the dune is more than 1 km long, 100 m wide and 30 m high (Grolier *et al.*, 1974). (US Geological Survey photograph by E. C. Morris.)

Fig. 5.13 (a)

Fig. 5.13 (b)

same places during the next wind episode in order for the dunes to extend and/or grow in the same position, which seems unlikely until they have grown sufficiently long to affect the position of the vortex cells. More likely, the spacing of longitudinal dunes is due to other factors including availability of sand from the source, and the strength and direction of the winds. No doubt, once a sand sea is established and the dunes are of significant size and extent, the dune shapes exhibit 'feedback' to the winds and affect the boundary-layer wind pattern. As Bagnold (1941) states, mean dune spacing (or wavelength) is probably a statistical effect dependent on the sand supply. The uniformity of dune fields is probably due to interaction between the secondary flows produced by dune geometry and the geostrophic wind patterns.

A second group of ideas on the formation of longitudinal dunes involves flow patterns resulting from multiple wind directions. Bagnold (1941) suggested that seif dunes evolve through the modification of barchan dunes, in which one horn is elongated by a secondary wind blowing diagonally to the prevailing wind which formed the initial barchan (Fig. 5.14). Similar ideas have been advanced by McKee & Tibbits (1964), McKee (1979*b*), and others.

Tsoar (1978) attempted to resolve the various conflicting ideas about longitudinal dunes by undertaking a study which involved: (1) concurrent wind measurements obtained from multiple anemometers arrayed on longitudinal dunes, (2) grain-size analyses, (3) use of tracers to determine sand and air movement, (4) analyses of internal structures, and (5) study of the changes, for different seasons, in shape and rate of advance of the dunes. He found that when winds strike the dunes obliquely and pass over the crest, the wind direction on the lee side is deflected and blows along the lee flank parallel to the crest. When the angle of the wind is 30°, or less, to the crest, the wind speed on the lee side is increased, causing net sand erosion. When the angle is 50–90°, the velocity on the lee side is decreased

Fig. 5.13. (*a*) High-altitude oblique photograph northward across the Moenkopi Plateau in northern Arizona. Fluvial sands from the Little Colorado River (foreground) are transported by the wind toward the plateau as various transverse dunes. As the sands encounter the scarp of the plateau, they form either echo dunes (*b*) or climbing dunes which feed into longitudinal dunes (arrow) on the plateau. (US Air Force photograph; courtesy of C. Breed.) (*b*) Echo dunes along the Moenkopi Plateau; sands moving from the left encounter the steep cliffs of the Plateau; separation of flow by winds striking the cliff causes a reverse flow which retards further advance of the dune. (Photograph by R. Greeley, 1980.)

Fig. 5.14. (*a*) Formation of seif-type longitudinal dune by the modification of a barchan dune subjected to bimodal wind directions. (From Bagnold, 1941.) (*b*) Diagram showing the formation of a longitudinal dune by the 'roller-vortex' method, developed in a unidirectional wind, as proposed by Folk (1971*a*, *b*). (Courtesy of C. Breed.) (*c*) Diagram of longitudinal dune formation, proposed by C. Breed (personal communication, 1981), involving multiple wind directions to explain dunes on the Moenkopi Plateau of northern Arizona.

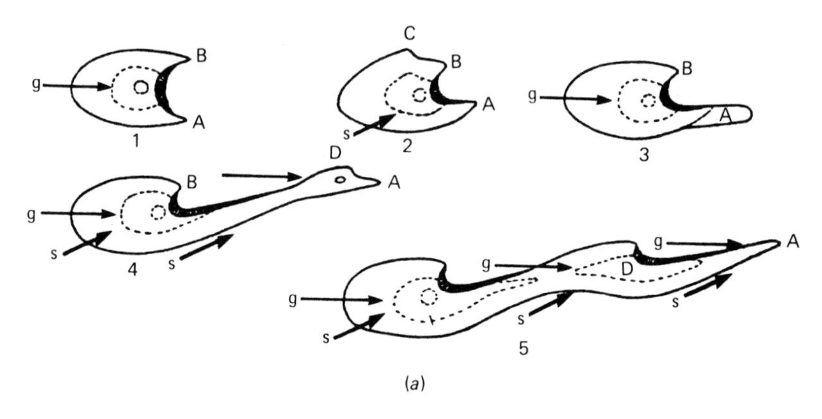

(*a*)

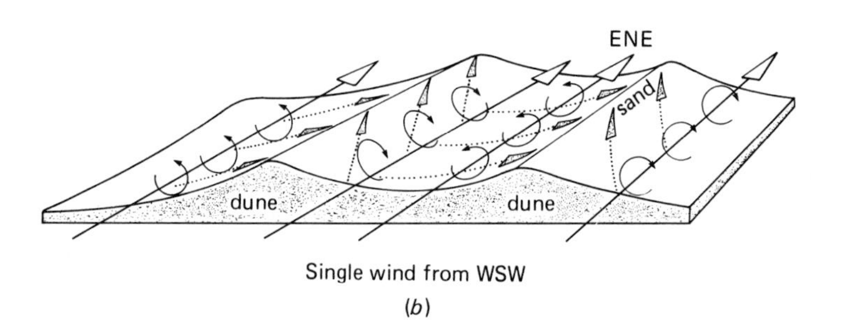

(*b*)

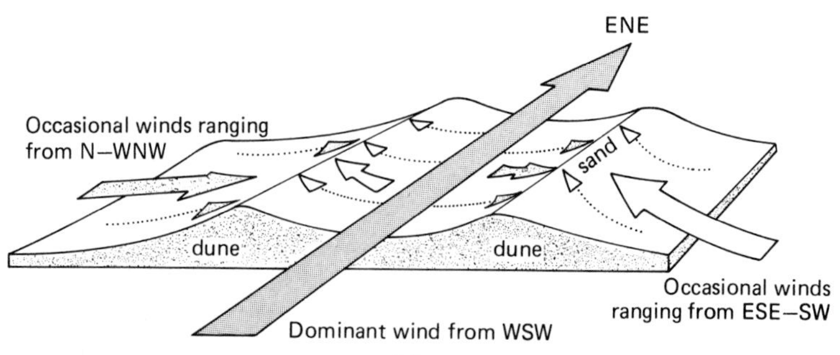

(*c*)

and can result in net sand deposition. Thus, the sinuosity of the dune through these critical angles causes alternating erosion and deposition, thus accounting for the 'peaks and saddles' along the crest line on some longitudinal dunes. Tsoar concluded that longitudinal dunes require a bidirectional wind regime where the two wind directions do not form an angle larger than 150–180°, and a sharp rim crest, so that flow separation can occur on the lee flank.

While the relationships noted by Tsoar can explain the maintenance and morphology of some longitudinal dunes (notably the seif variety), there is still the problem of initial formation of the longitudinal pile, upon which the wind can work. Tsoar considered the *zibar* dune (also called *granule-armored* dunes) to be an important element in the development of longitudinal dunes in essentially open, sandy terrains (i.e. not associated with topographical obstacles). Zibar is a Bedouin term for rolling, low-profile sand ridges which lack slip faces and have surfaces formed of well-packed coarse sands (Holm, 1960). Tsoar demonstrated that many longitudinal dunes appear to be spawned from zibar (Fig. 5.15). Sands being driven by winds diagonally across the zibar saltate easily because of

Fig. 5.15. Vertical aerial photograph of seif-type longitudinal dunes in the Negev Desert, showing zibar 'feeding' into the longitudinal dunes. Direction of dune movement is from left to right; the road indicates the approximate scale. (Photograph from Survey of Israel, Tel Aviv; from Tsoar, 1978, 1982.)

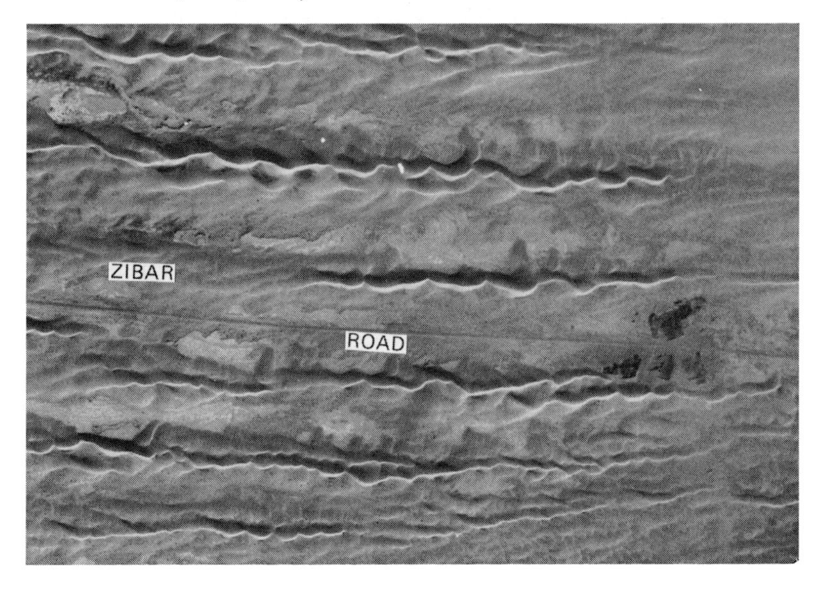

the coarse, tight-packed surface; as the sands move off the zibar, they are slowed enough so that there is net accumulation. Deflected winds (Fig. 5.16) then feed the accumulated sands into the longitidunal dunes.

The model developed by Tsoar appears to be consistent with various observations of, at least, the seif variety of longitudinal dunes. Whether it can apply to all longitudinal dunes, such as those of the Simpson Desert of Australia and of northern Arizona, must await further study.

5.3.4 *Parabolic dunes*

Casual examination of sand dunes on aerial photographs often leads to erroneous conclusions about wind directions. Dunes, which might appear to be a type of barchan (Fig. 5.17), upon closer examination are found to have slip faces on the side opposite to the 'arms'. These are the so-called parabolic dunes, which have their 'arms' open into the wind (Fig. 5.9). Although seldom approaching a mathematical parabola (some investigators prefer the term U-shaped dunes for this reason), the term 'parabolic dune' has come into general acceptance. Because their formation involves removal of sand from between the 'arms', parabolic dunes are also called 'blowout' dunes by some investigators (see Jungerius *et al.*, 1981, for discussion of blowouts).

Fig. 5.16. Model by Tsoar (1978) for the formation of seif-type longitudinal dunes; wind blowing diagonally to the dune axis 'overshoots' the crest, reattaches on the opposite flank and is deflected as a vector 'down-dune'; secondary flow directs some sand flow in other directions. (From Tsoar, 1978.)

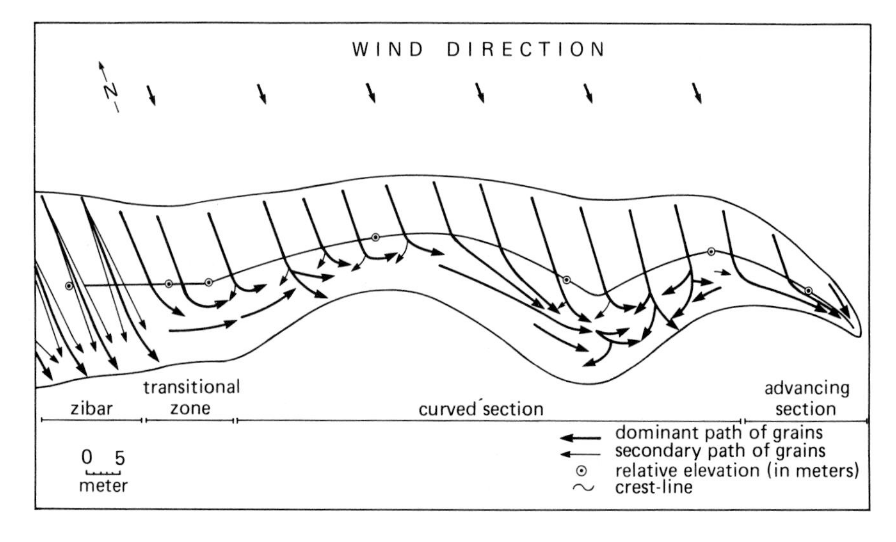

According to the Hack scheme (Fig. 5.8), parabolic dunes result from sand passage over vegetated regions, with the vegetation supposedly 'anchoring' the arms. Deflation of sand from between the arms allows the dunes to migrate downwind.

The largest parabolic dunes, reported by Breed & Grow (1979), occur along the Columbian–Venezuelan border, where arms of 12 km in length are common. In many areas of the world, passage of parabolic dunes through an area may leave longitudinal sand ridges parallel to the dominant wind (Fig. 5.18), with very little indication of the original dune(s) with which they were associated.

5.3.5 *Dome dunes*

Dome dunes are low, rounded mounds of sand which lack, or have

Fig. 5.17. Parabolic (or 'hairpin') dunes in the Snake River Plain, Idaho; direction of dune migration is toward the upper right. (From Greeley, 1977.)

poorly developed, slip faces (Fig. 5.9); some may be the same as the zibar dunes discussed in Section 5.3.3. Most simple dome dunes are less than a few hundred meters across, such as those in White Sands National Monument (McKee, 1966) or in Arizona. Larger dome dunes, as described in Saudia Arabia, may be 1 km or more across but they are generally compound or complex dunes.

5.3.6 *Star dunes*

Star dunes, also called *rhourds* and *oghurds*, are peaked mounds of sand having radiating ridges which from the air resemble pin wheels (Figs. 5.9 and 5.19). The arms have multiple slip faces and, unlike the controversy surrounding longitudinal dunes, there is almost unanimous agreement that star dunes involve multiple wind directions, although several different ideas about the nature of the multiple winds have been proposed.

Fig. 5.18. High-altitude vertical photograph of the Snake River Plain, Idaho, showing 'trails' of parabolic dunes which form long, curving ridges of sand, now mostly inactive; the Wapi lava field which overlies part of the dune field (upper part of picture) has been dated at 2270 ± 50 years before present. (NASA-Ames photograph 72–186, frame 5710, Oct. 1972; from Greeley, 1977.)

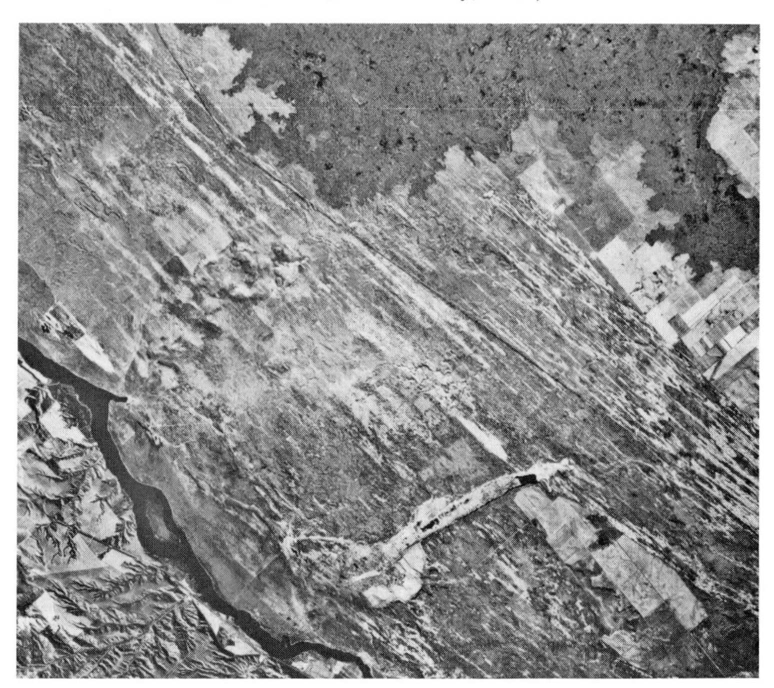

From their analysis of star dunes on Landsat images, Breed & Grow (1979) did not find a correlation between the spacing of star dunes and their size, a reflection of the idea that the dunes grow upward, not outward, although the lack of correlation may be due to incomplete growth of many of the star dunes analyzed. They may reach heights of 300 m (Carol Breed, personal communication, 1981) and measure more than 2 km across.

Star dunes occur in a variety of forms, including: (1) sharp-pointed, radiating-arm, pyramidal dunes typified by those in the Gran Desierto, Mexico, (2) rounded, compact stars with short arms, as found in the Grand Erg Oriental, Algeria, or (3) sharp-crested forms with arms elongated in a preferred direction, as in the Namib Desert. These variations apparently reflect differences in wind patterns and sand supplies.

Regardless of variety, star dunes are found singly or in chains, within or on the margins of sand seas, near the mouths of channels, or wherever multiple winds and a supply of sand occur. Unlike most other dunes, star dunes appear to be relatively fixed in place. If migration does occur, it takes place very slowly because of their large size and because they result from multiple wind directions.

5.3.7 *Other dunes*

The dunes described above constitute the basic types of simple dune forms. Each category of simple dune may also be expressed in many varieties; for example, the angle subtended by barchan 'arms' may vary from narrow to wide, and the crest may be simple U-shaped, or sinuous. In addition, compound dunes formed of two or more of the same dune type, such as small barchans superimposed on larger barchans, are common in most sand seas. Complex dunes are also common in most areas and intricate patterns are often the rule, rather than the exception, in dune fields.

Dunes that are subjected to two prevailing wind directions more or less 180° apart form slip faces that alternate from one side to the other and are called *reversing dunes* (Fig. 5.20). Seldom are the winds of equal strength; consequently, there is usually a dominant and a subordinate side with a net movement of the dune in the direction of the stronger wind.

Topography can have a significant effect on the size, shape, and orientation of dunes, regardless of type or variety, as illustrated by Smith (1982) and modeled by Tsoar (1983). Migrating dunes that encounter a cliff or a scarp may either 'climb' the cliff and are known as *climbing dunes* (Fig. 5.21) or, if the slope is too great, may stand off from the cliff and are then called *echo dunes*. Echo dunes result from flow separation and reverse flow to form a moat-like zone between the dune and the cliff. Dunes which approach a cliff or scarp from the upslope side may cascade over the slope as *falling dunes* (Figs. 5.22 and 5.23). Closed or semiclosed depressions can serve as a 'trap' for sands, forming enclosed sand dunes and dune fields (Fig. 5.24).

The term *draa*, defined earlier as used for very large dunes, also refers to *megadunes*. Nearly all draas appear to be compound and/or complex dunes. For instance, Cooke & Warren (1973) illustrate the Pur-Pur dune of Peru as an example of a draa. As shown in Fig. 5.25, the main dune mass has the general barchan form but has smaller barchan dunes on the windward slope of the mass, 'spawned' downwind from the 'horns' of the main mass. Over the past 20 years the main mass has moved at a rate of

Fig. 5.19. (*a*) Star dunes, Gran Desierto, Mexico. (Arizona State University photograph 2333-A, by D. Ball, 1983.)

(*a*)

about 1.0 m/yr (Grolier *et al.*, 1974), but the smaller dunes are moving at about 9.0 m/yr (Cooke & Warren, 1973).

5.3.8 *Internal structures in dunes*

For the past 30 years and more, Eddy McKee and his colleagues of the US Geological Survey have carried out an extensive program of studies of the internal features and sedimentary structures of sand dunes. This has been achieved primarily by trenching dunes from the upper surface to the lower base of the dune, and in directions parallel, transverse, and diagonal with respect to dune geometry. Coupled with laboratory simulations and study of fossilized dunes, McKee and his associates have gained unparalleled insight into the complexities of dune deposits.

Fig. 5.19. (*b*) Complex star dunes, eastern Rub al Khali, Saudi Arabia. (From McKee, 1979*a*.)

(b)

As reviewed by McKee (1979*b*), there are three types of sedimentary structures common to most sand dunes: (1) sets of cross-strata typically having foresets dipping downwind at angles, usually of 30–34°, which represent original angles of repose, (2) sets of tabular–planar cross-strata that often thin from the base upward, and (3) bounding planes between sets of cross-strata; the planes are mostly horizontal or dip leeward at low angles and often truncate more steeply dipping foreset beds. The arrangement of these structures, plus various others which seem to be unique to only one or two dune types, can be used to characterize dunes and wind directions, even in 'fossil' dunes.

Barchan and transverse ridge dunes result from unidirectional winds. Thus, the internal structure of these dunes (Fig. 5.26) is relatively simple in comparison to other dune types, and consists of sets of foreset beds typically truncated by horizontal or near-horizontal strata dipping gently to windward when viewed in cross-section (parallel to the wind). Barchanoid ridges are more complex, reflecting the intersection of multiple, scalloped slip faces.

Fig. 5.20. Reversing dune at the Great Sand Dunes National Monument, Colorado. (Photograph by J. Shelton, from McKee, 1979*a*.)

Longitudinal dunes viewed in cross-section (perpendicular to the wind) show two sets of foreset beds dipping in opposite directions away from the crest (Fig. 5.27). Bagnold (1941) noted that these beds grade laterally into tractional deposits which make up the broad, basal 'plinth' part of some longitudinal dunes. Results from trenching of a seif dune in Libya by McKee & Tibbits (1964) tend to confirm the basic structure described by Bagnold. More-recent work by Tsoar (1982) suggests a much more complex bedding structure for longitudinal dunes, as shown in Fig. 5.28.

Relatively few parabolic dunes have been dissected in order to study their internal structure. However, parabolic dunes at White Sands, New Mexico, were trenched by McKee (1966) and showed features which he thought might be unique to this type of dune: (1) many foresets are concave downward, which McKee (1979*b*) attributed to cross-winds that undercut the bases of the foreset beds as the dune moved forward, and (2) organic

Fig. 5.21. Climbing parabolic dunes approaching Juniper Buttes, Snake River Plain, Idaho; wind is from lower left to upper right. (From Greeley, 1977.)

accumulations that locally warp the strata, primarily from root growth.

McKee (1966) trenched a dome dune to determine its internal structure. He concluded that dome dunes begin as transverse-type dunes which are controlled by a dominant wind direction but are modified by very strong winds that bevel the windward face and transport sands in such quantities as to bury the leeward zone. In coastal dune fields, he (1979*b*) suggested that the shape of dome dunes may be controlled by moisture and vegetation.

Complex star dunes have never been trenched, but because of the manner in which they developed from multiple winds, McKee predicted that the internal structure would be complicated and would display a wide spread in bedding dip direction. The 'arms' for simple star dunes have been trenched (McKee, 1966) and were found to have a steep dip (31–33°) on one side and a more gentle dip on the opposite side, suggesting that seasonal,

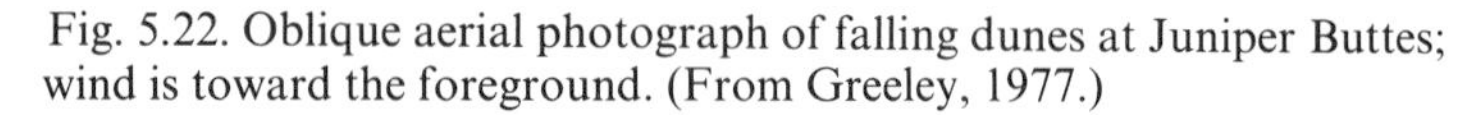

Fig. 5.22. Oblique aerial photograph of falling dunes at Juniper Buttes; wind is toward the foreground. (From Greeley, 1977.)

Fig. 5.23. Diagram showing development of the climbing and falling dunes at Juniper Butte, shown in Figs. 5.21 and 5.22. (From Koscielniak, 1973.)

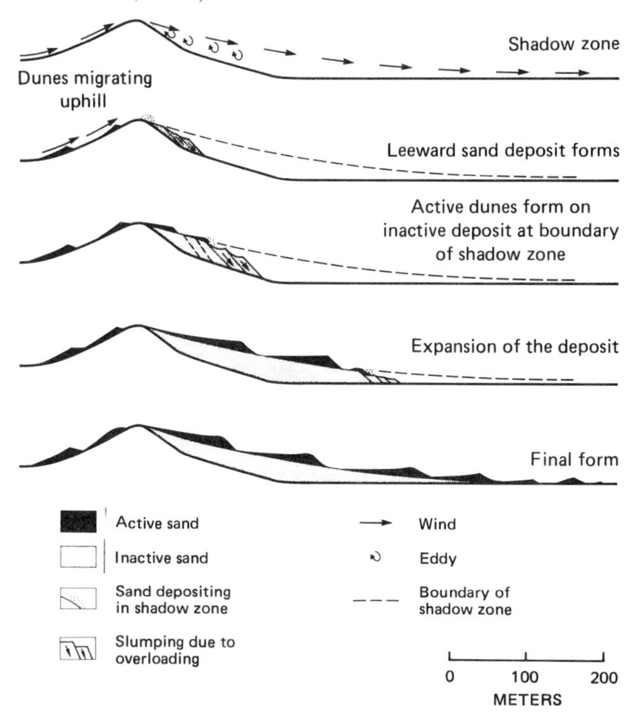

Fig. 5.24. Bruneau dune (arrow), in western Idaho, is a star dune more than 160 m high, formed within a semienclosed topographical basin ≈ 6 km across. Prevailing wind direction is from right to left. (Photograph by R. Greeley, July, 1971.)

rotating wind directions were responsible for establishing the position of the arms.

5.3.9 *Dune migration and control*

The migration rate of a transverse bedform is given by Wilson's (1972*a*) formula (modified after Bagnold, 1941):

$$c_r = (q_c - q_t)/h\gamma_p \tag{5.2}$$

where c_r is the advance rate, q_c is the mass transport rate at the crest, q_t is the transport rate in the trough, h is the bedform height and γ_p is the bulk density. If the bedform is large enough to have a slip face, such as a barchan dune, essentially all of the sand is trapped in the dune, q_t is zero, and the equation is identical to Bagnold's (Eq. (5.3)).

$$c_r = q/h\gamma_p \tag{5.3}$$

It is interesting to note that, in dimensionless form, Eq. (5.2) can be written (within the help of Eq. (3.20)) as

$$c_r/u_* = C_1(\rho/\gamma_p)\,(u_*^2/gh) \tag{5.4}$$

Where C_1 is a dimensionless coefficient and the parameter u_*^2/gh is a form

Fig. 5.25. Pur Pur dune in Peru; this complex dune is ≈1 km across (horn-to-horn) and ≈55 m high. (Photograph courtesy of US Geological Survey.)

Fig. 5.26. Cross-section of a transverse dune showing typical bedding structures. (From McKee, 1979*b*.)

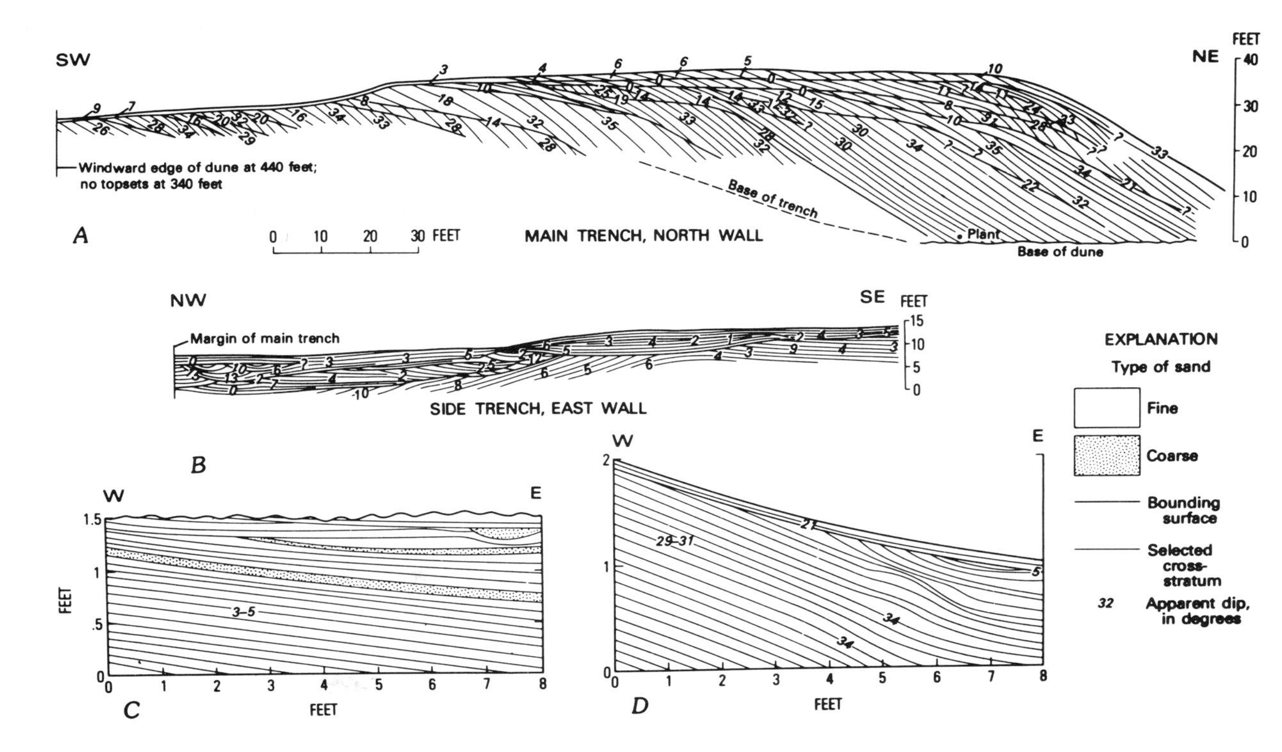

Fig. 5.27. Diagrams showing cross-sections and bedding in longitudinal dunes; (*a*) from McKee & Tibbits (1964), (*b*) from Bagnold (1941).

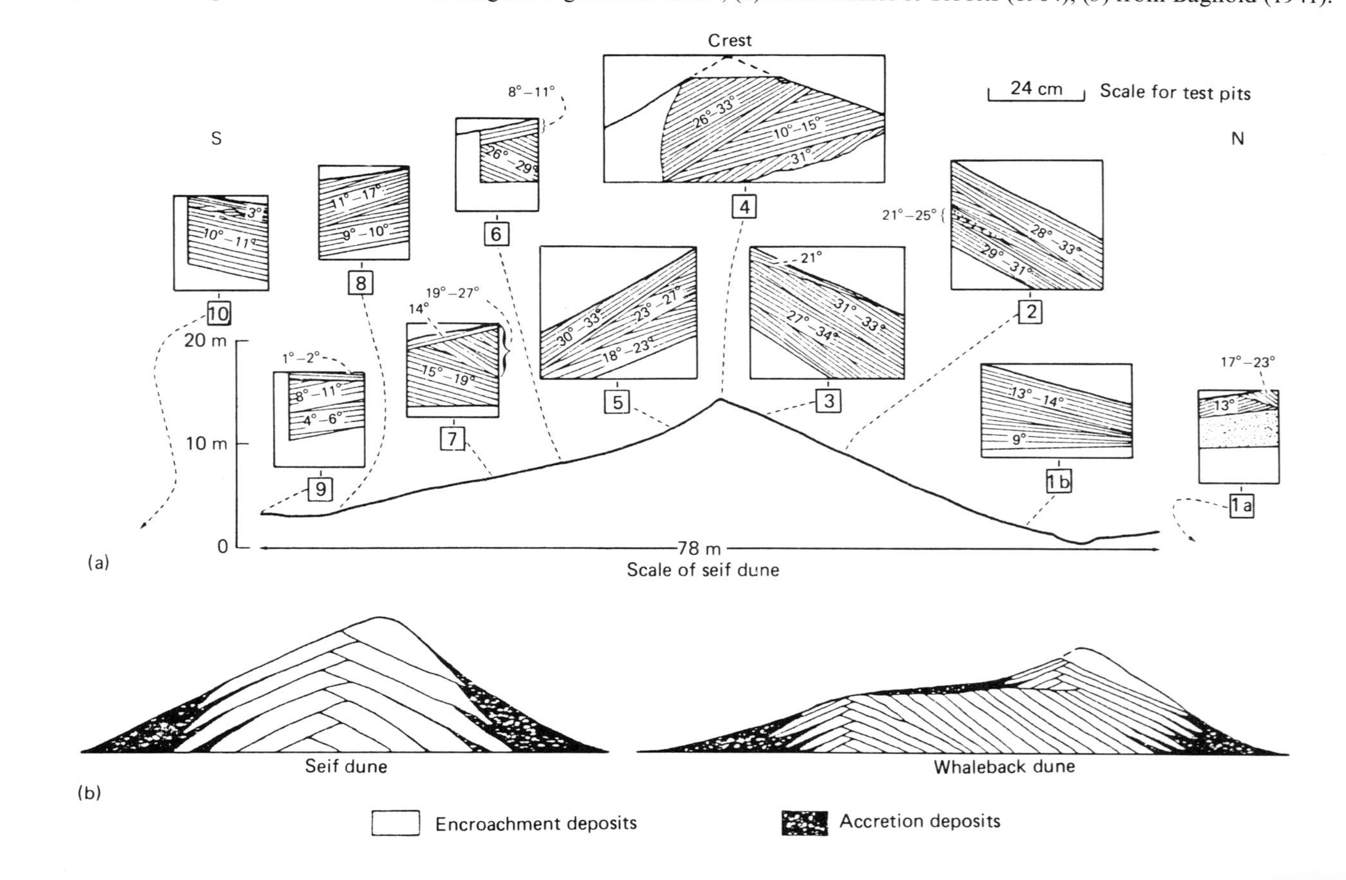

of the well-known Froude number. For sand-size particles, Bagnold's threshold coefficient, A (see Eq. (3.5)) is approximately constant, so that c_r/u_* becomes

$$c_r/u_* = C_2(u_*/u_{*_t})^2\,(D_p/h) \tag{5.5}$$

The particle diameter is D_p, C_2, another dimensionless coefficient, and u_{*_t} is the threshold value of the friction speed, u_*. Since the threshold friction speed on Mars is about ten times that for Earth, the advance rate during an active period ($u_*/u_{*_t} > 1$) would also be about ten times as large for the same dune height. The bedform advance rates on Venus would be correspondingly less than those on Earth.

Measured advance rates for bedforms on Earth of height ranging from 1 cm (ripples) to 20 m (barchans) are plotted as a function of bedform height in Fig. 5.29. The data show that the advance rate, c_r, does conform to Eq. (5.2) or Eq. (5.3) with greater values of c_r for either smaller bedform height or smaller bulk density (such as snow).

The control of migrating sand dunes is important in many areas of the world where agricultural or developed land is adjacent to active dunes. Jensen (1976) lists the following methods of dune stabilization:

(1) *Foredune barriers:* Artificial dunes are created with fences gradually built upon each other on the sand accumulated by

Fig. 5.28. Model of the internal bedding structure for a longitudinal dune, as derived from analyses of sedimentary laminae. (From Tsoar, 1974, 1982.)

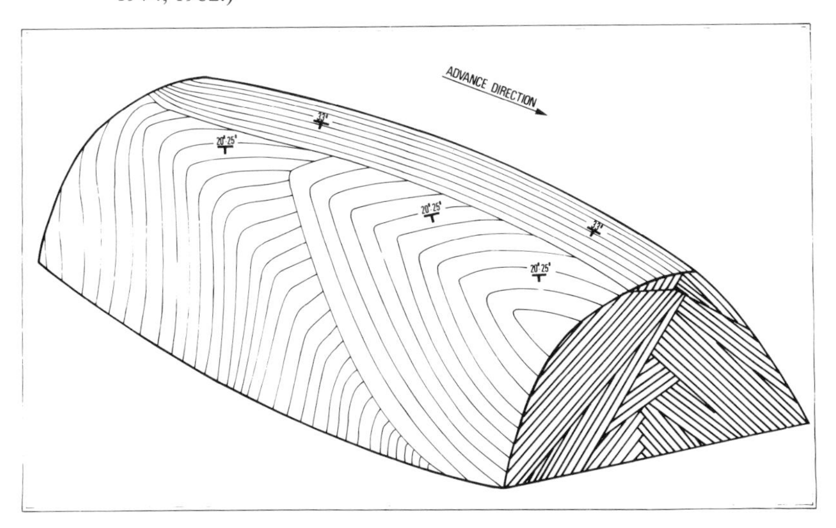

the preceding fence. This method is often used in front of a source of sand (e.g., desert sands or coastal beaches) to check the transport of fresh sand to the area to be stabilized (Fig. 1.8).

Fig. 5.29. Rate of advance, c_r, as a function of height, h, for aeolian bedforms. The smaller bedforms, such as ripples, were measured during short periods of continuous advance. The dune advances were measured over periods of years and thus were moving for only part of the measurement period. The ripples thus have a greater measured value of the product, $c_r h$. The snow ripples (and the snow barchan) exhibit higher advance rates, c_r, than the sand because of the lower bulk density.

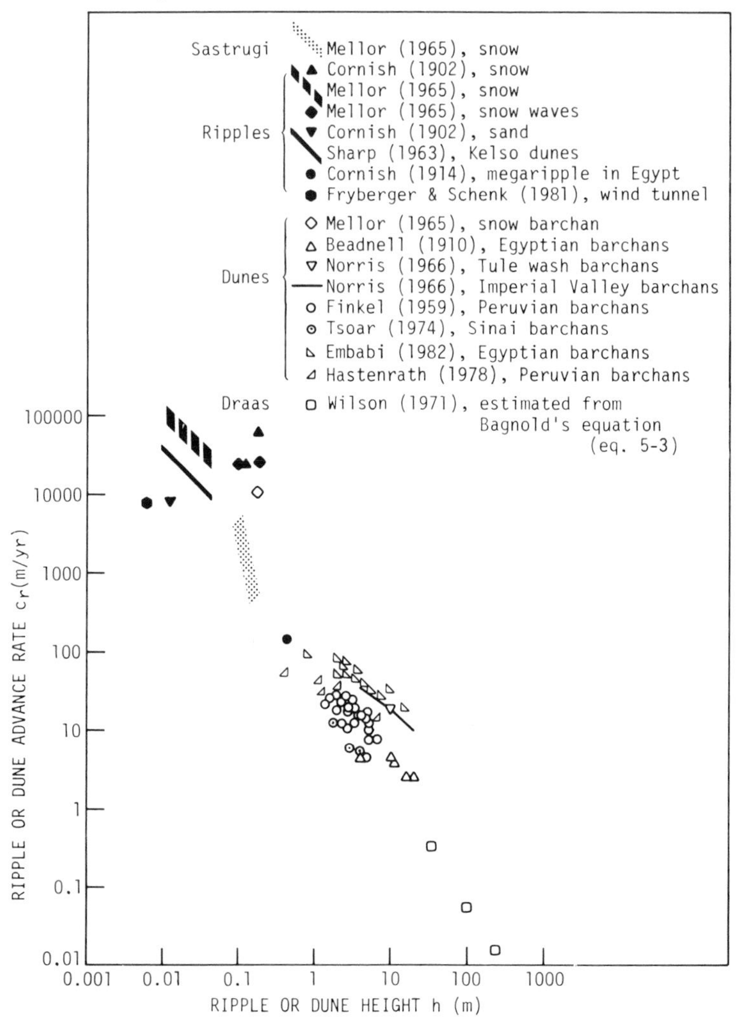

(2) *Microwindbreaks:* These are fences or 'palisades' about 0.5 m high, arrayed in parallel lines – if there is only one prevailing wind direction – or in checkerboard patterns if there are several wind directions. The distance between the fences varies between 4 and 20 m according to the fence height and the condition of the particular dunes. In arid and semiarid zones, only dead plant material should be used for the microwindbreaks to avoid competition later with the trees (where forestation is the objective). In humid tropics, on the contrary, dead plant material may house noxious animals, therefore living hedges are recommended.

(3) *Chemical mulching:* Because of the labor-intensive nature of making 'fences' and the difficulty in some areas of finding plant material for the microwindbreaks, some countries have made use of chemical mulching such as petrochemical sprays. These stabilizing materials have been developed from asphalt-type materials, latex and other chemical substances and several commercial products are available. Typically they are diluted with water and sprayed on the sands. However, the results from chemical mulching seem less successful than traditional microwindbreak type of fixation, although the results are largely dependent on application rates and methods. Combined with the seeding of grasses, chemical mulching has given excellent results in some areas. Whether chemical mulches are economical or not depends on the costs of these products in a given region in comparison with the savings in labor costs.

(4) *Revegetation:* Revegetation with grasses can be accomplished in many cases without previous stabilization by using rather dense slip-plantings. This is the method commonly used in the stabilization of coastal dunes in temperate zones. Experimental work has also been carried out for direct seeding in conjunction with mulching.

Control of blowing sand has a longer history in some countries. For example, in Denmark (Møller, 1980), experiments on sand dune control started in 1824 and today the formerly uncontrolled coastal dunes are under nearly complete control by means of vegetation and sand fences.

5.3.10 *Clay dunes*

Ripples, dunes, and sand drifts are composed mostly of sand-size particles transported by saltation, or moved by saltation impact. Windblown particles smaller than sand ($\leqslant$0.62 mm) ordinarily do not saltate, and so we would not expect to find sand-type bedforms composed of silt or clay. Dunes composed mostly of clay, however, occur in many places on Earth. First described by Coffey (1909) in the Gulf coast regions of Texas, clay and silt–clay dunes have also been found in the Mojave Desert, in Africa, and in Australia (reviewed by Bowler, 1973).

Analyses of clay-rich dunes in Texas and Australia show that the grain-size distribution may have a wide range, with clay content ranging from about 23% by weight to more than 77%. Sand and silt particles, however, were also found in all of the dunes analyzed (Bowler, 1973).

All of the clay-rich dunes reported occur in association with drying salt flats or on the margins of playas. As the playas or coastal lagoons dry, the silt and clay deposits crack, often developing mud curls (Fig. 5.30).

Fig. 5.30. (*a*) Mud cracks at Race Track playa, Death Valley, California; mud curls represent very fine (clay) grains which are the last to be deposited when playa is wet; curls are easily picked up by the wind. (Photograph by R. Greeley, 1982.) (*b*) Windblown mud curl chips, pellets, and aggregates are caught by a small bush and form a drift deposit. (Photograph by R. Greeley, 1982.) (*c*) Processes described in (*a*) and (*b*) can form dune-like features which have cross-bedding and other sedimentary structures similar to sand forms, shown here in this silt–clay dune at Rogers Lake playa, California (the metal plate is $\approx$8 cm long). (From Greeley, 1979.)

Fig. 5.30 (a)

Fig. 5.30 (b)

Fig. 5.30 (c)

Efflorescence of salts and the mechanical disintegration of the mud curls produce pellets, or aggregates of silt and clay which are easily picked up by the wind. These mobile aggregates were found by Huffman & Price (1949) to have modal diameters of 0.15–0.3 mm, typical of normal dune sands. Thus, the pellets are in the appropriate size range for saltation, and because of their somewhat lower density than holocrystalline sand grains, they are more easily moved by the wind. Such pellets are picked up from the surface of the playa or salt flat and transported to the margins where they are commonly trapped by vegetation (Fig. 5.30).

Fairly large dune forms may accumulate in this manner. Known as *lunettes* (Hills, 1940) because of their half-moon shape, crescentic clay dunes in Australia may be more than 30 m high and many kilometers across. Lunettes typically develop around the margins of the playa or salt flat which contributed the pellets. Thus, the 'horns' of clay-rich dunes often point upwind (toward the source), in contrast to normal barchan dunes in which the horns point downwind.

Because the pellets, or aggregates, of silt and clay are deposited from saltation transport, some of the internal bedding structures of normal dunes are also found in clay-rich dunes, including cross-bedding, as shown in Fig. 5.30. However, it seems unlikely that the dunes would migrate very far from their source; the aggregates are relatively weak and break apart with continued saltation. Moreover, percolation of water from rain tends to dissolve the aggregates, causing them to 'weld' together. Not only does this halt the saltation mechanism, but it also tends to transform the dune into a mass of clay, while preserving the original bedding structure.

Although the role of holocrystalline sand in the clay dunes is not fully understood, it seems reasonable that the presence of sand would enhance the saltation of the aggregates and would contribute to the formation of sand-type sedimentary structures, such as cross-bedding. However, clay dunes have not been studied to determine correlations between the percentage of sand and the form of the dune, or the presence and type(s) of sedimentary structures.

5.3.11 *Dunes on Mars*

Although science fiction writers have long incorporated ideas about martian deserts and sand dunes into their models for Mars, the first speculations based on observational data were made by Belcher and colleagues (1971) from their examination of Mariner 6 images obtained in 1969. They noted the similarities between crescent-shaped albedo patterns on Mars and similar patterns on terrestrial sand dune fields, and suggested that barchan or parabolic dunes were present on the 'Red Planet'. Poor resolution of the Mariner 6 images and lack of topographical detail made the suggestion extremely tenuous, however, and it was not until the Mariner 9 mission returned higher-quality images that 'acceptable' sand dunes were found (Sagan *et al.*, 1972).

The best-known field of dunes photographed by Mariner 9 occurs in a 200 km diameter crater in the Hellespontus region near 45°S, 33°W. At low resolution, the field is seen as a dark splotch about 60 by 30 km on the crater floor, similar to many other dark splotches in the area. High-resolution images, studied in detail by Cutts & Smith (1973) and Breed (1977), show a complex dune field of parallel ridges spaced at 1–2 km (Fig. 5.31). In general, the dunes are largest in the center of the field. Dunes on the margin of the field are small, some are isolated, and some are crescent shaped.

Fig. 5.31. Dune field within a ≈200 km impact crater in the Hellespontus region of Mars. (Mariner 9 frames MTC 4264-15, 19.)

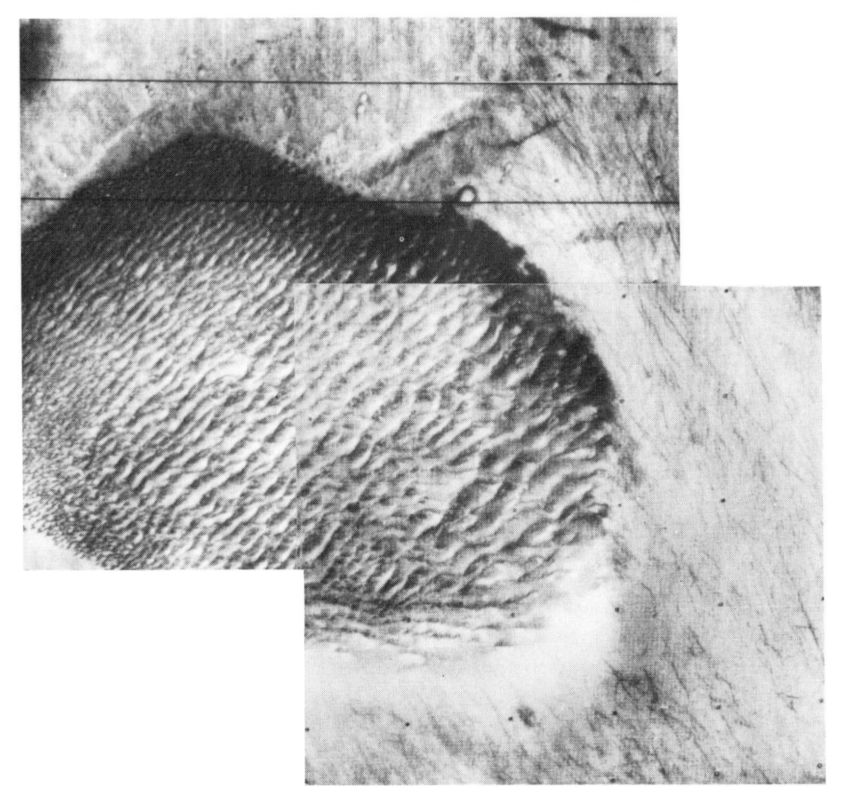

Comparison with other wind indicators in the area suggest that the dunes are of the transverse form.

The Viking mission has shown that many of the dark splotches seen on low- and moderate-resolution images are in fact dunes when observed at higher resolution. (Figs. 5.32 and 6.30). The dark appearance (i.e., low

Fig. 5.32. Small dune field, $\approx$25 km long, associated with an impact crater on Mars. (VO frame 571B53.)

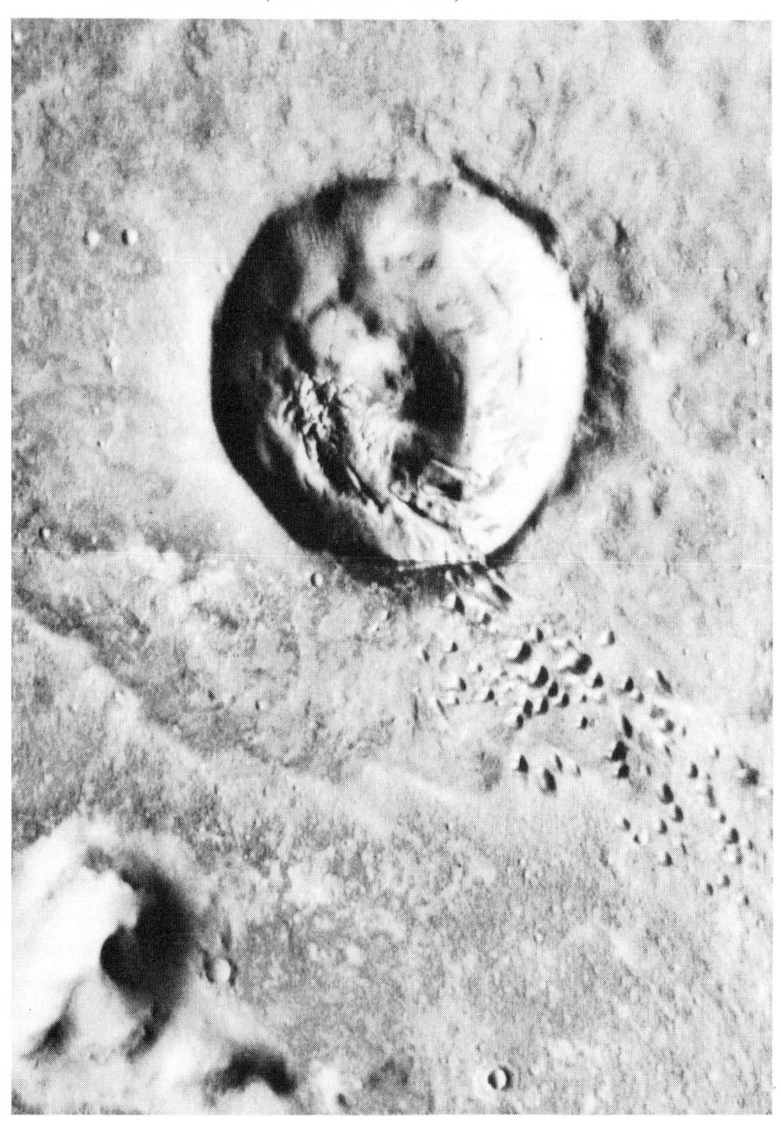

albedo) of the dunes suggests compositions that are silica-poor (Cutts & Smith, 1973) and, given the prevalence of basaltic volcanism on Mars (Greeley & Spudis, 1981), the grains may be basalt fragments.

Dunes are found in many areas on Mars, including the floor of the martian 'grand canyon' (Valles Marineris), in and around craters (Smith, 1972; Fig. 1.16) in the southern cratered terrain, and in valleys where they are topographically controlled.

The largest concentration of dunes on Mars occurs as an asymmetric circumpolar band in the north polar region. Although first hinted at by markings seen on Mariner 9 images acquired in 1971 (Cutts & Smith, 1973), the full extent of the field was not realized until the Viking mission (Cutts *et al.*, 1976). Mapped in detail by Tsoar *et al.* (1979) and Breed *et al.* (1979), the north polar sand sea covers 7×10^5–8×10^5 km^2 and is comparable in size to Rab'al Khali in Arabia, the largest active erg on Earth. Most of the north polar dunes are found between 100 and 220°W and between 77 and 85°N, although scattered dunes are found outside these boundaries as well.

The north polar dunes tend to occur in the zone marked by the so-called north pole dark collar, a low-albedo zone that has been observed telescopically for many years. This area was imaged by the Viking Orbiter 2 spacecraft during two martian seasons; orbits 56 through 120 corresponded to northern hemisphere summer and obtained images when the surface temperature was above the CO_2 frost point. Orbits 487 through 580 covered the martian spring when surface frost and ice blanketed most areas.

All of the dunes observed in the north pole area are transverse forms, including isolated barchans (Figs. 1.15 and 5.33). Some of the barchans are asymmetric and have one 'arm' or 'horn' elongated, suggesting multiple wind directions. Some margins of the sand sea show sets of 'framing' dunes – a band of larger dunes set apart from the main mass of dunes (Fig. 5.34). Tsoar and his colleagues attributed the 'framing' dunes to two possible factors: (1) the upwind plains are partly covered with sand and, as these sands are transported into the dune mass, they are deposited in the first set of dunes the sands encountered; hence, these dunes would tend to grow to a larger size than the subsequent dunes downwind, and (2) because the framing dunes are larger, they would travel at a slower rate and would lag behind the other dunes.

Breed and her colleagues studied the morphometry of the martian north polar dunes and compared them to sand dunes on Earth. They found that the width, length, and spacing of the dunes are comparable to larger transverse dunes on Earth. They also noted, as did Tsoar *et al.*, that longitudinal dunes are absent on Mars, which can be attributed to differences in geomorphic evolution, in wind regimes, or in sand supply.

Tsoar *et al.* (1979) mapped various wind-related streaks, slipface orientations of the dunes, and other indicators of wind direction in an attempt to determine near-surface wind circulation patterns in the north polar region. Although image resolution is marginal for detection of slip faces, and illumination of the dune slopes can make the interpretations difficult, useful data can be derived. Predominant winds in the summer appear to be toward the southwest from the pole, with a westward turning

Fig. 5.33. Barchan dunes in the northern polar region during the northern winter; the white material is seasonal CO_2 ice; the black streamers from 'horns' may be active grains blown from the dune onto the ice surface; the area of photograph is $\approx$35 by 40 km. (VO frame 544B05; from Tsoar *et al.*, 1979.)

in response to Coriolis forces. From latitudes 75 to 80°N, winds are either southerly (through longitude 130–30°W) or westerly (35–130°W). These winds apparently result from the strong thermal gradient that is generated between the permanent cold ice cap and the warm, thermally absorbent dark zone surrounding the cap. Orbiter images show this zone to be a spawning ground for cyclonic winds (Fig. 5.35) which undoubtedly generate sand-moving winds on the surface. Winter winds are considered to be southerly and westerly.

This analysis of wind directions raises the question of whether or not the north polar dunes (or any of the dunes on Mars) are currently active. Tsoar and colleagues (1979) suggested that some of the north polar dunes are active, based on a combination of several lines of evidence: (1) some barchan dunes have 'streamers' from their horns, which appear to be superimposed on seasonal ice deposits (Fig. 5.33), (2) the dunes do not have a degraded appearance, but seem to be 'fresh' – terrestrial dunes quickly lose their sharp form, (3) observed active cyclonic winds (Fig. 5.35) in the summer and predicted strong thermal winds should result in surface shear stresses of ample strength to cause sand saltation, and (4) some dunes have a suggestion of reversing slipfaces from one season to the next. Unfortunately, each of these lines of evidence have uncertainties: the possibly reversing slipfaces could be an illusion resulting from different viewing

Fig. 5.34. 'Frame' dunes (arrow) in the martian north polar erg; the area shown is ≈20 by 30 km, centered at 78°N, 89°W. (VO frame 519B36; from Tsoar *et al.*, 1979.)

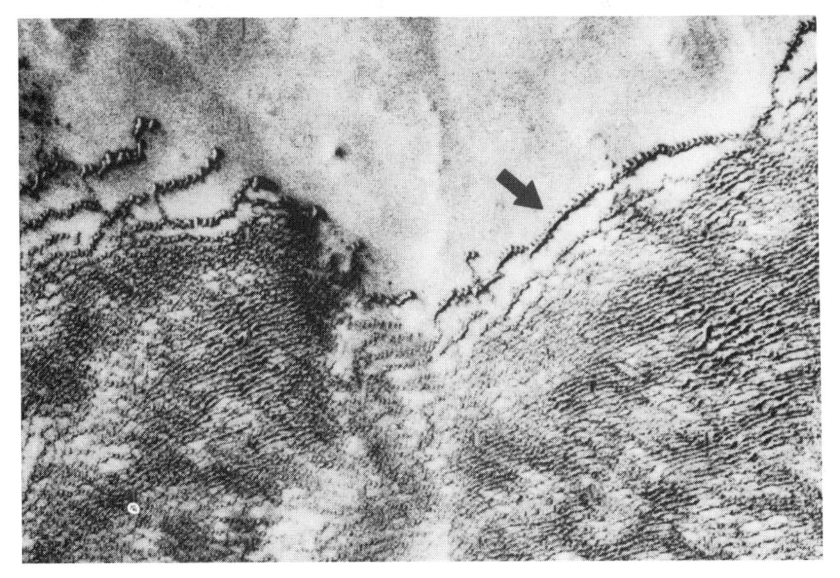

perspectives and differences in illumination in the Orbiter pictures; resolution of the images may not be adequate to define the geomorphic 'crispness' of the dunes – and in any event, it may not be possible to assign ages or geomorphic activity on Mars based on terrestrial experience; and finally the streamers seen in association with the barchan horns could be explained as relict features. Nonetheless, there is enough uncertainty to leave open the question of dune activity.

Another uncertainty about the north polar dunes is the source and composition of the particles that make up the dunes. Given the suggested wind direction and the relationships of the framing dunes, Tsoar and associates considered the plains of the northern latitudes to be the primary source for particles of the north polar erg. These plains are considered to be volcanic (probably basaltic) in origin – which could account for the low albedo of the dunes – and to be subjected to various periglacial processes which might weather the rocks to produce sand-size grains capable of transport by the wind.

By analogy with Earth, where most dune sands originate from fluvial deposits, Breed *et al.* (1979) suggest that the enormous channel systems on

Fig. 5.35. High-altitude view of the martian north polar cap showing ice (bright areas), clouds (faint white patches in the upper part of the picture) and cyclonic storm (arrow). VO frame 814A02; from Tsoar *et al.*, 1979.)

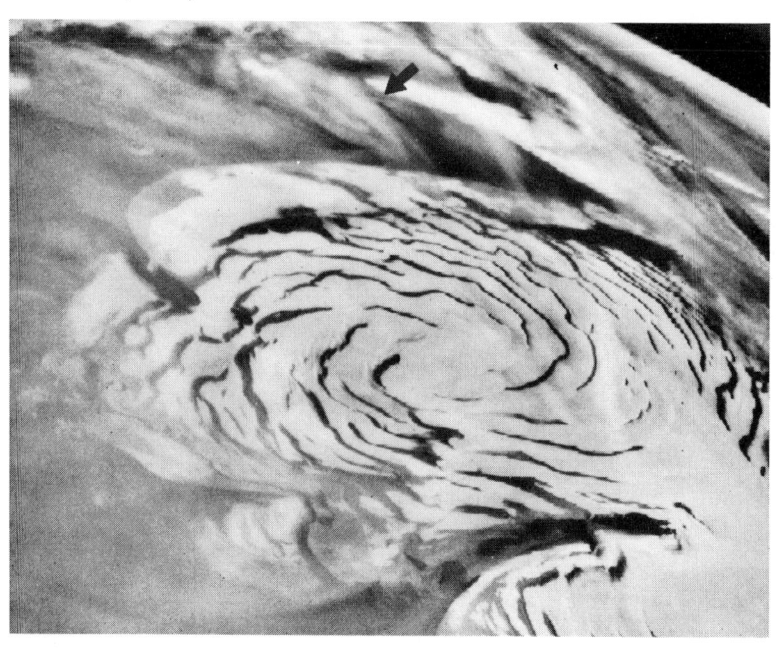

Mars may contribute particles to the north polar erg. Most of the major outflow channels on Mars empty into the northern latitude plains and undoubtedly contributed substantial volumes of sand-size grains to the general area.

On the other hand, some investigators have suggested that the primary source of sands is from deposits of the north polar cap and that the primary migration of dunes is southward. Thomas (1982) and Howard *et al.* (1982) propose that the sand dunes are migrating from the erosional scarps in the layered terrain (Fig. 7.12), which is considered to be alternate layers of ice and wind-deposited particles. These deposits may shed sands or sand-size aggregates as the terrain is eroded. Although Thomas and his colleagues suggest that the primary transport of sands was to the south, they note that the sand and dunes may be trapped by seasonally reversing winds.

Some of the wind-deposited particles in the layered terrain are thought to be dust grains which acted as nuclei for frost, with the aggregates settling out of the atmosphere and accumulating in layers. However, once in place, it is difficult to see how sand-size aggregates could be released. Furthermore, it is doubtful that such aggregates could survive the multiple saltation impacts required for formation of the vast dunes in the north without disintegrating to subsand-size particles. Unfortunately, unless some new ideas emerge from examination of currently available data, the question of source and composition of the north polar ergs will have to await further missions to Mars.

5.4 Sand shadows and drifts

Bagnold (1941) described accumulations of sand that result from fixed obstacles, such as bushes, rocks, or cliffs, as sand shadows and drifts. Sand shadows form in the lee of obstacles to air flow, where the carrying capacity of the wind is reduced. The zone of deposition is governed primarily by the size and shape of the obstacle.

Sand drifts form where winds are funneled between two objects and then fan outward. Deposition generally does not occur in the zone of funneling because the wind tends to be accelerated as it passes between the objects, but rather, where the wind spreads out. Similar drifts can develop in the lee of a stream bank, or along a cliff.

We discuss the influence of obstructions and wind erosion and deposition in more detail in the next chapter.

5.4.1 *'Drifts' on Mars and Venus (?)*

The Viking Landers returned magnificent views of the martian surface for two localities on the 'Red Planet'. Although the two areas differ in their geological setting, Lander images for both sites reveal abundant patches of material, which have been described as aeolian deposits. During the excitement of the first few days of the mission, these deposits were referred to as dunes, but, upon closer inspection, most of the forms are seen to have concave surfaces and give the appearance of having been sculpted. Consequently, the Lander imaging team (Mutch *et al.*, 1976) has used the term 'drifts' for these deposits and ascribes their origin to wind deposits that have been eroded.

Use of the term 'drifts' for the martian features is somewhat unfortunate because the original meaning of the term, given by Bagnold, is genetic, and the origin of the martian deposits is not known. Fine bedding structures can be identified in the martian material, some of which resembles cross-bedding. Sharp & Malin (personal communication) have studied these deposits and suggest that at the Viking Lander 1 site, there may be remnants of a much more extensive mantle which may have completely buried most, if not all, of the rocks in the field of view. Thus, the 'drifts' would simply be deposits which are protected from erosion in the lee of rocks, or which are still in the process of being eroded.

Interpretation of possible windblown deposits on Venus are even more uncertain than on Mars. Examination of the Venera images (Fig. 1.19) reveals deposits of fine grains which could be attributed to aeolian processes, but the resolution of even the latest images from Venera 13 and 14 is marginal for the detection of bedding structures.

5.5 Sand sheets and streaks

Sand sheets and streaks consist of patches of sand which lack slipfaces; sand sheets, as defined by Breed & Grow (1979), are also referred to as sand plains by some workers and occur in most sand seas. Sand streaks, also known as sand stringers, sand streamers and, possibly zibar dunes, have sharply defined edges and are typically associated with dominantly unimodal winds. From their study of various sand streaks in Egypt, El Baz and colleagues (1979) noted that many light streaks consist of coarse grains of highly reflective quartz, which form a lag deposit resting on finer sands.

Many sand sheets and streaks are readily apparent on orbital images of Earth and appear to be abundant in many regions on Mars. We discuss these in more detail in Chapter 6.

6

Interaction of wind and topography

6.1 Introduction

It is possible, as we have seen in the last chapter, for a sea of windblown sand to create its own topographical features in the form of sand dunes. In this chapter we discuss the interactions which occur between windblown particles and topographical features such as hills, craters (particularly as related to craters on Mars), vegetation, and other obstructions to the boundary layer. In order to understand these interactions, it is necessary to discuss atmospheric circulation and the effects that obstructions have on wind flow near the ground.

Some of the primary indicators of topographical influences are various types of wind streaks. Mariner 9 and Viking images of Mars show a remarkable array of bright and dark surface markings that result from interactions between the wind and topography (Fig. 1.3(*a*)). Many of these streaks serve as surface wind direction indicators. The differences in streak geometry and their variability with time have sparked much discussion and controversy on their origin and evolution. We review the streaks of Mars and discuss their possible origins, as well as present some possible terrestrial analogs that may shed light on the atmospheric–topographical interactions involved in streak formation.

6.2 Atmospheric motions

General atmospheric circulation results primarily from planetary rotation combined with the variation of incident solar flux with position on the planet (see Chapter 2). On a smaller scale, atmospheric circulation is influenced by topographical features such as mountain ranges, and on Earth is further influenced by the oceans through the exchange of heat and water vapor with the atmosphere. The boundary-layer characteristics discussed in Section 2.4 are appropriate for flat areas with no surface irregularities; the introduction of topographical features, such as hills and craters, results in significant deviations in wind flow patterns, which in turn cause variations in aeolian erosion and deposition. It is necessary to understand these flow field effects at both large and small scales in order to interpret wind-formed features.

6.2.1 *General circulation and rotational flows*

Ideally, if we consider only atmospheric motion on a planetary scale and disregard friction and planetary rotation, we would expect a north–south circulation pattern to develop in which warm air would rise at the equator, travel poleward at high altitudes, and then return to the equator close to the surface. Because of surface friction and planetary rotation, there are three idealized circulation cells on Earth, in both the northern and southern hemispheres – one corresponding to the easterly trade winds near the equator, one developed by westerly winds in the mid-latitudes, and one formed by polar easterly winds. In reality, this three-celled model is much too simple because of topography, the asymmetric distribution of land and water, and seasonal variations due to the axial tilt of Earth, but it serves to illustrate the fundamental pattern on a planetary scale.

On a somewhat smaller scale, secondary circulation patterns often develop around high- and low-pressure centers. Among those which produce strong winds are monsoons, hurricanes (called typhoons in the Far East and cyclones in the Indian Ocean) and extratropical cyclones. *Monsoons* – which develop most strongly in Asia – are seasonal winds that form in association with thermally produced continental high-pressure zones during the winter and low-pressure zones in the summer. *Hurricanes* typically form in low latitudes and are large, axisymmetric vortices that derive their energy from heat released by condensation of water vapor. Hurricanes, with diameters of several hundred kilometers, can have surface winds exceeding 35 m/sec over land and can develop even higher winds over the ocean.

The *extratropical cyclone* is produced either by mountain ranges which act as barriers to large-scale circulation, or by the interaction of air masses along large storm fronts. The storms are often accompanied by destructive winds generated by the interaction of cold, dry, polar air masses mixing with warm, moist, tropical air.

The general circulation on Mars has some resemblance to that on Earth because of the similarities in axial tilt and rotation rate (Table 1.1). However, great differences between the two planets result from the lack of oceans on Mars. Computer models of general circulation patterns on Mars predict strong, thermally driven winds at the solstices, with flow toward the poles at low levels during the summer, rising in the summer subtropics, return flow toward the winter poles at high altitude, and descending flow at winter mid-latitudes (Leovy, 1979). Although computer modeling is not readily verifiable because of limited observations, wind directions indicated by various wind streaks on the surface (Fig. 6.1) and measurements of atmospheric pressure by the Landers support the results to some extent.

In general, the strength and direction of winds near the surface are highly dependent on topography and have strong diurnal and semidiurnal amplitudes. Large-scale topography (≈ 10 km) generates thermally induced diurnal winds which move downslope in the morning and upslope in the evening.

Circulation in the atmosphere of Venus is quite different from that of both Earth and Mars because of the relatively slow (243-day) axial rotation. The lack of significant Coriolis accelerations indicates that the general circulation of Venus is, to first approximation, a *Hadley cell* – i.e., warm air rises at the equator and sinks at the poles (Chamberlain, 1978). Measurements made by the Venera and Pioneer spacecraft on Venus

Fig. 6.1. Rose diagram showing light- and dark-streak orientations derived from Viking orbiter images in Chryse Planitia on Mars, and a wind hodograph for the Viking Lander 1 site. The dark streaks are oriented approximately parallel to the predominant wind directions. (From Greeley *et al.*, 1978.)

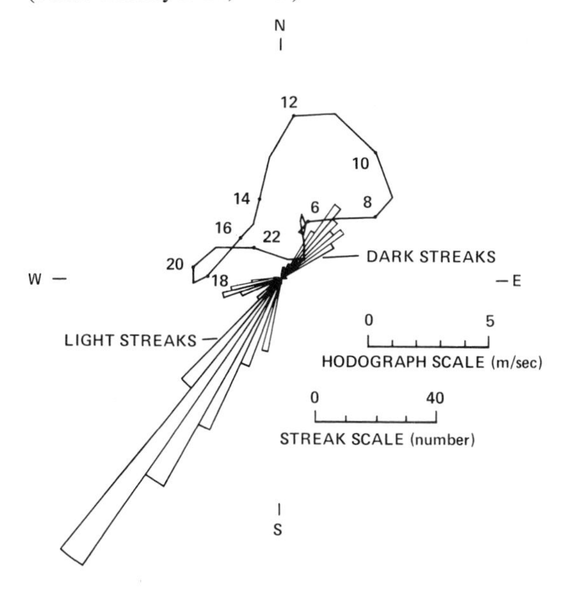

yielded wind speeds of about 1 m/sec near the surface of Venus, while considerably higher easterly speeds of up to 200 m/sec were detected at high altitudes within the cloud layer (Donahue, 1979).

6.2.2 *The vortex*

The vortex – a rotating column of fluid – is ubiquitous in fluid flows. *Vorticity* (mathematically the curl of the wind vector) is a measure of the angular velocity of a parcel of fluid (the magnitude of vorticity is twice the angular velocity). Fig. 6.2 illustrates the tangential velocity profile in a viscous, two-dimensional vortex and the corresponding vorticity. Vorticity, created through friction, manifests itself either in a shear layer (as in the boundary layer) or in a vortex. In the absence of friction, the laws of motion predict the absence of vorticity, and so the vorticity in the vortex is nearly zero outside of the radius for maximum tangential speed where friction is small, and vorticity is highest in the center where the effect of friction is largest.

Vorticity in hurricanes is generated primarily from the interaction between atmospheric pressure gradients and Earth's rotation. The central region of the hurricane (called the 'eye'), where winds are light, corresponds to the area on the left side of the graph in Fig. 6.2 where the tangential wind speed falls to zero at the vortex center. The hurricane structure is more complex than in Fig. 6.2, however, because there is also vertical circulation within the hurricane vortex.

The vorticity created in an extratropical cyclone can give rise to smaller vortices, such as tornados. A tornado is a vortex of considerably smaller

Fig. 6.2. Pressure, vorticity and speed ratios within a viscous vortex; P_{AMB} = ambient pressure, P_{MIN} = minimum pressure.

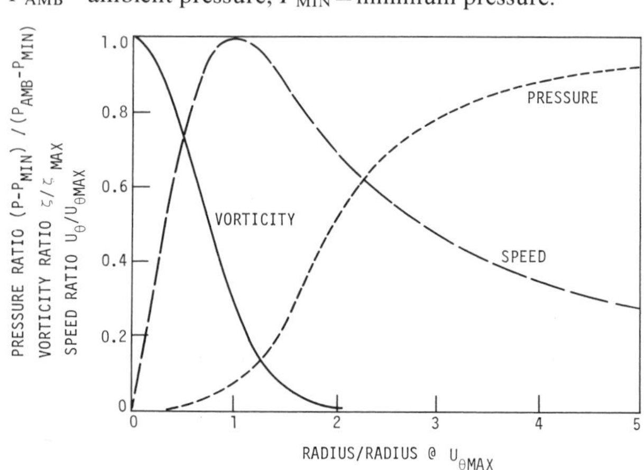

size than a hurricane (tornados typically range in diameter from several meters to perhaps 2 km). Tornados are often visible because of water vapor condensation in the low-pressure core. Maximum wind speeds in tornados are probably higher than those in hurricanes and can be of considerable destructive force.

The dust devil is another atmospheric vortex and is of smaller size than the tornado. Dust devils (discussed in more detail in Chapter 7) are not associated with extratropical cyclones, however, but are formed in unstable air moving over hot surfaces. The air over an especially hot portion of the surface rises and colder air rushes in to take its place. If atmospheric vorticity is generated, as can occur by shear from a boundary-layer obstruction, a dust devil may be formed. The low pressure in the vortex associated with the air rotating at high speed can raise considerable dust from the surface.

Hurricanes, tornados, and dust devils all involve vortices in which the primary axis is perpendicular to the surface. Vortices parallel to the surface are also common and can form in several ways, including instability in the Ekman layer (see Section 2.4.5), which causes parallel vortices of alternating rotational sense. Some investigators have linked such rows of vortices to the formation of longitudinal dunes. Such parallel vortices can undoubtedly exist when the wind is blowing over existing longitudinal dunes, but they probably are not involved in the initiation of such dunes (see Section 5.3.3).

6.3 Topographical effects on surface winds

6.3.1 *The effects of mountains*

Air rising over a mountain crest which has lost its moisture due to precipitation may experience sufficient adiabatic compression on the downslope side, after crossing the crest, to become a high-temperature *foehn* or *chinook wind.* If the rise in temperature is too low (such as where a steep slope separates a cold plateau from a warm plain), the potential energy of the still, cold air is converted into kinetic energy and a strong, cold wind results. This type of wind, called a *bora*, often occurs along the northeast coast of the Adriatic Sea.

Other types of winds associated with large topographical features include *valley winds* and *mountain lee waves*. Valley winds are caused by air being forced to flow between mountains or hills. From consideration of conservation of mass flow, as the stream cross-sectional area decreases, the wind speed must increase at the same rate. A typical example is the wind flowing from the cool coast of southern California through the San

Gorgonio Pass onto the desert near Palm Springs. These winds often generate intense sand storms and have led to one of the better-known localities for ventifacts (see Chapter 4).

Lee waves often form where mountain ranges are located upwind from flat plains (Fig. 6.3), as in the Boulder, Colorado area east of the Rockies. When air flows over the crest of a range of mountains, it must rise in order to do so. In turn, it descends when traveling over the leeward slope (unless the flow separates from the surface at the crest). The momentum gained by the air as it travels down the slope carries it past an equilibrium condition, forcing it to 'bounce' back up again. The momentum gained on the upward motion again carries it past equilibrium and the air starts downward a second time. The standing wave pattern thus formed can exist for several oscillations as shown in Fig. 6.3. The wavelength can be related to the extent of stable stratification in the boundary layer. Conditions for mountain lee waves to occur are most favorable when the wavelength predicted from the values of wind speed and degree of stability coincides with the size of the mountain range. If conditions are not favorable for production of standing waves, separation will occur on the lee slope and lee waves are not formed. Lee waves can also occur downwind of isolated mountains but are confined to a leeward wedge-shaped region with the wedge apex at the mountain peak. Some lee waves are easily seen because clouds occur at the upper portions of the waves and form in a *cloud street* (parallel rows of clouds). Lee waves and cloud streets have been observed on Mars in association with some topographical obstacles (French *et al.*, 1981).

6.3.2 *Separated flows*

Most boundary-layer obstructions, such as hills, are not streamlined, but have sharp edges which cause *flow separation*. Flow is said to be 'separated' when the air streamline leaves the surface, either because the air

Fig. 6.3. Schematic cross-section of mountain leewave streamlines.

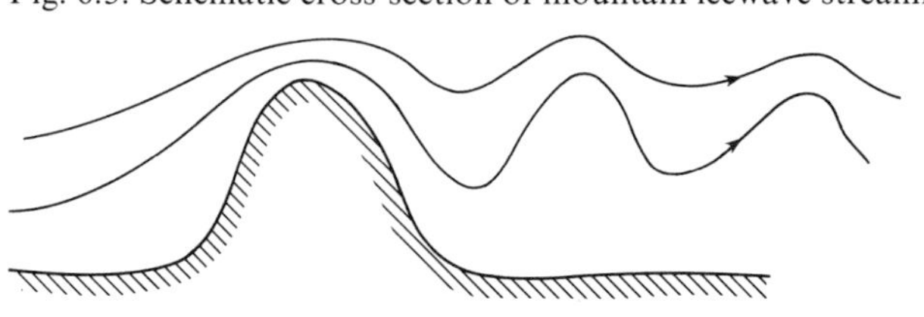

is encountering increasing pressure on its downwind journey or because its momentum is too high to follow the curvature of the surface.

Examples of 'two-dimensional' flow separation are shown in Figs. 6.4–6.8. If the windward edge of a cliff or escarpment is sufficiently steep, separation will occur upwind and rotational flow is generated, as shown in

Fig. 6.4. Schematic cross-section of streamlines with separated flow upwind of a windward facing escarpment.

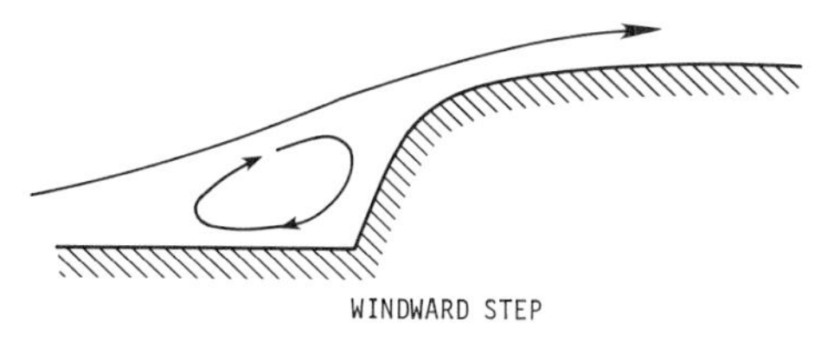

Fig. 6.5. (*a*) Flow over a two-dimensional cavity. (*b*) Snow drift patterns formed with two-dimensional cavities in a flat plain. Note the periodic character of the drift patterns. Wind direction is top to bottom. (Photograph by Iversen, 1979.)

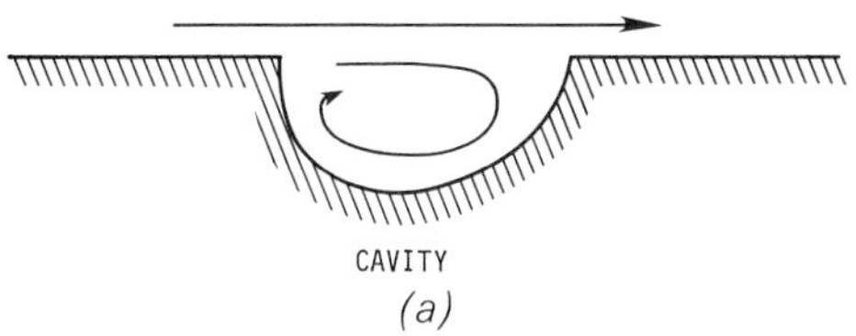

(*a*)

(*b*)

Fig. 6.4. Notice that the flow near the surface at the base of the escarpment is in the upwind direction. This type of *reverse flow* can also occur within cavities (Fig. 6.5(*a*)). A 'two-dimensional' topographical feature does not necessarily result in a two-dimensional flow; Fig. 6.5(*b*) shows a snow drift associated with a two-dimensional cavity. The periodic drift formation is probably related to the three-dimensional cellular flow in a two-dimensional cavity, as discovered by Maull & East (1963). Reverse flow can also occur downwind of a leeward-facing escarpment (Fig. 6.6).

On a streamlined surface, the point of separation depends on the Reynolds number and the degree of upwind turbulence. A sharp edge, however, fixes the separation line, as illustrated in Fig. 6.7. The flow is

Fig. 6.6. (*a*) Cross-section of separated flow streamline downwind of a leeward-facing scarp. (*b*) Example of leeward separation at the Kelso dunes, California. Some of the sand particles in the leeward separation zone on the right could be observed traveling upwards and to the left near the surface. (Photograph by Iversen, 1978.)

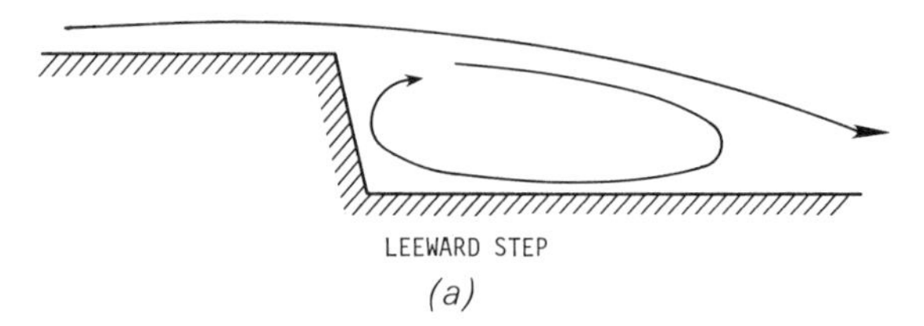

(*a*)

(*b*)

upwind near the surface, both on top of the plateau and just windward of the cliff base. The separated flow on top of the cliff will reattach on the upper surface of the cliff if it is long enough, as shown in the diagram. A *stagnation point* is shown just below the cliff top on the windward edge. Below this line strong downdrafts occur adjacent to the cliff surface. The separated region on the leeward side of the rounded ridge, pictured in Fig. 6.8, is smaller than for a blockier shape. Notice the convergence of streamlines at the ridge top. This is indicative of the increase in wind speed and speed gradient and the resultant increase in surface friction speed at the top. The logarithmic wind speed profile (Section 2.4.2) is not maintained in any portion of the flow field near boundary-layer obstructions such as those shown, except perhaps in relatively thin layers very near the surface.

Flows around three-dimensional objects are even more complicated. Fig. 6.9 shows a highly simplified diagram of the horseshoe vortex pattern which develops around some three-dimensional objects. The upwind boundary layer, because of the velocity gradient which starts from zero speed at the surface, contains vorticity, the orientation of which is perpendicular to the wind and parallel to the surface. As the air sweeps around the obstruction, the vorticity in the boundary layer is bent around it and is manifested in the form of the horseshoe vortex, which may trail downwind for many diameters. Although more than one vortex may form, one is usually much stronger than the other(s). The surface shear stress directly beneath the vortices can be very high. Maximum shear stress occurs directly upwind of the object if the height of the object is

Fig. 6.7. Separated flow pattern caused by a sharp-edged cliff.

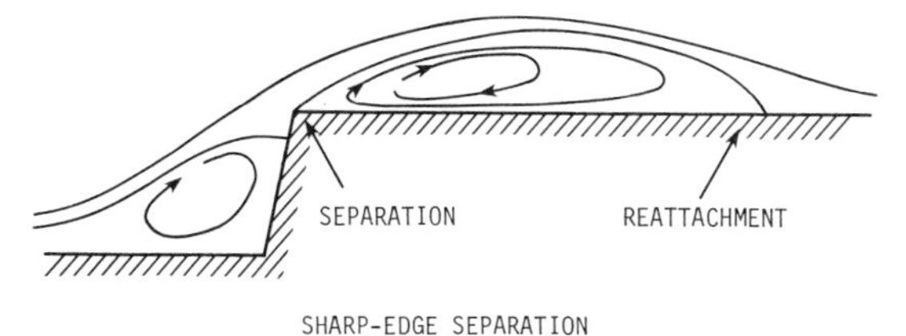

Fig. 6.8. Streamline pattern over a smoothly contoured ridge. Separation will occur on the leeward side if the hill is sufficiently steep.

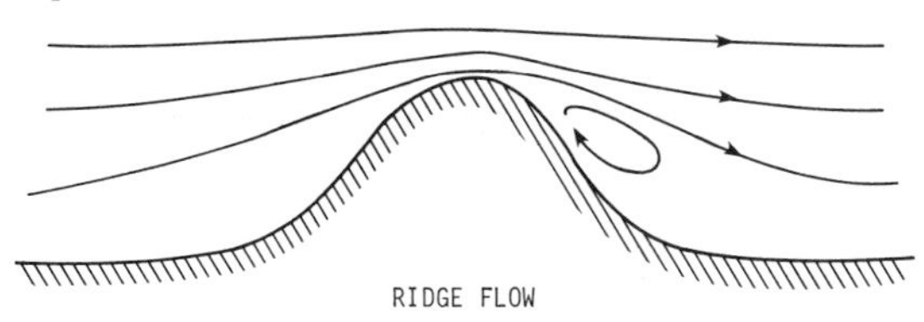

significantly greater than its breadth. For a low, broad object, however, such as a raised-rim crater, the maximum shear stress occurs under the downwind trailing vortices. The pattern illustrated in Fig. 6.9 can exist through a large range of scales. For example, a horseshoe vortex trailing from a large isolated mountain was blamed for tearing off most of the vertical tail fin of a large Air Force B-52H airplane. The wind speed prior to the encounter was measured at 36 m/sec and the aircraft was flying at a speed of 180 m/sec. The violent response of the aircraft upon penetrating the vortex (at 4000 m altitude, about the same height as the mountain peak) from smooth air lasted 9.6 sec. The airplane landed several hours later after burning off most of the fuel load and after experimentation with methods of control by differential thrust, flaps, and air brakes (Fisher, 1964).

Additional complexity in flow past three-dimensional shapes is illustrated in Fig. 6.10. The plan-view of the flow over the prism shows only the outline at the surface of the main separated flow boundary, the trace of the horseshoe vortex, the outline of the interior separated wake, and the cross-sections of the vertical portions of a vortex loop embedded within the separated wake. The diagram does not illustrate the three-dimensionality of the flow field above the surface, however, including vortices shed from

Fig. 6.9. Horseshoe vortex system formed by a boundary-layer obstruction.

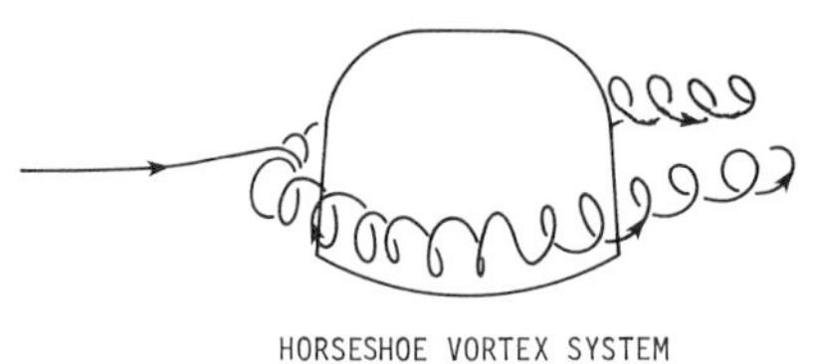

Fig. 6.10. Plan-view of the surface features of flow over a rectangular prism situated on a flat plain.

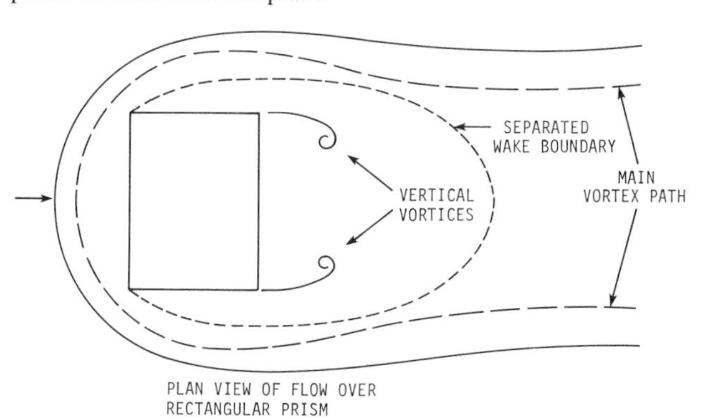

the corners. The strongest vortices shed from an upper surface occur when the wind vector bisects the surface edges (Fig. 6.11). The shear stresses associated with these vortices are again very high, a fact noticeable when roof damage occurs on buildings with flat roofs. All the flows represented in Figs. 6.4–6.11 are time averaged. The actual flows are unsteady due to boundary-layer turbulence and the natural instability of some of the flow patterns.

6.4 Wind streaks

Wind streaks are patterns of contrasting albedo that can be seen on the surface as a result of various aeolian processes. Wind streaks are found on Earth and are very common on Mars. They may possibly occur on Venus and Titan. Understanding how, where, and why they form is important not only for understanding modifications of surfaces in the geological context, but also for providing insight into meteorological phenomena.

6.4.1 *Modes of formation*

Most wind streaks on Earth and Mars form in association with topographical obstacles, although some are also found on flat plains. Changes in wind flow – as might result from flow over a hill – can be manifested on the surface in a wide variety of ways, any of which could lead to the formation of wind streaks. There are at least three characteristics of aeolian surfaces that lead to albedo contrast and, therefore, to wind streaks: grain size, grain composition (color), and aeolian bedform.

Let us first consider grain size. Some wind streaks result from differences in grain size within the streak zone, in comparison to grains outside the streak zone. It is well known that the albedo (reflection of light from surfaces) is influenced by grain size; patches of small grains generally have a higher albedo (i.e., brighter) than patches of large grains. Differences in grain size between wind streaks and the surrounding surface can come

Fig. 6.11. Vortex formation on top of a sharp-cornered prism.

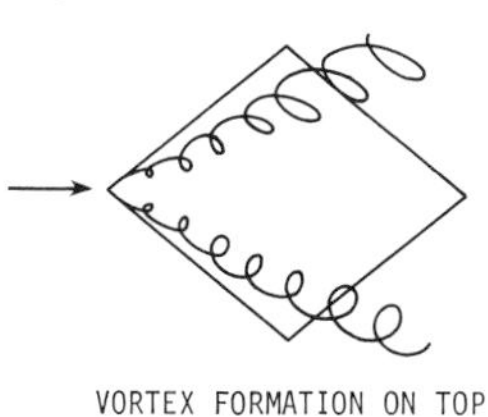

about either as lag deposits or as transported deposits. Fig. 6.12 shows the wind threshold curve for grains of different sizes (qualitative, but based on Fig. 3.17), subdivided in three general sizes: (1) *small* (high albedo, or bright), (2) *medium* (moderate albedo), and (3) *large* (low albedo, or dark). Keeping in mind the fact that the 'medium' size grains are those most easily removed by the wind (lowest threshold), then if a surface covered with grains of medium to large sizes were subjected to winds above threshold speeds, the medium-size grains would be removed, leaving the large ones behind as a lag deposit; the net result would be a generally darker surface.

On the other hand, if the original surface were blanketed with small- to medium-size grains, and subjected to wind, then, in principle, the medium-size grains might be removed, leaving the small grains behind, and the resulting surface might appear brighter. It is more likely, however, that saltation of the medium grains would also set the fine material into motion and thus this case of streak formation is probably impossible to achieve.

Wind deposition of different grain sizes may also lead to the formation of wind streaks. For example, fine-grained volcanic ash erupted from a volcano may be carried by the wind to form wind streaks. Or, material may

Fig. 6.12. Schematic diagram relating wind threshold curve as a function of particle diameter to the effective albedo of a bed of particles. In general, fine grains (bright) and coarse grains (dark) require higher wind speeds for saltation than do medium-size grains.

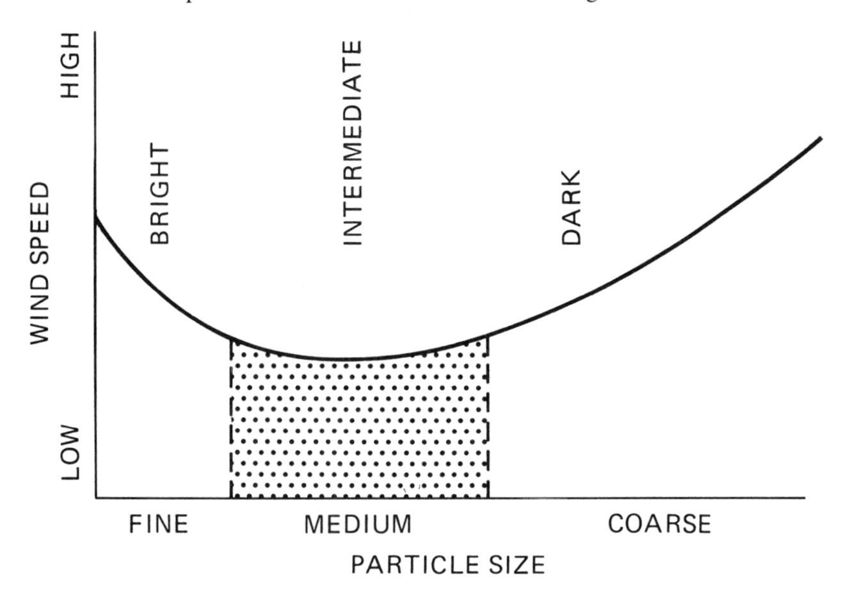

be eroded by the wind, from a hill, stream bank or crater, and drift downwind to form a streak.

Differences in composition may also lead to wind streaks. For example, if the volcanic ash, or the eroded material described above is of a different composition than the surrounding plain, then it may show up as a different color and/or albedo, regardless of grain size. Composition may also affect threshold wind speed; dense grains – such as basalt – are more difficult to move than low-density grains, such as quartz, simply because they are heavier. Thus, if there were a bed of like-size grains, but of mixed compositions, the low-density grains might be deflated, leaving the high-density ones behind. Of the common minerals on Earth, high-density grains are rich in iron and magnesium and are dark-toned, whereas low-density grains are silica-rich and are light-toned. Thus, at least on Earth, we would expect the wind-deflated surface to be darker after the light-toned, low-density particles were removed.

Bedforms such as ripples and dunes may also show as light and dark surface patterns, at least from distances of orbiting spacecraft. As shown in Fig. 6.13, because of the asymmetry of these bedforms, depending upon the lighting geometry, areas containing bedforms may appear to be darker than smooth surfaces – or the albedo may depend on ripple wavelength and/or ripple geometry.

Given the three conditions of the surface, outlined above (grain size, composition, and bedform), which can lead to the formation of albedo differences, let us now consider where we might expect these patterns to develop. Earlier in this chapter we discussed how wind flow patterns are affected by topography, and how the surface shear stress – the ability of wind to move particles – can change in response to these patterns and to vorticity. There are specific zones where the shear stress is high, and those where the stress is low related to the surface roughness and geometry of the topography. If we combine this fact with the various surface conditions, it is possible to generate both bright and dark streaks in specific areas related to the topographical obstacle.

Thus far, we have considered wind streaks in theoretical terms. In the sections below we consider streaks on Earth, in the laboratory, and on Mars, and then discuss the possibility of wind streaks on Venus and Titan.

6.4.2 *Streaks on Earth*

Wind streaks occur in many areas on Earth, but they are often difficult to recognize, especially from the ground.

Recent images of Earth obtained from various orbiting spacecraft, however, show that streaks are common in many desert regions (Fig. 6.14). Farouk El Baz and his colleagues (1979) found that at least part of some dark streaks result from irregular chips a few centimeters in size (the coarse size contributes to the low albedo). They also found a dark streak extending from a 15 m high inselberg. The dark streak surface is a reddish soil veneered by a lag pavement consisting of 5–7 mm pebbles eroded from the hill. An erosive vortex wake system caused by the hill has deflated the streak to a depth 2 m lower than the surrounding plain which is covered by a sand sheet.

Conventional air photos also reveal wind streaks and bedforms associated with a wide range of topographical features. As discussed in Section 5.3, some dunes and dune fields are topographically controlled;

Fig. 6.13. Diagrams showing some possible origins of light and dark streaks: (A) the entire surface is covered with fine dust; dust is later removed in the wake of the crater as a result of preferential erosion (see Fig. 1.3) to expose a dark substrate; (B) the surface is initially covered with a mixture of fine-grained (high-albedo) and medium-grained (low-albedo) particles; medium-grained particles are easier to move by the wind than fine-grained ones, thus, the medium-grained particles are removed from the wake zone at the crater, leaving the fines that appear as a high-albedo zone; (C) the crater is a source for the dark particles that are carried by the wind downstream and deposited as a plume; these are probably rather rare on Mars; (D) under very high winds, the 'shadow zone' in the immediate lee of the crater may lengthen; it would consist of medium- and fine-grained particles (low-energy zone) and would appear relatively dark in comparison to the surrounding terrain where the medium-grained particles would have been removed. Particle sizes are defined as follows (based on threshold curves conducted under martian surface pressures): (A) fine, <40 μm, (B) medium, 40–600 μm, (C) coarse >600 μm. (From Greeley *et al.*, 1978.)

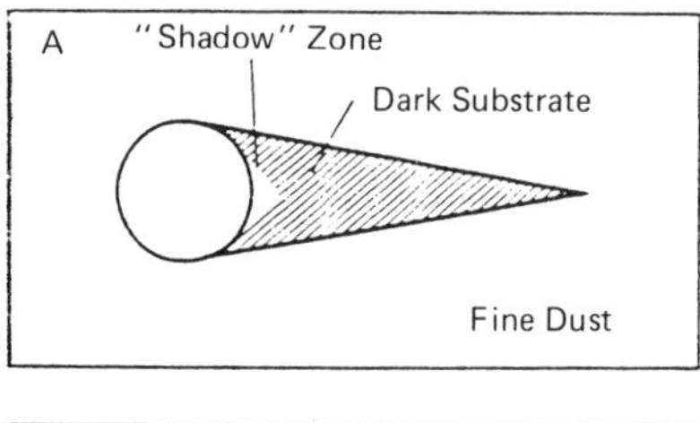

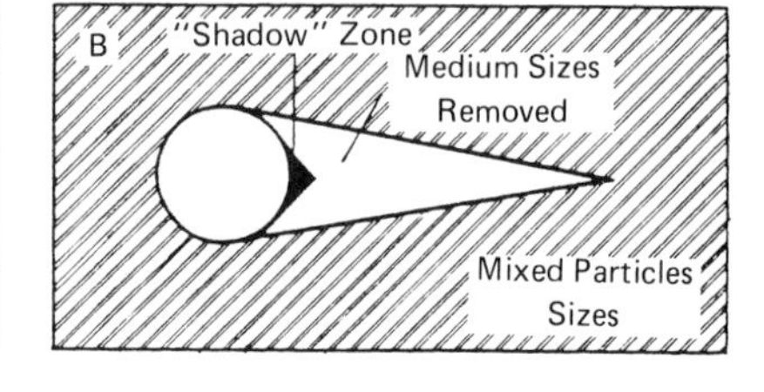

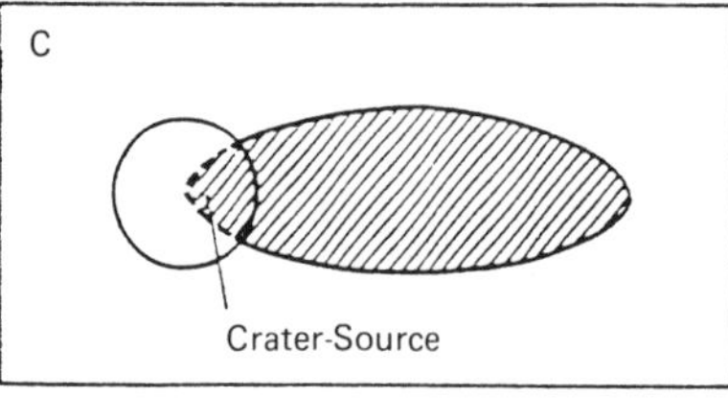

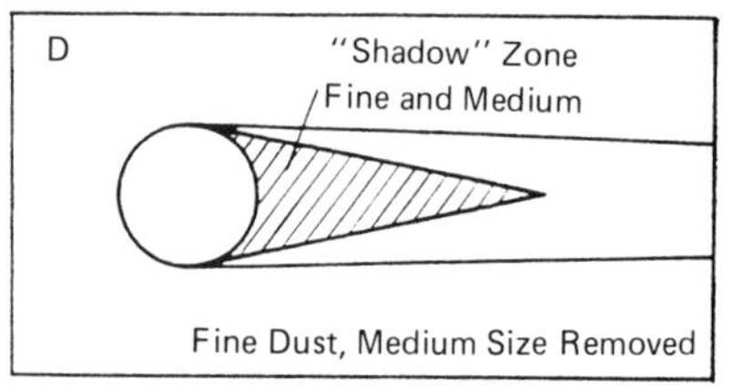

Grolier and his colleagues (1974) of the US Geological Survey found particularly striking examples in Peru.

Craters are particularly effective landforms as generators of wind streaks. Although relatively few craters on Earth occur in aeolian environments, those that have been found are important analogs to martian features. Fig. 6.15 shows some terrestrial cases in which there are distinctive zones of wind erosion and deposition.

Fig. 6.14. Light and dark streaks in the Borkou Lowland of Chad. The dark areas are sandstone ridges with sand and lakebed remnants between the streaks. (From Mainguet, 1972, courtesy of the Institut Geographique National, France.)

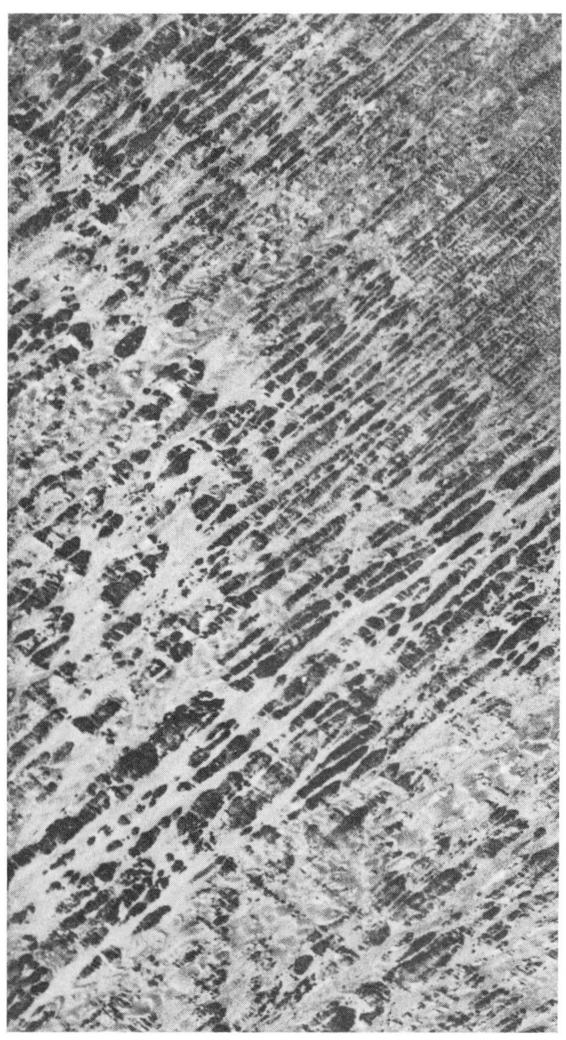

Fig. 6.15. (*a*) Sketch map of the Wolf Creek impact crater in Australia, showing zones of erosion (quartzite) and deposition (dunes) associated with flow over the crater. Wind is from the right (from McCall, 1965). (*b*) Low-altitude, oblique aerial photograph of the Aouelloul Crater in Mauritania, western Sahara, showing a similar pattern of erosion and deposition as that around the Wolf Creek crater. (Photograph courtesy of Smithsonian Institution.) Both craters show patterns which correspond to the flow field portrayed in Fig. 1.3.

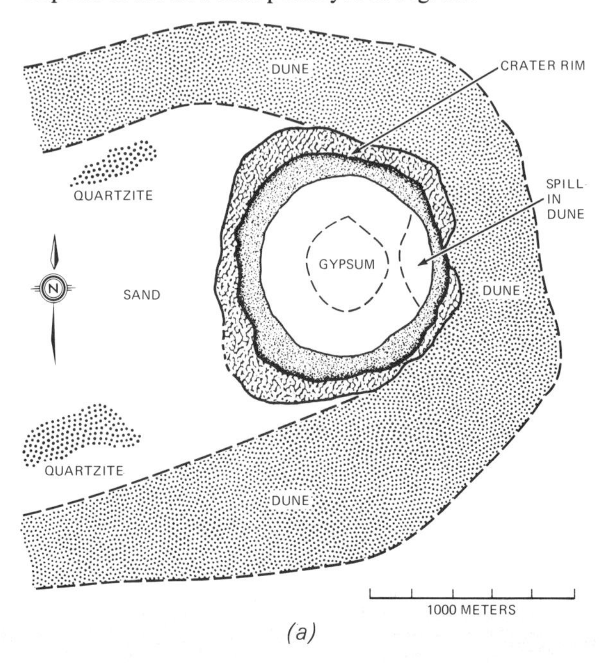

(*a*)

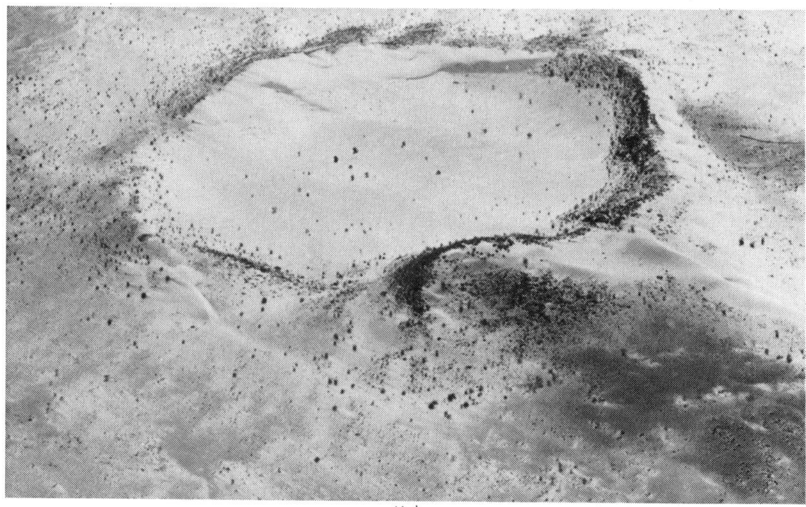

(*b*)

Streaks in snow

Wind streaks can also develop in snow. Fig. 6.16(*a*) shows twin dark streaks associated with a small crater blanketed with snow. The hay bales in Figure 6.16(*b*) show depositional regions (bright) outside of a central deflational streak. Those in Fig. 6.16(*c*) show a short deposit region in the lee of the bale and a larger deflational, or scour, area on the windward side, on the flanks, and trailing downwind for many diameters. These features in drifting snow have been described by Cornish (1902) and are called 'current crescents' by Allen (1965). Larger deflational streaks caused by farm buildings are illustrated in Fig. 6.16(*d*). A depositional area of snow caused by artificial roughness elements is shown in Fig. 6.16(*e*). Here

Fig. 6.16. (*a*) Twin deflation streaks (dark areas) in snow deposited leeward of a 5 m diameter raised-rim crater. Wind direction is from lower right to upper left. Note the sastrugi bedforms. (Photograph by Iversen, 1972.) (*b*) Trailing streaks formed by snow blown past hay bales lying in a flat farm field. Cylindrical bales are ≈ 1 by 1m. Note the snow deposits on either side of the bale wakes. Wind is from upper left to lower right. (Photograph by Iversen, 1979.) (*c*) Another field of scattered bales and associated deflation streaks. Note the central deposit ridges which are absent in (*b*) and the wave patterns in the snow. (Photograph by Iversen, 1979.) (*d*) Large deflation streaks trailing from farm buildings after a snow storm. (Photograph by Iversen, 1979.) (*e*) Effect of surface roughness on deposition; meter-high stakes have caused snow to accumulate in this field.

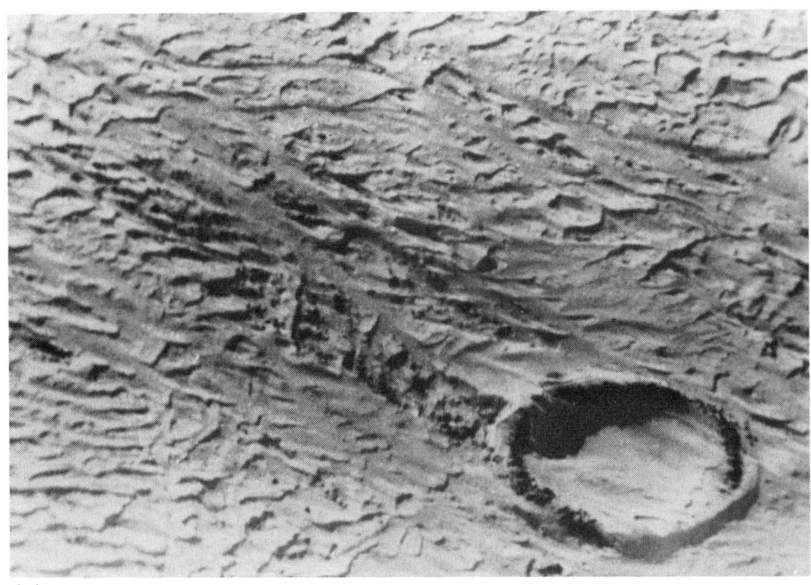

(*a*)

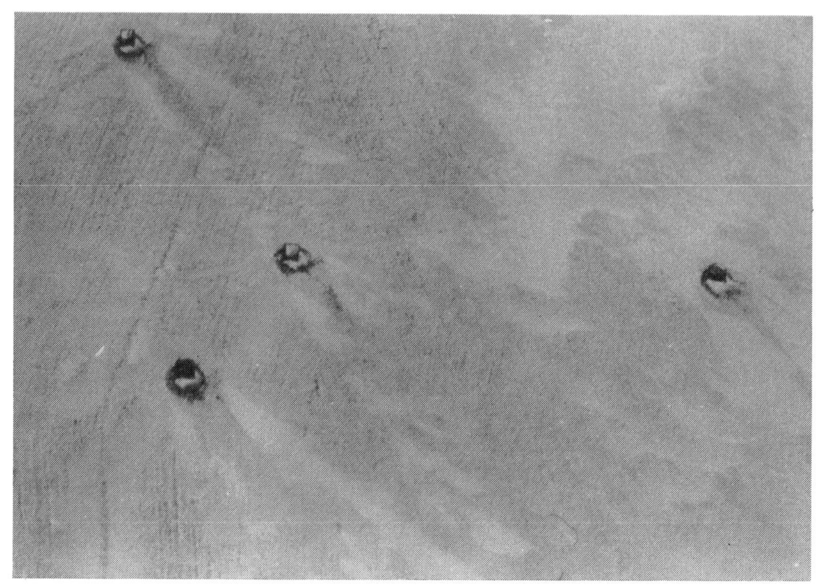

Fig. 6.16 (b)

Fig. 6.16 (c)

Fig. 6.16 (d)

Fig. 6.16 (e)

the roughness elements increase the threshold friction speed and net deposition of the snow results (see Section 3.3.4).

6.4.3 *Wind tunnel streak simulation*

The wind tunnel is a useful means for simulating large-scale aeolian phenomena, as well as for investigating the fundamental physics of particle motion. The similitude requirements for scaling wind tunnel models to full-size features are complex and difficult. We have studied and reviewed these requirements (Greeley *et al.*, 1974*a*; Iversen, 1980*a,b*, 1981, 1982; Iversen *et al.*, 1975*a*, 1976*c*) and have concluded that dimensionless mass transport rate, particle-fluid density ratio, Froude number, u_*^2/gh, and particle terminal speed-to-threshold friction speed ratio, U_F/u_{*_t}, are the most important parameters, which, for perfect simulation, should all have the same values at model scale as at full scale. Although for many simulations, especially for Mars cases, it is not possible to satisfy all the parameters, qualitatively many similarities exist between the wind tunnel results and observations of full-scale features.

A series of experiments was carried out to determine the characteristics of wind erosion and deposition associated with raised-rim craters. The erosive effect of the vortex wake is illustrated in Figs. 6.17 and 6.18, in which streaks formed by the erosion of material are shown. In these experiments, the floor of the wind tunnel test section was initially covered with particles to a uniform depth. The development of eroded wake area was recorded photographically as a function of time (Fig. 6.18). The differences among the wake patterns developed are due to differences in crater size, wind speed, and particle characteristics. Although these differences are significant qualitatively, collectively the data correlate fairly well (Fig. 6.19); some of the scatter in the data can be accounted for by changes in Reynolds and Froude numbers (Iversen *et al.*, 1976*c*).

Similar experiments were carried out to assess the formation of wind streaks through depositional processes. In Fig. 6.20, a rimless crater model was filled with fine particles and the material was deflated from the crater interior and deposited on a rough surface (rough sand paper) downwind. In another experiment, a streak resulted from deposition of material injected into the air stream to simulate an active source of particles, such as a volcano. From these and other experiments, a general model of streak formation was developed, which takes into account the flow field and zones of erosion and deposition associated with crater-shaped objects (Fig. 1.3).

Erosion and deposition rates

Aeolian erosion and deposition are strongly controlled by topography, as demonstrated by the experiments described above. Let us first consider a curtain of saltating sand blowing over a flat, uniform surface containing scattered topographical features, such as hills or craters. Where the curtain encounters topographical irregularities, there will be places where the local wind speed and shear stress decrease from the average value. In those places, net deposition will occur until a transport rate is reached which is in equilibrium with the surroundings. Conversely, net erosion will take place where the local wind speed and shear stress are increased above the average value outside the zone of topographical influence.

It is possible to predict where net deposition and deflation may occur from an analysis of the flow field and considerations of the particle threshold wind speeds. In regions of separated flow, wind speeds are generally lower than the average winds (except near the centers of vortices);

Fig. 6.17. (*a*) Iowa State University wind tunnel simulation of a crater wake dark streak. The deflation streak is dark because the wind tunnel floor is black and the particles are white. Crater rim height is 3 cm and its diameter is 30 cm. The material on the floor is sugar, with an average diameter of 393 μm. Terminal speed-threshold friction speed ratio is 10.1. (From Iversen *et al.*, 1973.)

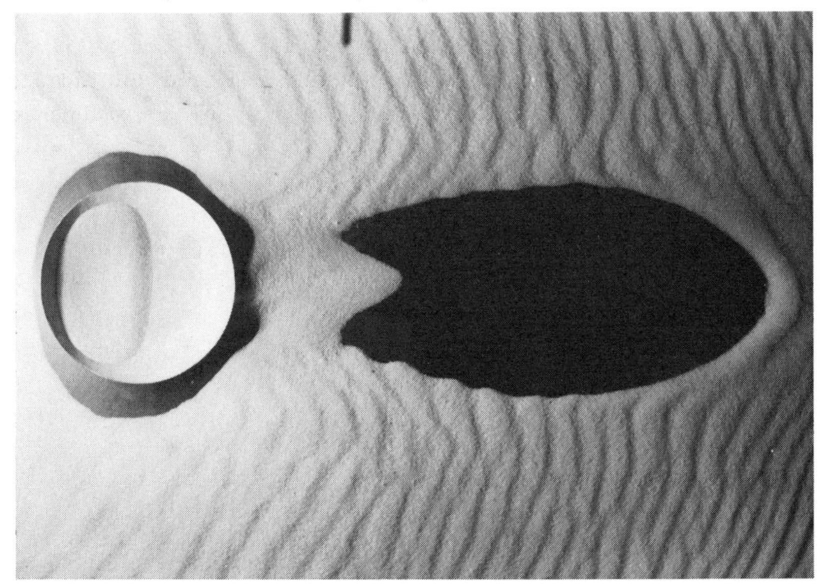

thus, deposition will occur just to the lee of most obstacles and in areas immediately in front of obstacles where the wind speeds are low. Conversely, deflation will occur beneath the cores of vortices, as shown in Figs. 6.9–6.11. These relationships can be combined with Eq. (3.27) to establish an erosion rate similitude to analyze wind streaks associated with craters (Iversen *et al.*, 1975*a*). Small models of craters were placed in a wind tunnel, and partly blanketed with a bed of loose sand. The model and the bed were then subjected to wind and the area that was deflated to a constant depth, d, was recorded as a function of time; this, in turn, was related to the mass transport rate using Eq. (3.27)

$$\frac{dA}{dt} = \frac{\text{area}}{\text{mass}} \times \frac{\text{mass}}{\text{time}} \propto \left(\frac{1}{\rho_p d}\right) q D_c \propto \frac{\rho u_*^3 D_c}{\rho_p g d}\left(1 - \frac{u_{*t}}{u_*}\right) \tag{6.1}$$

Or as volume rather than area,

$$\frac{\mathrm{d}V}{\mathrm{d}t} \propto \frac{\rho u_*^3 D_c}{\rho_p g}\left(1 - \frac{u_{*t}}{u_*}\right) \tag{6.2}$$

Fig. 6.17. (*b*) Simulation of a deflation streak in the Iowa State University wind tunnel; crater diameter is 20 cm; glass particles are 200 μm in diameter. Terminal speed–threshold friction speed ratio is 14.3. Dark material within and blown from the crater is composed of fine copper particles. (From Iversen *et al.*, 1973.)

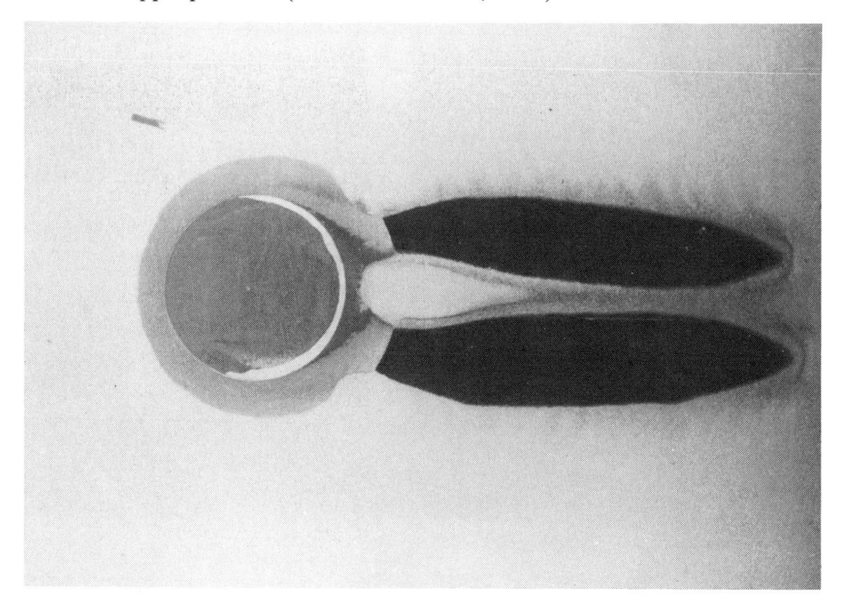

Fig. 6.18. (*a*) Sequence of the development of a deflation streak in the Iowa State University wind tunnel. Crater diameter is 20 cm; glass particles are 200 μm. Terminal speed–threshold friction speed ratio is 5.8. (From Iversen *et al.*, 1976*c*).

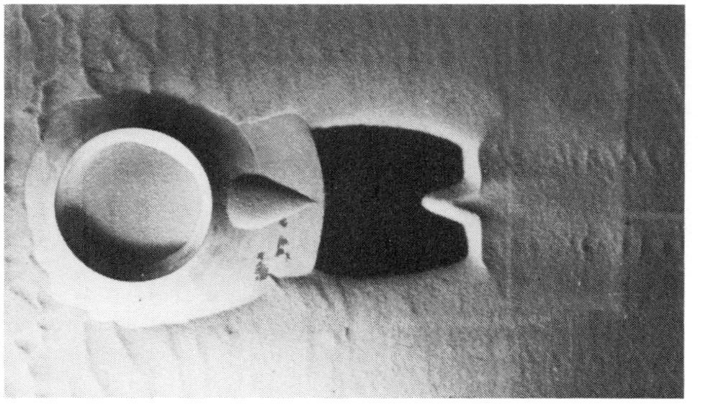

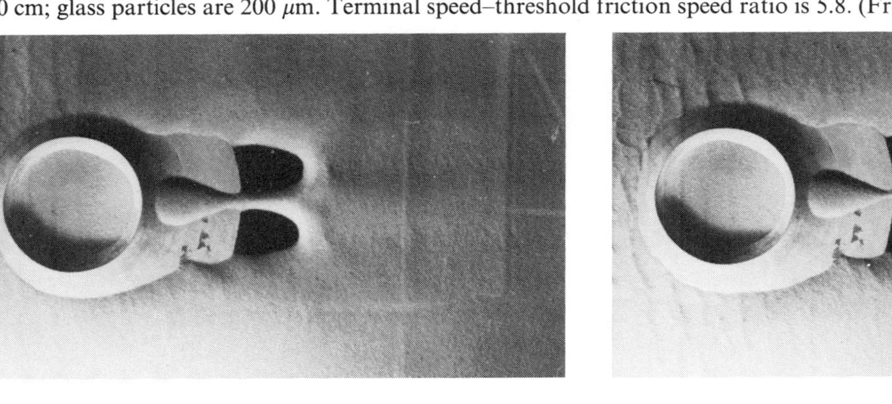

Fig. 6.18. (*b*) Deflation sequence; crater diameter is 30 cm; glass particles are 91 μm. Terminal speed–threshold friction speed ratio is 2:4. (From Iversen *et al.*, 1976*c*.)

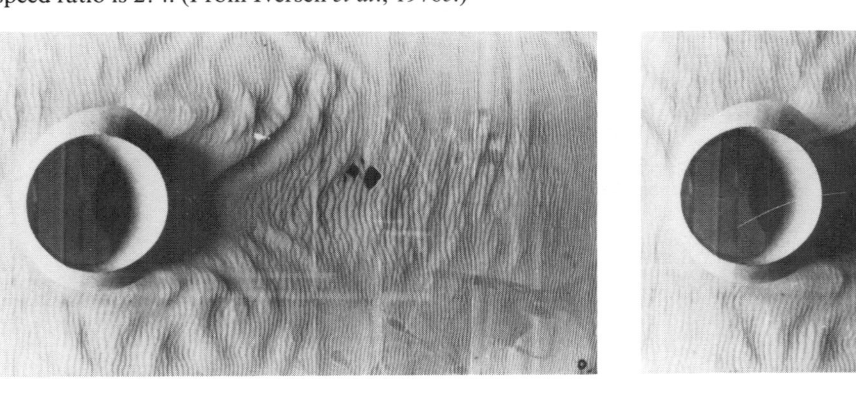

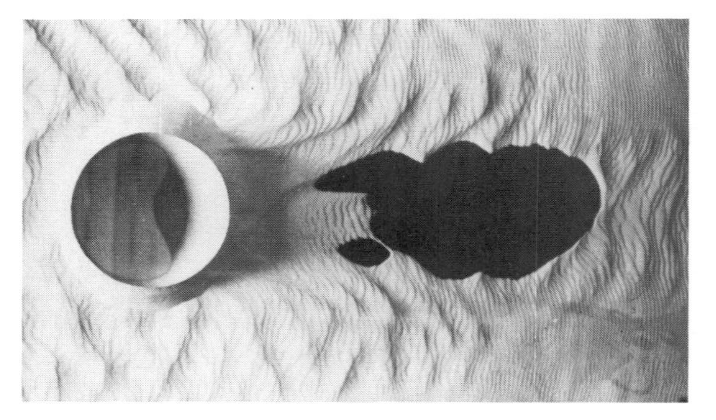

where D_c is a characteristic horizontal dimension (e.g., crater diameter). The results for deflated streak area as a function of dimensionless time are shown in Fig. 6.19 (using a form of Eq. (3.29) and wind speed instead of friction speed). Net deflation takes place where there is a local region of increased surface shear stress. Because mass transport increases with the third power of wind speed, only a small change in speed is needed to increase the deflation rate significantly. In the case of crater streaks, a horseshoe vortex system in the crater wake accounts for the increase in surface shear stress (Iversen *et al.*, 1975*a*, 1976*c*). The deflation rate in these wind tunnel tests was greater, by about a factor of 10, than the mass

Fig. 6.18. (*c*) Deflation sequence; crater diameter is 10 cm; glass particles are 200 μm. The differences among the patterns of the development are due to differences in wind speed, particle properties and, perhaps, the values of the Froude and Reynolds numbers.

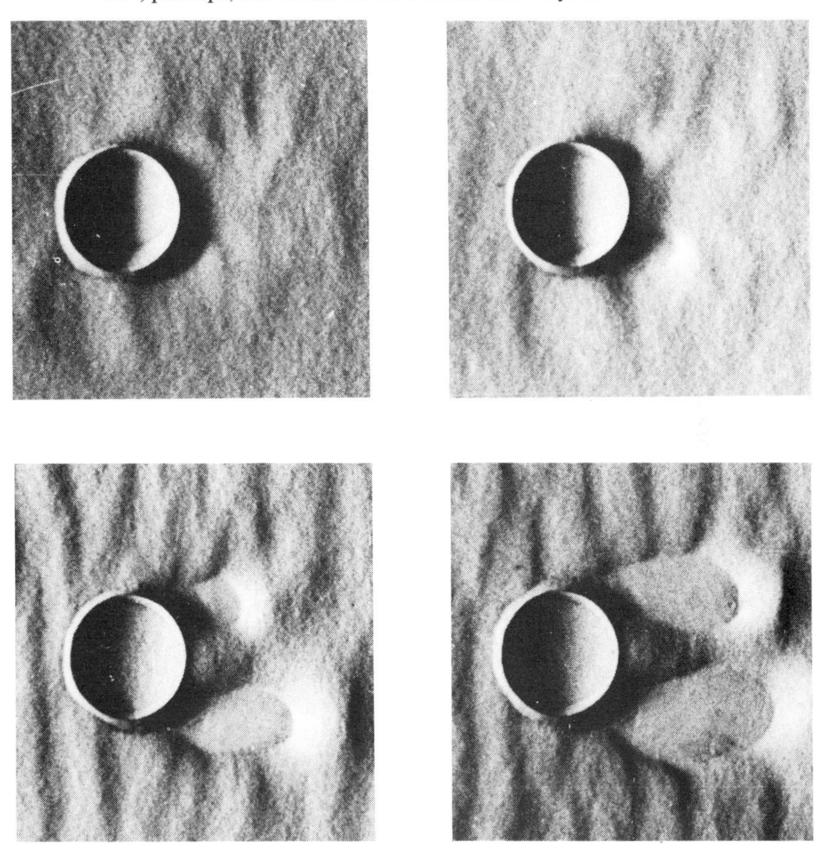

transport rate that would be expected in the absence of the crater model.

Eq. 3.27 has also been used to analyze rates of deposition (Iversen, 1980*a*, 1981). The deposition or deflation rate for a particular geometry, however, does not represent the total mass transport over the entire surface, nor are the rates constant fractions or multiples of the average transport rate as the wind speed changes. Rather, from the deposition data obtained experimentally, it was found that the mass deposition rate was a function of a densimetric Froude number, i.e.,

$$q_{\mathrm{d}} = \frac{\rho u_*^3}{g}\left(1 - \frac{u_{*\mathrm{t}}}{u_*}\right) f\left(\frac{\rho u_*^2}{\rho_{\mathrm{p}} g h}\right) \tag{6.3}$$

The mass deposition rate was found to decrease with increases in densimetric Froude number.

6.4.4 *Amboy field experiment*

Wind tunnel experiments have the advantage that certain parameters can be controlled and isolated for study. In the case of the formation of wind

Fig. 6.19. Deflated streak area as a function of time. The area cleared by the horseshoe vortex wake system was recorded for a variety of wind speeds, particle properties, and crater diameters in the Iowa State University wind tunnel (see Fig. 6.18). (From Iversen *et al.*, 1975*a*.)

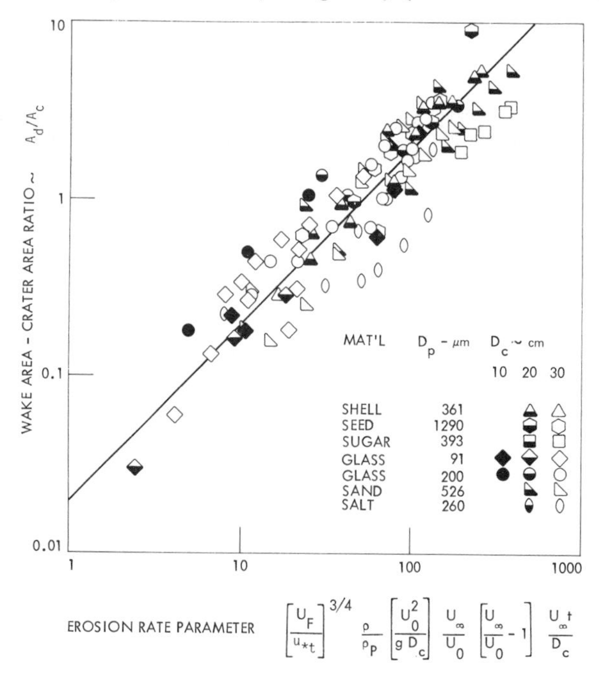

streaks, a number of interesting results were found in the simulations. But how realistic are the results in a natural aeolian environment? After searching for possible full-scale features, the Amboy lava field of southern California (Figs. 6.21 and 6.22) was selected for study of the formation of wind streaks (Greeley & Iversen, 1978, 1983; Iversen & Greeley, 1978).

The Amboy field consists of a cinder cone surrounded by basaltic lava flows and is situated in a valley. The prevailing wind is down the valley from the northwest and carries fine sand which drifts across the lava. In addition to a predominant wind streak around the cinder cone, numerous smaller streaks, ventifacts, and other aeolian features are found on the field.

The authors have performed a variety of experiments at Amboy, including measurements of wind-speed profiles and determination of wind-flow patterns over the field (Figs. 2.7 and 2.8), determination of particle sizes and compositions across the field, and experiments involving the genesis of desert pavement. In addition, they have conducted wind tunnel experiments to simulate flow over the Amboy lava flow for comparisons with data collected in the field.

Fig. 6.20. Depositional streak; the material placed in the interior of the rimless crater prior to the experiment in the Iowa State University wind tunnel was 79 μm ground nut shell. The ratio of terminal speed–threshold friction speed is 1.1.

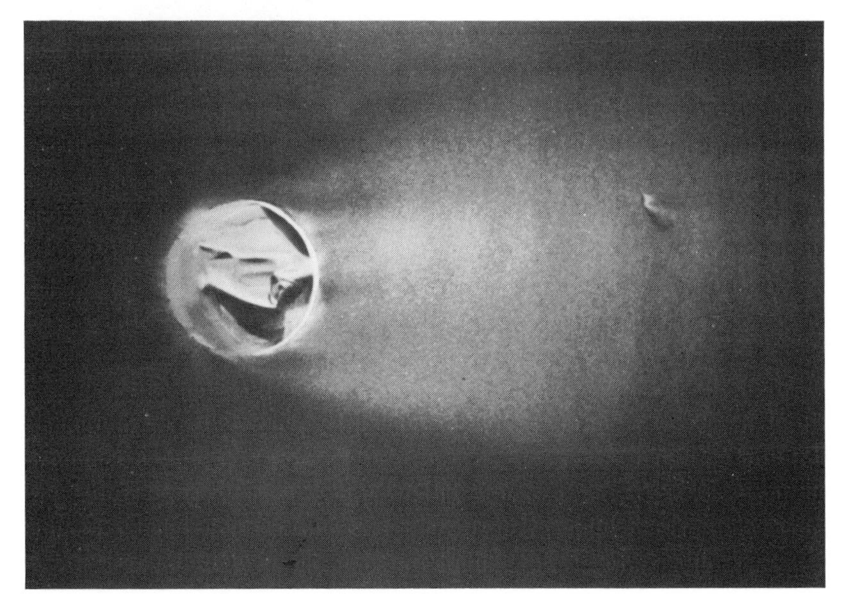

The cinder cone rises about 75 m above the surrounding lava flows and has a diameter at the base of about 460 m. A dark streak extends toward the southeast from the cone for about 4 km and is 0.5 km wide. The area of the streak consists primarily of exposed lava and patches of desert pavement, with virtually no ash or cinders derived from the cone. Within the zone of the streak, bedrock exposures and desert pavement make up about 90% of the surface area, while outside the streak only 20–30% lava surface is exposed, the remainder being mantled with windblown sediments.

Desert pavement occurs throughout the lava field and is found both inside and outside the streak zone. Pavement surfaces consist of gravel-size basalt pebbles underlain by fine silt (Fig. 3.12(*a*)). If the dark pebbles are pushed down into the brighter particles underneath, only one or two strong winds are needed to reestablish the pavement, but if the pebbles are removed, the surface remains bright for a much longer period of time. Both the bare rock and dark pavement areas represent areas where the wind exerts high surface shear stresses in which sands are rapidly deflated. The only exception is that in the near-leeward region of the dark streak adjacent

Fig. 6.21. High-altitude, aerial photograph of the Amboy lava field. The Amboy cinder cone is at upper right center. (NASA Ames photograph; from Greeley & Iversen, 1978.)

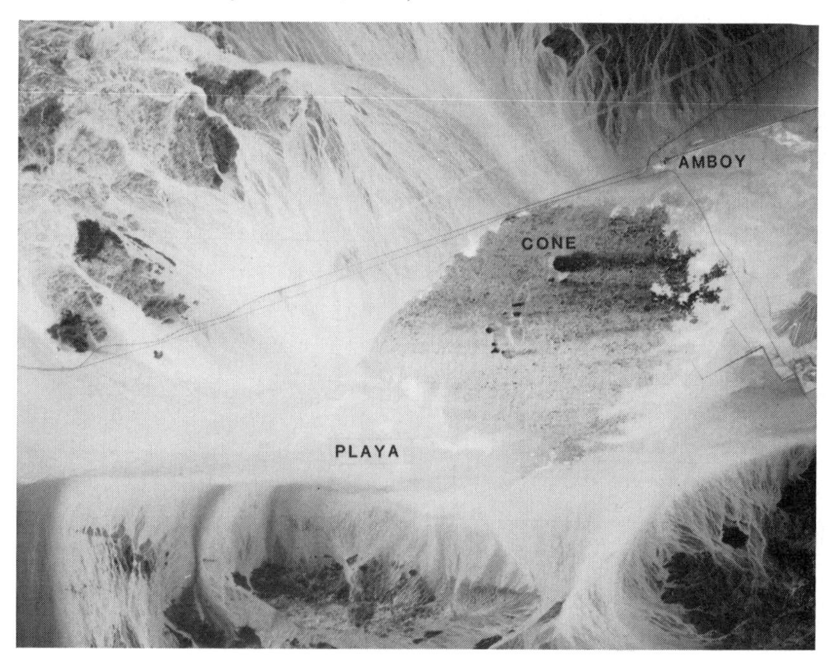

to the cone base, where wind speeds are low in the separated wake area. The low-speed area is the shape of an equilateral triangle with the apex pointing downwind.

In order to determine the changes in wind flow over the smooth valley floor to the lava field and around the cinder cone and crater, several 15 m towers were erected at critical locations and arrayed with anemometers to determine the wind velocity profiles. In addition, a scale model of part of the lava flow was constructed and placed in a boundary-layer wind tunnel.

Fig. 6.23 compares wind tunnel simulations with an aerial photograph of the cinder cone and its streak. In the simulation, sand (light-colored) was introduced at the wind tunnel inlet and allowed to drift across the model (dark surface), which was initially devoid of loose material. The various

Fig. 6.22. Oblique aerial view looking upwind at the Amboy crater, showing the locations of meteorological towers. (From Greeley & Iversen, 1978.) See Section 2.4.2 and Figs. 2.6 and 2.7. (Photograph by Greeley, 1976.)

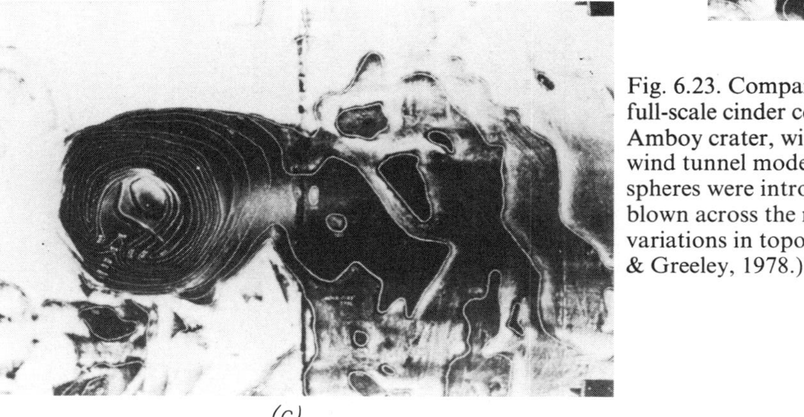

(a)

(c)

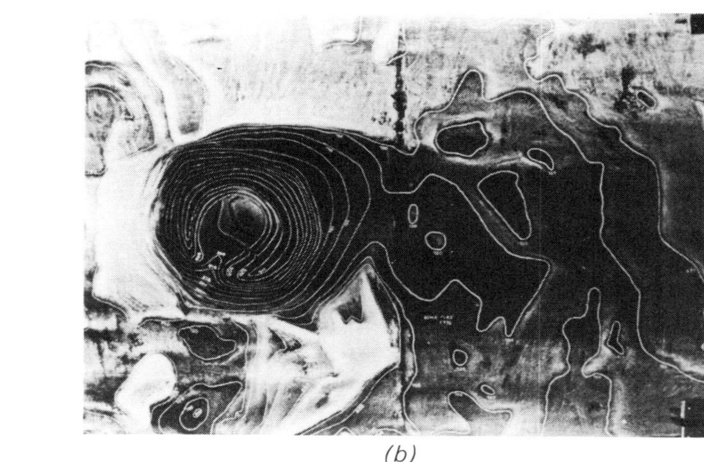

(b)

Fig. 6.23. Comparison of a wind tunnel model simulation with the full-scale cinder cone at the Amboy lava field; (*a*) Aerial views of the Amboy crater, wind direction left to right. (*b*, *c*) Iowa State University wind tunnel model of the Amboy crater (scale 1:1000); 90 μm glass spheres were introduced upstream of the test-section and were then blown across the model. The irregular edge of the dark streak is due to variations in topography for both model and full-scale. (From Iversen & Greeley, 1978.)

dark spots outside the streak zone are topographically high areas and the bright areas on either side of the cone are sand deposits which fill topographical depressions. The light and dark areas on the model agree fairly well with the aerial photograph and demonstrate, at least qualitatively, the validity of the wind tunnel model.

From field measurements and wind tunnel results, we have found the following:

(1) Both the relatively smooth alluvial plain upwind of the lava field and the field itself are relatively homogeneous aerodynamically (i.e., with constant roughness), so that, during strong winds with near-neutral temperature distribution, the wind profiles for the first 15 m above the surface are logarithmic (see Eq. (2.11) and Fig. 2.7).

(2) The equivalent roughness height of the lava surface is almost five times that of the alluvial plain, causing a surface friction speed about 20% greater for the rougher surface than for the smoother alluvial plain upwind of the lava flow.

(3) The average surface shear stress within the streak, calculated from measured wind tunnel speed profiles, is 23% higher than the area outside of the streak. The shear stress was minimum in the leeward separation zone adjacent to the cone base and reached a maximum on the lateral flanks of the cinder cone.

Fig. 6.24 illustrates several other wind-related surface features on the Amboy flow field. The dark areas on the plateau face into the prevailing wind and are swept free of deposited particles. The darker areas on the upper surface of the plateau are desert pavement. Fig. 6.24(*b*) is a close-up view of one of the knobs on the plateau; the dark areas on the flanks are caused by the strong surface shear stresses associated with the vortex system wrapped around the mound. A small sand deposit lies to the leeward of the mound. Measurement of the small sage-brush deposit streaks show convergence of the wind on the lateral sides of the mound wake. A small, nearly rimless crater near the leeward edge of the plateau is shown in Fig. 6.24(*c*). Note the bright deposits downwind of the leeward steps in both the crater and at the plateau edge. Fig. 6.24(*d*) illustrates several aspects of the aeolian environment on the ground at Amboy. Although the surface in the immediate foreground of the photograph contains an assortment of large dark rocks (several centimeters), from an aerial perspective the area is brighter than the surroundings because the rocks constitute a rough surface which traps the sand (as the snow is

Fig. 6.24. (*a*) Lava plateau at the Amboy lava field. The plateau is elevated ≈10 m above the plain and is ≈450 m wide by 1 km long. Fig. 7.29 (*b*) is the small knob just to the left of the photo center (arrow) and the crater in Fig. 7.29 (*c*) is above the center adjacent to the plateau leeward edge. The lava channel shown in Fig. 3.12 (*b*) is at the upper right center at the end of the plateau. (Photograph by Iversen, 1976.) (Greeley & Iversen, 1978; Iversen & Greeley, 1978.) (*b*) The knob is ≈2 m high. The darkest area around the knob is due to lag deposits of basaltic pebbles. The light areas on the lee side are sand deposits. (Photograph by Iversen, 1976.) (Greeley & Iversen, 1978; Iversen & Greeley, 1978.) (*c*) A rimless crater on the lava plateau. The bright areas are sand deposits; the dark areas are solid lava protusions or lag deposits of basaltic pebbles. (Photograph by Greeley, 1976.) (Greeley & Iversen, 1978.) (*d*) Desert pavement at the Amboy lava field. (Photograph by J.D. Iversen, 1976.)

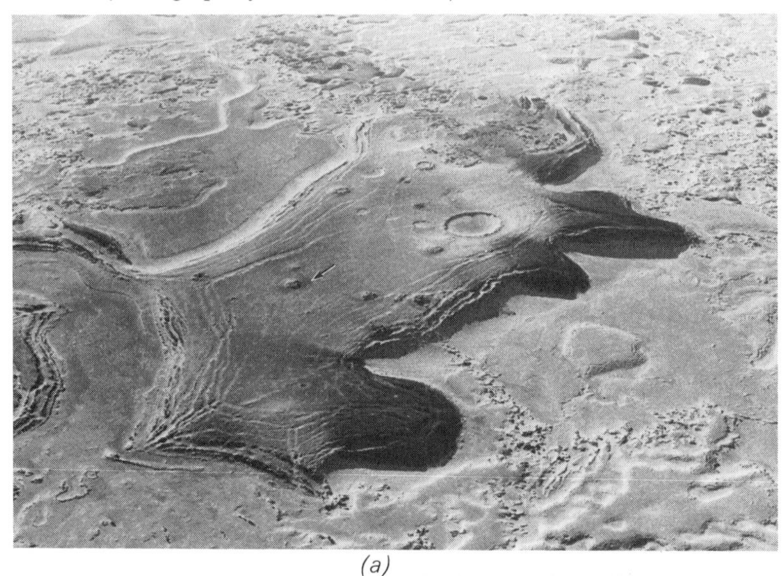

(a)

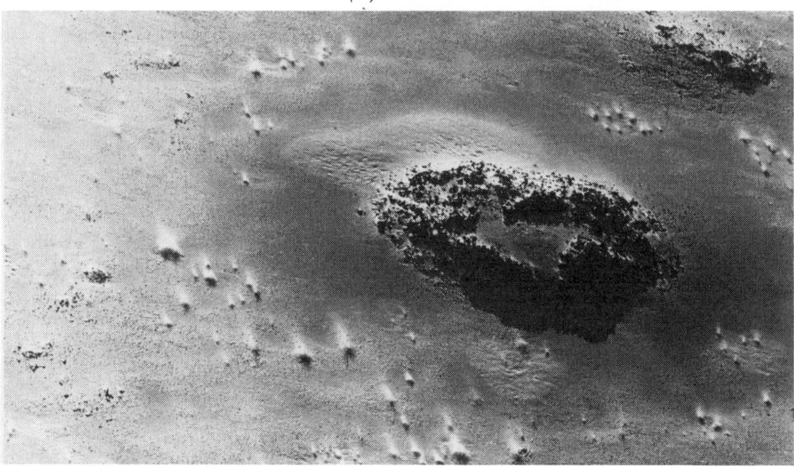

(b)

(c)

(d)

trapped in Fig. 6.16(*e*)). On the darker surface in the background, there are no large rocks and thus the sand is blown downwind, leaving the dark pebble veneer or pavement. The footsteps are bright in color because the dark veneer pebbles have been pushed down into the sand. Finally, the porous desert brush plants have trapped the sand, forming bright streamers downwind. The effective porosity of the plants causes a different kind of deposit geometry than does a solid barrier.

6.4.5 *Wind streaks on Mars*

One of the most striking discoveries to come from the highly successful Mariner 9 mission to Mars was the existence of albedo patterns on the surface, which changed their size, shape, and position as a function of time. Termed *variable features* by the imaging science team (Sagan *et al.*, 1972), these patterns seemed to be linked to the great dust storm of 1971 and were correctly attributed to aeolian processes.

Beginning in 1976, the Viking results provided a major increase in the data base for variable features by providing increased image resolution, near-global coverage by images at a fairly uniform scale, and information on wind speeds and directions measured at the Lander sites for coupling orbital data with the ground.

In the decade since their discovery, several planetologists have studied variable features intensively in order to establish a classification scheme and to determine such factors as origin and evolution. From these studies it has been possible to shed light on several important aspects of Mars, including: (1) derivation of patterns of near-surface winds both on a global scale and as a function of season, (2) characterization of martian surficial materials as a function of terrain and location, and (3) evaluation over geological time on the modification of the martian surface by the wind.

Peter Thomas and his colleagues at Cornell University have classified martian variable features and provide an excellent review (Thomas *et al.*, 1981). We have slightly modified their classification scheme to remove some of the genetic implications (Table 6.1). There are 6 basic types: (1) dark streaks, (2) bright streaks, (3) mixed-tone streaks, (4) splotches and related streaks, (5) dune shadow streaks, and (6) frost streaks. The first four categories are the most abundant types of variable features on Mars and are based on the albedo of the features in relation to the background. It must be noted that the surface of Mars is rather low contrast and that the light and dark patterns seen on most images, although real, have been drastically contrast-enhanced by image processing techniques, so bright areas appear much brighter and dark areas darker. The next category refers

Table 6.1. *Classification of martian wind streaks (after Thomas* et al., *1981)*

Type	Obstacle-shape[a]	Shapes	Length (km)[b]	Occurrence	Variability
Bright streaks	Crater, hill, scarp	Teardrop, tapered, parallel, fan	5–25	Low, mid-latitude	Minor, southern hemisphere summer
	Trough	Serrate, linear, irregular	10–100	Syria Planum, Labyrinthus Noctis, Coprates	Rapid, southern hemisphere summer, fall
	Sheet	Serrate, linear, irregular	10–100	Syria Planum, Labyrinthus Noctis, Coprates	Rapid, southern hemisphere summer, fall
Dark streaks	Crater, scarp	Fan, tapered	10–30	25–40°S, flatter areas	Form ≈100 days after dust storms
	Coalesced	Irregular	5–150	Regional slopes in Tharsis, Syrtis Major	Same as crater–scarp; mostly after dust storms, some during rest of year
	Linear	Linear	5–100 long 0.1–4.0 wide	Scattered	Insufficient data
Mixed tone	Crater	Teardrop, tapered	5–20	Syrtis Major, Tharsis	Insufficient data
Splotches and related streaks	Crater splotch	Parallel, tapered	5–100	Oxia Palus, 40–60°S, Cerberus	Sporadic, south, summer
Frost	Craters	Teardrop, tapered	5–30	Polar caps, 55–70° lat.	With CO_2 caps
Dune shadow	Craters	Teardrop	5–20	North polar dunes	
Frost–sediment	Craters, ejecta	Arrowhead	5–30	North polar region	

[a] Streaks within each type are arranged by the kind of associated obstacle, or, by streak shape if the obstacle is not visible.
[b] Typical lengths in kilometers; not the complete range of lengths.

to patterns of dunes relative to topographical obstructions. The last category consists of patterns that are presumed to involve frost.

Both bright and dark streaks vary greatly in geometry, in terms of 'parent' crater size, streak length–diameter ratio, and streak shape. Streak shapes can be tapered (i.e., widest part at the crater lee rim), parallel-sided, fan-shaped (narrowest part at the lee rim), or oval. Although there are large statistical variations, dark streaks tend to have slightly smaller streak-length-to-crater-diameter ratios (Arvidson, 1974; Greeley *et al.*, 1978), slightly greater tendency to be fan-shaped or oval (Veverka *et al.*, 1978*a*; Gifford *et al.*, 1978), greater tendency toward ragged downwind margins (Thomas *et al.*, 1981), and greater tendency to asymmetry (Greeley *et al.*, 1978).

Dark streaks

Many dark streaks are highly variable, having been observed to appear or disappear completely in a matter of weeks. The most rapid change that has been observed occurred during the first 100 days after the decay of the 1977 dust storm. Thomas and his coworkers define three types of dark streaks (Table 6.1), with the crater- and scarp-associated forms being the most common (Fig. 6.25). The shapes of dark streaks vary from tapered to parallel to fan-shaped. Dark streaks are typically 5–30 km long, but can be up to 80 km. Many of the smaller forms are asymmetric with respect to their 'parent' crater. Those associated with scarps emanate from the base of the scarp, not the upper edge.

Most dark streaks associated with craters, scarps, or ridges (Sagan *et al.*, 1974) occur in either the Tharsis region (Fig. 1.14(*a*)), or in the zone between 25 and 40°S, although many streaks occur outside these areas as well. Dark streaks are best developed in smooth, flat plains and on broad, regional slopes. Although they appear to be relatively unaffected by the topography which they cross, many show gaps where topographically low areas, such as those across channels, occur (Thomas *et al.*, 1981).

Dark streaks would seem to be best explained as erosional (deflation) zones associated with topographical obstacles. Any three-dimensional boundary-layer obstruction will generate a horseshoe vortex system (Section 6.3.2 and Fig. 6.9). The strong winds developed beneath streamwise vortices in the crater wake cause deflation of particles from the surface, as demonstrated in wind tunnel simulations (see Section 6.4.3). The erosive power of the wind to the lee of obstacles has also been noted in Earth's atmosphere (see Sections 6.4.2 and 6.5.4). This erosion (probably in the form of deflation) may remove fine, loose material to expose darker,

Fig. 6.25. (*a*) Dark streaks associated with craters. The area shown is ≈140 by 160 km (VO frame 603A25.) (*b*) Dark streak associated with large (17 km in diameter) crater having rugged apron of ejecta, showing that dark streaks can form on both rugged terrain and smooth plains. (VO frame 459A79.) (*c*) Dark streak formed in association with a scarp (the area of the picture is ≈170 by 190 km). (VO frame 579A56.) (*d*) Dark linear streaks; mechanisms to explain these and similar features have not been suggested. The area of the image is ≈30 by 40 km. (VO frame 580B20; from Thomas *et al*., 1981.)

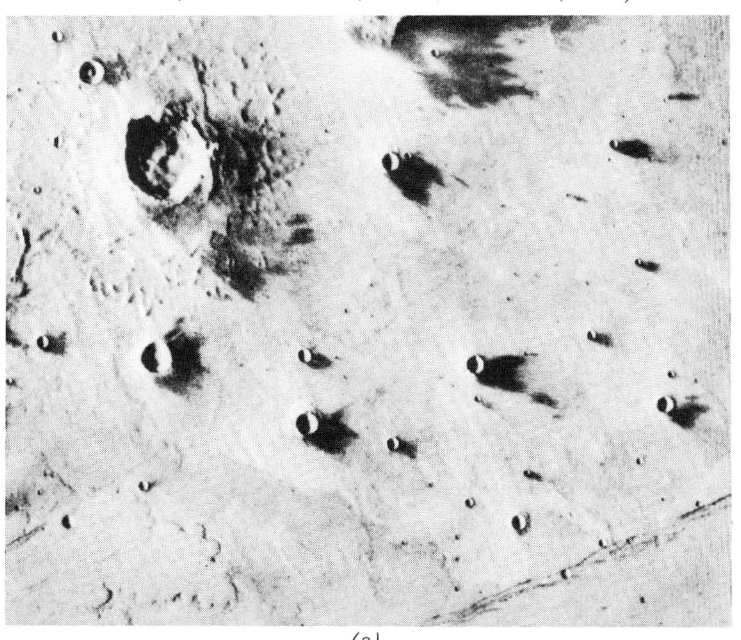

(a)

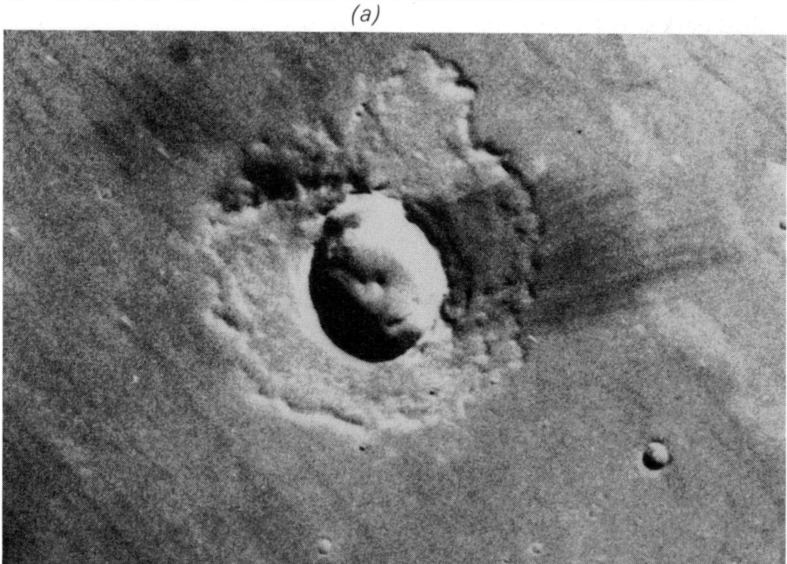

(b)

Fig. 6.25 (c)

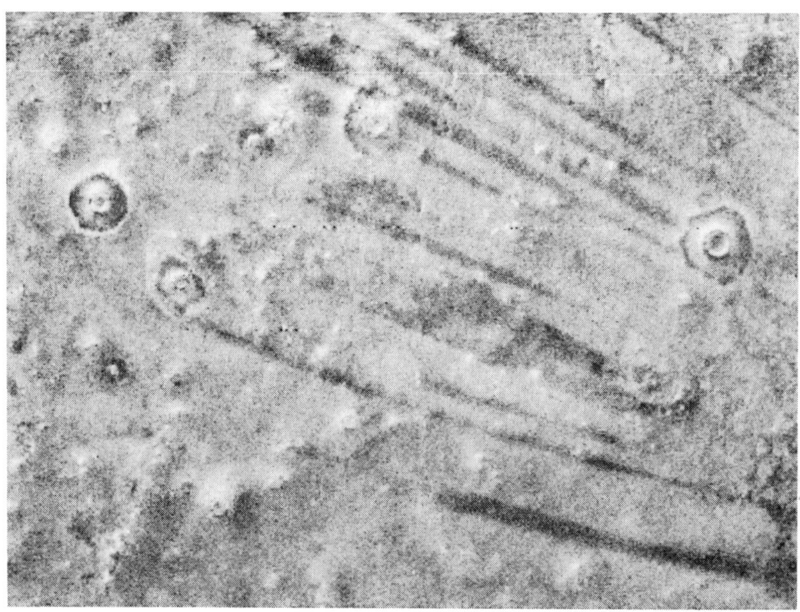

Fig. 6.25 (d)

underlying 'bedrock', or may remove easily transported grains to leave a lag deposit in the wake zone or, as we see at Amboy (Fig. 6.22), it may involve a combination of both exposed bedrock and lag deposits.

Thomas and coworkers consider the dark streaks to be erosional features involving the removal of previously deposited thin layers of dust to expose a darker substrate, probably composed of larger particles. Because the threshold wind speeds for such fine grains are extremely high on Mars, they suggest that the dust is removed by saltation impact from sand grains, which are more easily dislodged by the wind. The dust particles, once airborne, are suspended by turbulence and carried far downwind. Alan Peterfreund (1981) has analyzed thermal inertia data for the surface, obtained from the Viking mission, to assess particle sizes and has found areas of dark streaks to have an abundance of sand-size grains, thus lending credence to the idea of saltating sand grains as a 'triggering' mechanism for the removal of dust.

Kuzmin (1978) has proposed that all dark and light streaks are depositional and are composed of bedform features such as barchan dunes or barchanoid ridges. He proposes that streaks are light or dark depending upon whether the lee slopes or the downwind slopes of the dunes are illuminated. Veverka *et al.* (1978*b*), however, refute Kuzman's proposal simply because both light and dark streaks have been observed under identical conditions of viewing and illumination. There is a possibility, perhaps, that ripples formed on the surface – too small to be resolved by the orbital cameras – cause some of the albedo differences. The higher surface shear stresses in the crater wake could presumably cause longer wavelength ripples (or a smoother surface) than the surrounding area and thus perhaps cause the difference in albedo.

Dark linear streaks (Table 6.1) have approximately parallel sides and can be up to 100 km long and 4 km wide. Generally, they are not associated with topographical obstacles (Fig. 6.25(*d*)) and their origin is not known.

The authors have measured some of the changes that occurred in dark crater streaks in the Daedalia region of Mars during the Mariner 9 mission (Greeley *et al.*, 1974*c*; Iversen & Greeley, 1984). According to Eq. (6.1), if the depth of erosion is a constant for all craters, then the time-rate of change of streak area, $\mathrm{d}A_\mathrm{d}/\mathrm{d}t$, should be proportional to crater diameter. The rate of change of dimensionless streak area, $\mathrm{d}(A_\mathrm{d}/A_\mathrm{c})\mathrm{d}t$, would be inversely proportional to crater diameter, as suggested in Fig. 6.26. Variations from the straight line are due to differences in elevation, crater geometry and, probably, local topography. Using the coefficient of 2.85 (Fig. 3.21(*b*)) and the result from Fig. 6.26 (Iversen *et al.*, 1975*b*),

$$d = 1.2(10)^{-6}\left(\frac{\rho u_*^3}{\rho_p g}\right)\left(1 - \frac{u_{*t}}{u_*}\right)\Delta t \tag{6.4}$$

The increases in streak area recorded in Fig. 6.26 took place in a time interval Δt of only 912 hr. Assuming, for example, reasonable values for the variables in Eq. (6.4) ($\rho/\rho_p = 6(10)^{-6}$ and $u_{*_t} = 150$ cm/sec), the expression for deflation depth d becomes (in centimeters)

$$d = 0.2(1 - r_t)/r_t^3 \tag{6.5}$$

where $r_t = u_{*_t}/u_*$. If the friction speed is assumed constant and is 1.5 times the threshold value in the crater wake, for example, the value of depth, d, becomes about 2 mm. Thus, the rapid changes in dark streaks can indicate a deflation of only a small depth of surface material.

The difference between light and dark streaks is not necessarily due to either crater shape or surface material. Fig. 6.27 illustrates a group of craters which exhibit both bright and dark streaks oriented in nearly opposite directions. Clearly, the difference in this case is meteorological

Fig. 6.26. Increases in dark streak area in the Daedalia region of Mars during 38 spacecraft orbits. (Measured from Mariner 9 images, from Greeley *et al.*, 1974*c*; Iversen *et al.*, 1975*b*.)

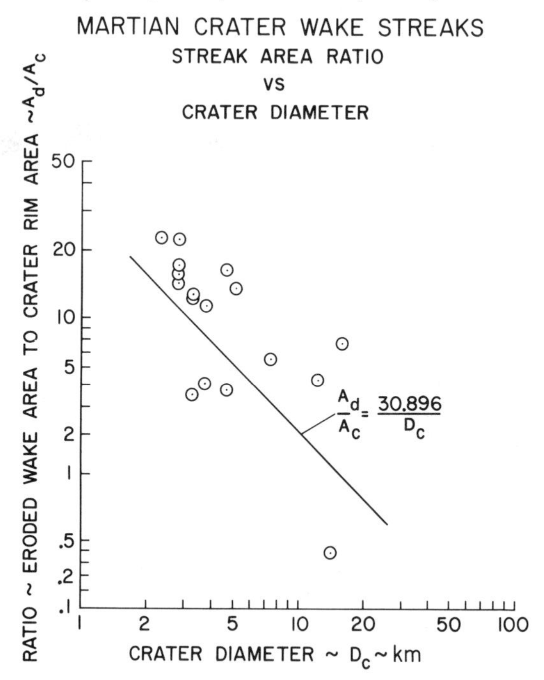

rather than topographical (Veverka *et al.*, 1981), but whether the difference is due only to wind speed or to atmospheric stability, as they suggest, is uncertain.

Bright streaks

Thomas *et al.* (1981) recognize three types of bright streaks, depending upon the feature with which they are associated (Table 6.1). Studies by Arvidson (1974), Veverka *et al.* (1978*a*), and Greeley *et al.* (1978) show that more than three-quarters of all streaks are bright, i.e., they exhibit a higher albedo than the background on which they occur. Most bright streaks extend from craters, hills, and other topographical features, including some troughs (Fig. 6.28). They are seen to cross all types of terrain, including fairly rough surfaces, without significant interruption. The shapes of bright streaks include teardrop, tapered, parallel, and fan, all with boundaries that tend to be diffuse. Although the typical length is 5–20 km, some bright streaks have been found up to 90 km long (Thomas *et al.*, 1981).

Fig. 6.27. Viking Orbiter image showing craters having both bright and dark streaks; the occurrence of these streaks suggested to Veverka and his colleagues (1981) a model of streak formation involving meteorological factors, rather than factors related to surface properties or crater geometry. The area is $\approx$170 by 220 km. (VO frame 553A54.)

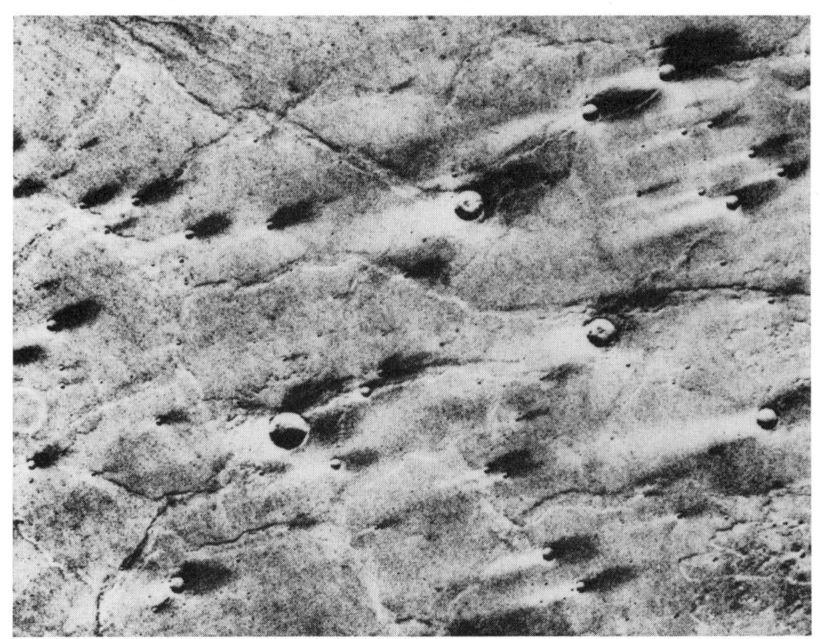

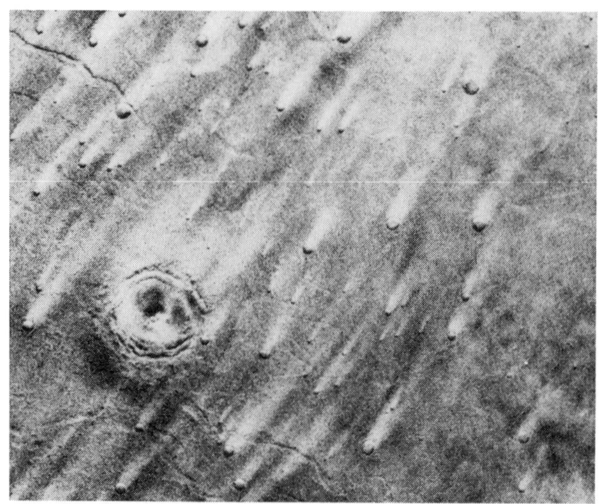

Fig. 6.28 (a)

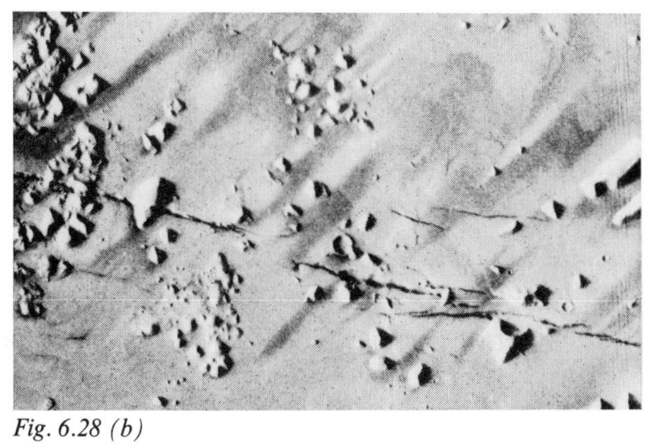

Fig. 6.28 (b)

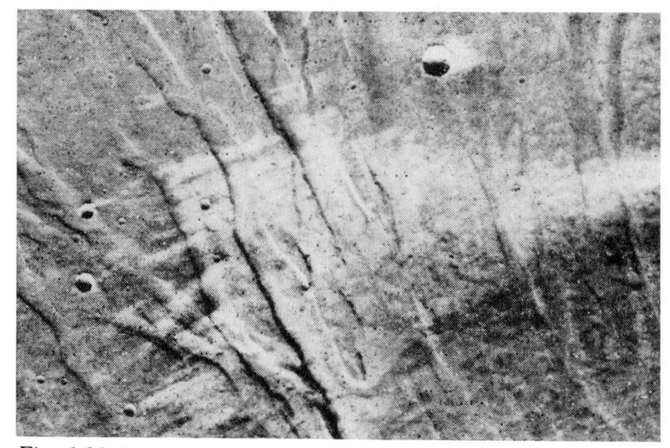

Fig. 6.28 (c)

Bright streaks are the most stable of all variable features. During the operation (which lasted for over one year) of Mariner 9, no changes were detected. However, Veverka and his colleagues (1974) examined Mariner 6 and 7 images which had been taken two-and-a-half years earlier and concluded that some bright streaks and patches had developed by the time of the Mariner 9 mission. Comparison of bright streaks seen on Mariner 9 images (1971–72) and the later Viking images (1976–80) showed that some streaks changed slightly in their length, orientation, or sharpness (Thomas & Veverka, 1979).

Most bright streaks occur in the low- to mid-latitudes and in areas that are regionally smooth but which may have locally rugged features, such as craters and hills.

Most investigators agree that bright streaks represent areas of fine, dust-size particles, which are associated with martian dust storm activity (Sagan *et al.*, 1972; Arvidson, 1974; Greeley *et al.*, 1974*a*; Veverka *et al.*, 1974; Thomas *et al.*, 1981). The mechanism of formation, however, is not well defined. As outlined in Section 6.6.1, bright streaks could be deposits of light-colored minerals, or they could represent fine grains of material derived from sources associated with the topographical features from which the streaks emanate.

Many craters have both dark streaks and light streaks (Fig. 6.27) associated with them, and it is difficult to define a mechanism that takes both into account. Veverka *et al.* (1981) note that bright streaks apparently form during global dust storms and, because the crater morphology seems to have no control over whether light or dark streaks form, they attribute the mechanism of formation to meteorological conditions. Consequently, Veverka and coworkers derived a model related to atmospheric stability. They presume that bright streaks form during storms when the dusty atmosphere is heated by the sun. The surface may also become warmer, but the variation in temperature through the atmosphere is reduced. The smaller difference in temperature results in a stable boundary layer in which vertical motion is inhibited. More of the flow thus goes around the craters, rather than over them – compared to the dust-free neutral or unstable case – and the development of the trailing vortices is thus inhibited. The

Fig. 6.28. (*a*) Crater-related bright streaks in the Hesperia Planum region; area shown is 300 by 350 km. (VO frame 245B31.) (*b*) Bright streaks associated with small hills in Cerberus; the picture area is ≈ 80 km wide. (VO frame 883A03.) (*c*) Bright streaks crossing various troughs; the area is ≈ 100 km wide. (VO frame 922A04; from Thomas *et al.*, 1981.)

presumption is that the winds are 'blocked' in the crater wake and preferential deposition of the finer dust in the atmosphere occurs in this 'dead' zone. One problem with this hypothesis is that the flow begins to approximate a two-dimensional flow and a system of moving vertical vortices can occur (called the Karman vortex street). Such vortices have been photographed from Earth orbit of cloud formations streaming downwind of islands in stable conditions (Bayliss, 1976), but it is not known whether such vortices would preclude deposition of dust.

In the Veverka *et al.* model, the atmospheric conditions favorable for deposition occur in the latitude zones primarily between 30°S and 30°N where the equatorial Hadley circulation would intensify dust storms. They also note that, as the dust storm clears, the surface would begin to heat, leading to an unstable atmosphere and subsequent loss of the 'dead' zone. Wind flow would then revert to the generation of leeside vortices and resulting erosional surfaces typified by dark streaks, described above.

Mixed streaks

Mixed streaks consist of light and dark patterns associated with the same crater. They may involve bilateral symmetry (Fig. 6.29) or they may have different orientations. Those with bilateral symmetry might be explained by a flow field model which takes zones of wind erosion (dark) and deposition (light) into account (Greeley *et al.*, 1974*b*). Veverka *et al.* (1981) ascribe those representing bimodal winds to their model, which takes atmospheric stability into account and which would involve multiple wind directions.

Splotches and related streaks

Dark, irregular zones – called 'splotches' – are associated with some topographical depressions on Mars. Although most of the depressions are large, often degraded, craters, a few splotches have also been found associated with troughs and canyons. Many of the splotches are distinguished by tongue-like streaks which may extend over the crater rims onto surrounding plains. Some of these tongues are quite large; Thomas *et al.* (1981) note that such streaks may be up to 250 km long. Some splotches and related streaks, which show simply as dark patches on low-resolution images, are seen to be dunes in high-resolution pictures (Fig. 6.30). Other splotches, however, do not show obvious bedforms, at least at the limit of resolution ($\approx$20 m).

The Oxia Palus region is noted for its occurrence of splotches (Fig. 6.31) and has been the subject of several studies. Alan Peterfreund (1981) analyzed thermal inertia data (providing clues to the grain size) and spectral data (providing compositional information) for correlation with results from photogeological studies and concluded that the Oxia Palus features involve a complex interplay of erosion and deposition of multilayered (stratified) deposits.

In general, investigators agree that most splotches and related streaks represent deposition of material derived from within the source depression. Like all variable features, they change with time, but there is no clear pattern of the change being related to season.

Dune shadow streaks

The vast sand sea of the north polar region contains thousands of dunes. In a few places, the pattern of the dunes defines a streak-like zone associated with some craters, as shown in Fig. 6.32. These have been classified by Thomas *et al.* (1981) as *dune shadow streaks.*

Frost streaks

Streaks composed of frost, or possibly snow, have been identified

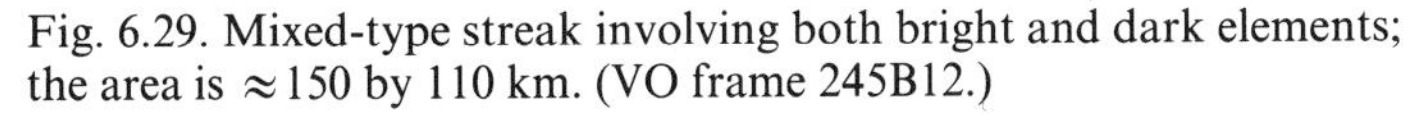

Fig. 6.29. Mixed-type streak involving both bright and dark elements; the area is $\approx$150 by 110 km. (VO frame 245B12.)

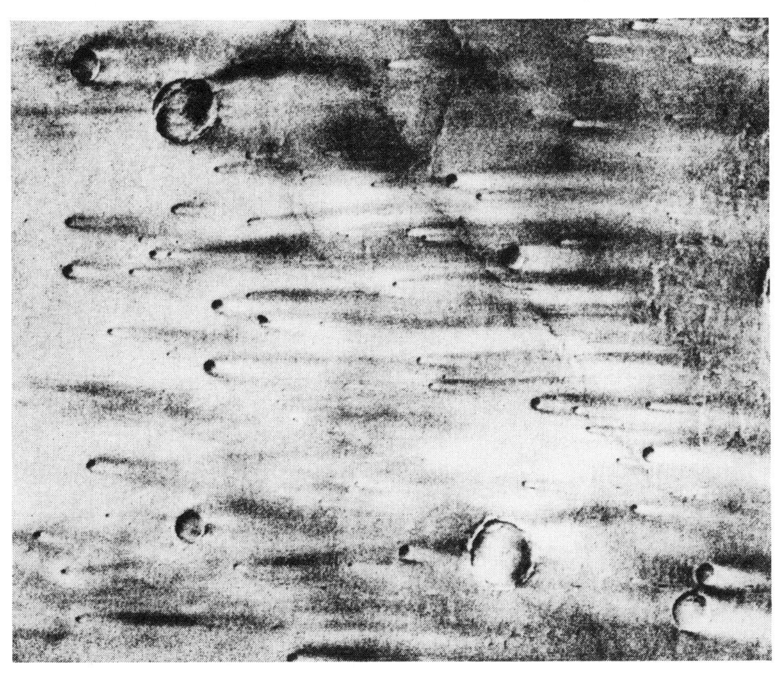

Fig. 6.30. (*a*) Low-resolution image of 150 km diameter crater in Noachis; the smaller crater (arrow) on the floor of the larger one has a dark streak draped over its rim. (VO frame 380B06.) (*b*) High-resolution image of small craters seen in (*a*), showing that the dark streak consists of climbing dunes; the area shown is ≈40 by 45 km. (VO frame 571B53; from Veverka *et al.*, 1981.)

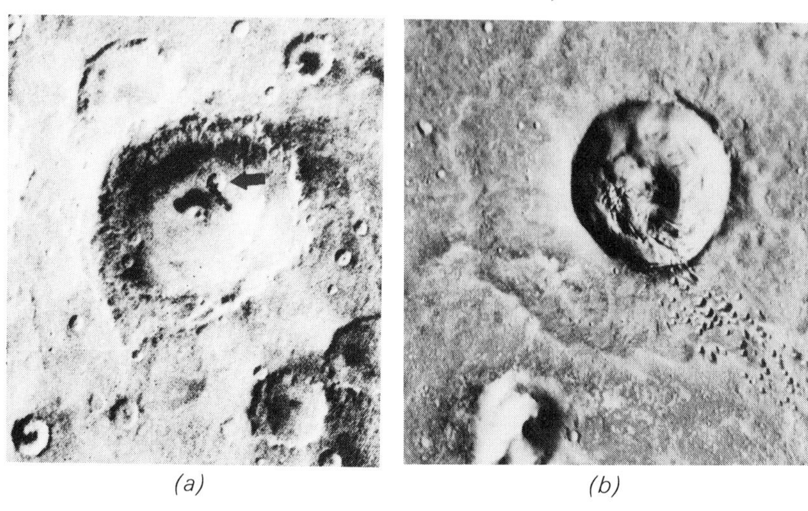

Fig. 6.31. Splotch-type streaks in Oxia Palus; the area shown is ≈500 by 800 km. (VO frame 669A50.)

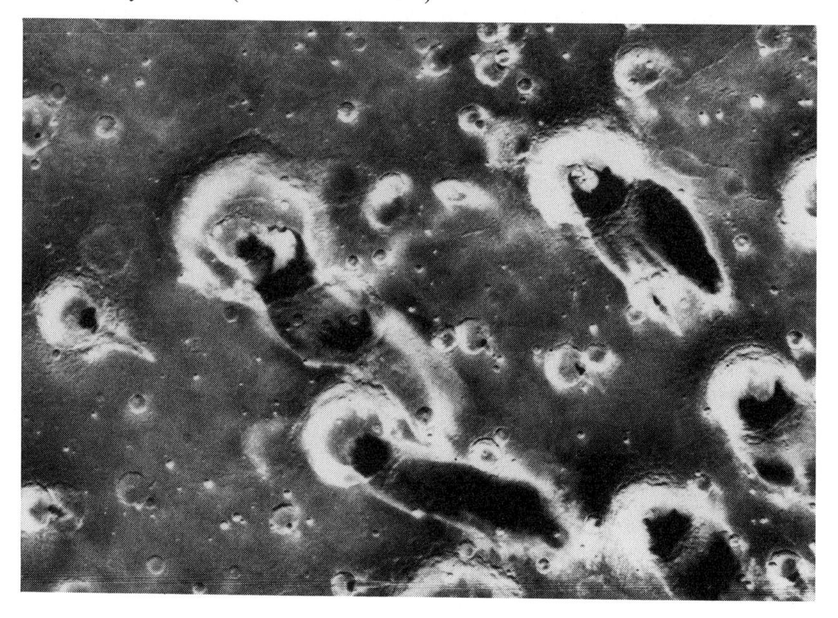

in both polar regions (Fig. 6.33) by Thomas & Veverka (1979). They are bright, tapered patterns which generally follow the seasonal CO_2 frost cap; some remain after the retreat of the CO_2 ice cap in the spring and may represent excess accumulations of CO_2 frost in the lee of some craters. Based on calculations of the thermal balance between the environment and the surface and the longevity of several frost streaks, Thomas & Veverka estimate that the frost streaks could be as thick as 3 m. Thomas and coauthors (1981) have analyzed frost streaks in detail and conclude that some are probably mixtures of frost and sediment.

Mapping wind directions

Several investigators have used the orientation of variable features (primary crater-related streaks) to map the patterns of near-surface winds on Mars. Using Mariner 9 images, Arvidson (1974) and Arvidson *et al.* (1974) devised a computer-based system to map streaks, splotches, and associated craters which could then reduce the data to show global wind-flow patterns. They found that the trend of bright streaks implied flow of dust into the southern hemisphere and suggested that dark streaks are relict features representing the transfer of sediment from polar regions toward the equator.

Fig. 6.32. Dune-shadow streaks in the north polar region; the zone in the lee of craters such as these, and other topographical obstructions to the wind, remain free of dunes, forming a general streak-like pattern; prevailing wind is from the left; the area shown is 50 by 60 km. (VO frame 71B64.)

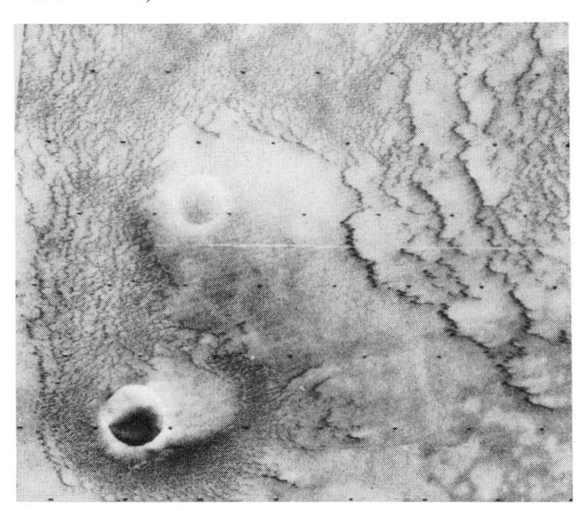

Thomas & Veverka (1979) and Thomas (1981) used the expanded-image data set provided by Viking to obtain a more uniform global pattern of streak orientation than was available from Mariner 9 images. Their maps (Fig. 6.34) show that in the middle and low latitudes the near-surface winds blow toward the southwest at about 30°N, then swing southeast at about 30°S. This pattern, defined primarily by bright streaks, is consistent with flow predicted for southern hemisphere summer by the meteorological models of Leovy & Mintz (1969) and Sagan *et al.* (1973). Thomas (1981) confirmed the earlier speculations of Arvidson concerning the general movement of sediments from the poles toward the equator and, based on analyses of various crater splotches, that such activity is much more extensive in the southern hemisphere than in the northern and reflects the general asymmetry between climate in the north and south.

Mapping by Thomas & Veverka (1979) also shows the strong topographical influence exerted by the large shield volcanoes in the Tharsis and Elysium regions. In addition, they found that seasonal patterns on a local scale are not repeated from year to year, especially in regard to the localities

Fig. 6.33. Frost streaks in the south polar region; the white zone represents CO_2 frost–snow preserved in the lee of the crater rims; the area shown is ≈100 by 150 km. (VO frame 527B13.)

of dust storm development and severity of dust storms. However, they also note that the global wind patterns are repeatable from one year to the next.

Both light and dark streaks occur in Chryse Planitia, site of Viking Lander 1 (Fig. 1.14). Streak orientations have been mapped and compared to wind directions and strength by Greeley and coworkers (1978), who found a good correlation between dark streaks and the prevailing winds

Fig. 6.34. Global distribution of wind streak directions, showing average local trends of streaks. Also included are some large, isolated streaks, most of which are of the splotch-related types. (*a*) Viking data, post-dust 1977 storm. (1) Bright streaks. (2) Dark streaks. (3) Dark, splotch-associated streaks. (*b*) Mariner 9 data (1971–72); bright streaks are shown by simple arrows; dark streaks have superposed dots. Areas that were imaged by Viking before the global dust storms are marked. Only in the cross-hatched area were significant differences from the Mariner 9 data observed. (From Thomas *et al.*, 1979.)

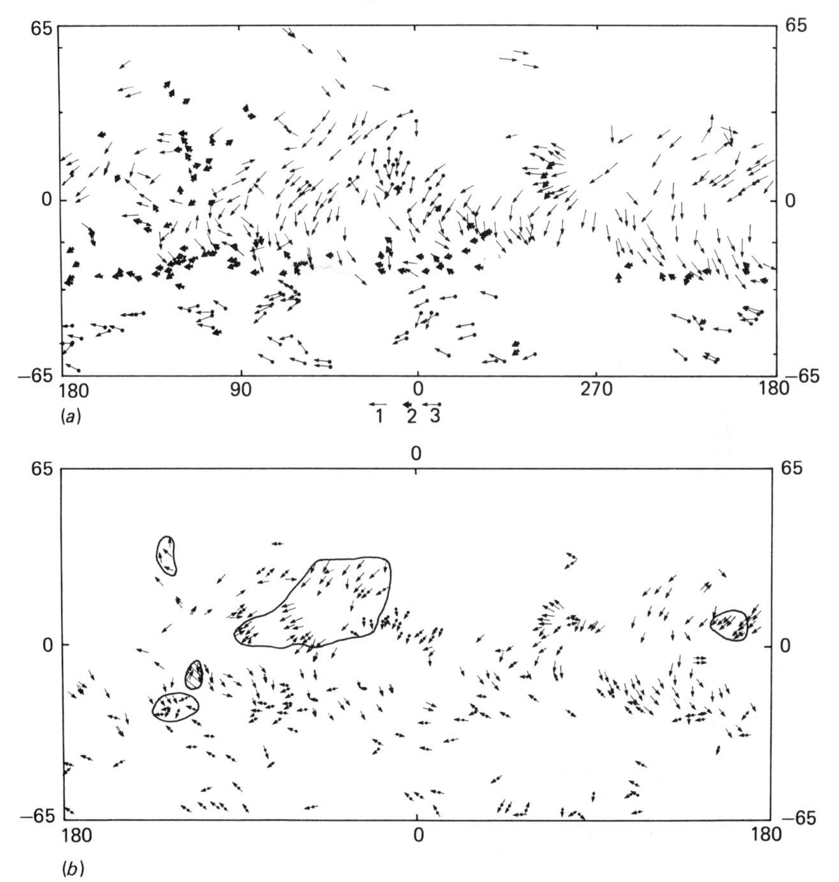

from the southwest (Fig. 6.1). Bright streaks were found to be oriented in the opposite direction and to be approximately parallel with the orientation of the sediments in the lee of various rocks seen in the Lander pictures.

6.4.6 *Wind streaks on Venus and Titan*

Because of dense cloud cover, the surfaces of Venus and Titan cannot be seen telescopically or by conventional camera systems from orbit. Thus, if wind streaks exist, they cannot be seen by conventional means. Although radar images of the venusian surface have been obtained from Earth-based observations and from the Pioneer–Venus mission, the resolution, at best, is many kilometers, and it is unlikely that wind streaks could be seen. Future missions to both Venus and Titan may carry radar imaging systems capable of resolving objects as small as a few meters.

The question is: 'Would radar be able to "see" wind streaks?' The answer is 'Maybe'. If dunes were the size of the resolution elements, or larger, then they may show up quite well, as has been documented on radar

Fig. 6.35. Shuttle Imaging Radar (SIR-A) photograph showing wind-related streaks in the Altiplano of Bolivia. The dark streaks, detectable by the radar system, exceed 14 km in length. (Image from DT 31, courtesy C. Elachi, Jet Propulsion Laboratory.)

images taken from Earth orbit. But streaks related to ripples, to lag deposits, or to compositional differences are more problematical. The 'signature' seen on radar images is a function of many complex parameters including the radar wavelength versus the size and shape of the objects on the ground, the angle of the radar beam subtended from the spacecraft and the ground, and the dielectric constant (includes composition) of the objects. Experiments have shown that lag gravels can be distinguished, and that some compositions are discernible on radar images flown over desert terrains (Elachi *et al.*, 1982).

Results from the Shuttle Imaging Radar Experiment (SIR-A), flown in 1981, show that some wind-related streaks are radar 'visible', as shown in Fig. 6.35. Thus, there is the possibility that wind streaks could be identified on radar images under some conditions. If they exist on Venus and Titan, they could be an important data base for understanding surface modifying processes and for deriving models of near-surface winds, just as they have been used on Mars.

7

Windblown dust

7.1 Introduction

Dust. It gets in our eyes, our shoes, even in our lungs, sometimes causing disease to those who breath it. Dust storms cause visibility problems on highways, resulting in many accidents and deaths each year. And soil erosion is a major worldwide problem where surfaces are disrupted through cultivation, overgrazing, mining, construction, vehicular traffic, or other activities which disturb the surface or destroy vegetation cover.

Dust storms are common on both Earth and Mars. Although many storms on Earth reach sizes that can be seen on photographs taken from orbit, they do not become global in extent, as some do on Mars. Nonetheless, the great North American dust storm of 12 November 1933, covered a region greater than the combined areas of France, Italy, and Hungary (Goudie, 1978), and for the people involved in the dust storm, it might just as well have been global.

The dust storm of November 1933 is an example of a storm caused by an area of extreme high pressure and strong barometric pressure gradient. It covered the region southward from the Canadian border, stretching from Montana to Lake Superior, and to the western Ohio and lower Missouri River valleys (Hovde, 1934; Miller, 1934). That year a severe drought was experienced over the northern and central Great Plains, and crop and pasture grass failures left the soil exposed to the wind. Maximum recorded wind velocities on 12 November were 25 m/sec at Bismarck, North Dakota, and at Davenport, Iowa. At Sioux City, Iowa, the visibility and ceiling were recorded as zero at 2.30 p.m., when objects greater than 15 m distant could not be seen. The top of the dust cloud at Omaha, Nebraska, was reached by airplane at 2500 m above the surface. By noon the next day the dust cloud reached Chattanooga, Tennessee, with visibility reduced to 300 m. Dust particles measured at Buffalo, New York, ranged from 5 to 500 μm, with the average about 20 μm. Extensive damage was reported throughout the region to buildings, fences, vegetation, and, of course, to farmland, by wind erosion of soil.

Many estimates have been made for the amount of dust injected each year into the atmospheres of Earth, Mars, and Venus. On Earth the most common methods of dust injection are by dust storms, volcanic eruptions, and industrial processes. Junge (1971) estimates the total mass of atmospheric dust and aerosols to be 2.5×10^{15} g/yr, of which about 20% is mineral debris, or about 5×10^{14} g/yr. For Mars, Pollack and colleagues (1977, 1979) estimate atmospheric dust loading to be about 2.9×10^{15} g/yr. Although very little is known about Venus and the dust–aerosol content of its atmosphere, a rough estimate of 3.2×10^{14} g/yr can be made, based on data from Knollenberg & Hunter (1980). This mass includes H_2SO_4 droplets and various contaminants, and it is not possible to determine with available data how much might be mineralogical dust. If we normalize these amounts for Earth, Mars, and Venus per unit surface area, we find that Mars is about 20 times 'dustier' than Earth, whereas Venus may be about 30% less dusty than Earth, at least near the surface.

In this chapter we discuss the characteristics of dust storms on Earth and on Mars, the nature of windblown dust particles, and we consider the effects of dust storms on soils and climate. We then discuss dust deposits and their significance on Earth and Mars.

7.2 Dust storms on Earth

Dust storms are common, especially in arid and semiarid parts of the Earth, as shown in Fig. 7.1, which illustrates the advancing front of a dust cloud in the American southwest (see also Fig. 1.1). Dust from deserts is often transported hundreds of kilometers, although many of the finest particles can be carried thousands of kilometers. The principal directions of dust transport from the world's deserts are illustrated in Fig. 1.6. Major areas where aeolian dust originates are the arid regions of North Africa (the Sahara), the Arabian peninsula, the lower Volga and North Caucasus in the USSR, the Canadian prairies, the Great Plains of the US, the Sonoran–Mojave Desert, and the drier parts of Argentina, Australia, and Afghanistan. Views obtained from orbiting satellites frequently show dust streaming more than 4000 km from the west coast of Africa, derived from the Sahara Desert.

Although strong winds causing surface erosion usually account for most of the dust in the atmosphere, significant quantities are at times injected by volcanic eruptions, as demonstrated by the 1982 activity of El Chichon volcano in Mexico. Because of the high-energy explosions, ash particles ≈ 1 μm in diameter were injected to altitudes exceeding 14 km, well above most clouds. Thus, the grains have no way of settling out of the atmosphere other than by gravitational settling, which would require weeks to occur.

Much of the wind erosion occurring in semiarid regions is due to cultivation and overgrazing. Minor amounts of dust are also injected into the atmophere by dust devils, or rise as 'fugitive' dust from activities such as mining and traffic. It has been estimated that windblown dust from soil erosion contributes approximately 500×10^6 tons of particulates to the atmosphere each year (Peterson & Junge, 1971). Many of the data on the characteristics of dust storms have been compiled by Goudie (1978). He presents tables on dust storm frequency, location, dust concentration levels, deposition distribution, deposition rate, and particle characteristics and concludes that in some areas deposition of aeolian dust is great enough to equal the loss of soil by fluvial erosion.

7.2.1 *Dust storm development and effect on climate*

According to Jackson *et al.* (1973), a special set of meteorological conditions is necessary to initiate dust storm activity and to enable long-range transport. These conditions include intense surface heating,

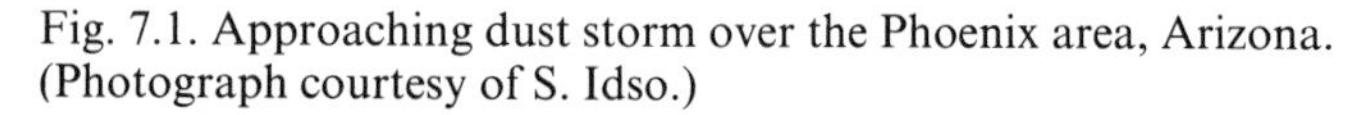

Fig. 7.1. Approaching dust storm over the Phoenix area, Arizona. (Photograph courtesy of S. Idso.)

upper-level atmospheric momentum, such as the jet stream, an increasing horizontal pressure gradient, and a large-scale convergence producing ascending winds. These conditions occur most frequently in the spring season. In some cases, there is a positive feedback in which airborne dust causes greater surface heating and a resulting atmospheric instability, with stronger winds supplying still more dust.

It has long been thought that the effect of atmospheric dust on world climate was to reduce temperature; i.e., the more dust in the air, the cooler the climate, due to greater reflection of incoming solar radiation. However, it may be that because of the relative absorptive capacities and reflection characteristics of the dust particles, dust aerosols may act to either cool or heat Earth (Idso, 1976, 1981). Low-level dust can have a warming effect on climate. Although incoming solar energy to the surface is reduced because of the dust, the accompanying decrease of thermal radiation outward from the surface can overcompensate and a net warming of the air near the surface can occur. For stratospheric dust, however, such as occurs from violent volcanic eruptions, the backscatter of solar radiation prevails over thermal blanketing to produce net cooling at the Earth's surface.

7.2.2 *Characteristics of aeolian dust*

Fine grains ('dust'), when alone, are quite difficult for the wind to set into motion, as shown in the wind threshold curve (Fig. 3.17). Bagnold (1960) describes a powdery surface on the Anatolian Plateau, which can lie undisturbed by relatively strong winds. He noticed great clouds of dust, however, kicked up by the feet of passing flocks of sheep. Thus, once disturbed, dust can readily be entrained and carried aloft by fairly gentle winds. Ordinarily, dust particles are intermingled with larger particles or clumps and thus become airbone by collisions from saltating sand. In order to be transported for long distances, the particles must be small enough so that the terminal or fall speed of the particles is less than the characteristic vertical speed component of the wind-induced turbulence. The smaller the particle, the smaller its terminal speed, so the smallest particles travel the greatest distance. For soils of large particle size, Gillette (1976) found that particles sampled in the lowest 1.3 cm above the surface show essentially the same particle-size distribution as the surface material. At 1 m above the surface, however, he found the proportion of particles less than 0.02 mm to be much larger than for the surface material. Particles collected in the atmosphere at a distance of 800 km from the dust storm source mostly ranged from 0.0002 to 0.02 mm (0.2–20 μm) in diameter. Particle sizes collected in various locations are illustrated in Fig. 7.2.

Fig. 7.3 plots the size distributions for sand, sand storm aerosols, loess, and airborne dust, compiled by Junge (1979) and shows schematically how the major fractions are derived from an original source by successive stages in a process controlled by the interaction between airborne transport and gravitational settling. The aerosol during a sand storm depicts a smaller average particle diameter than the sand itself because the largest particles are moving by creep. The most extensive loess deposits are found near the Yellow River in China and in the so-called loess hills downwind (east) of the Missouri River in the US. Most loess deposits occur fairly close to their source and thus the mean particle diameter is larger than that of the long-range dust particles blown from the Sahara and other arid regions.

7.2.3 *Erosion of agricultural land*

The most serious human problem caused by blowing dust is its effect on the ability of land to support crops and domestic animals. According to Woodruff & Siddoway (1965), the erodibility of land surfaces (primarily agricultural land) by wind is governed by 11 primary variables. The *soil erodibility index*, I, is a soil property based on soil cloddiness. *Knoll erodibility index*, I_s, is a factor related to windward slope. The *mechanical stability of the surface crust*, F_s, is important in determining threshold, but once erosion has begun on tilled soil, due to saltating particles, the crust is usually soon broken. Additional factors include *soil ridge roughness*, K_r, the

Fig. 7.2. Grain-size characteristics of windblown dust samples. (From Goudie, 1978.)

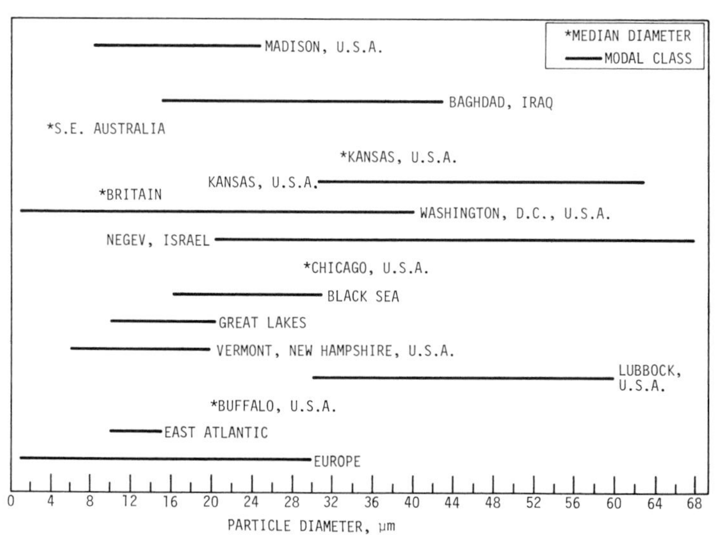

wind speed, U, and *soil surface moisture*, M. There are two distance factors, the distance across a given erodible surface, D_f, and a distance associated with the presence of soil erosion barriers, D_b. The remaining three factors deal with vegetative cover, i.e., quantity, R', type, S, and orientation, K_o.

Woodruff & Siddoway group these factors in order to obtain a six-parameter equation, i.e.,

$$E = f(I', K', C', L', U) \tag{7.1}$$

The resulting soil erodibility, E, is thus found in terms of mass per unit area per annum as a function of soil erodibility index, I', soil ridge roughness, K', climatic factor, C', field length, L', and wind speed, U. They present a series of equations and charts from which erodibility can be determined. Eq. (7.1) is used by the US Soil Conservation Service to estimate annual wind erosion losses from farm fields, under given conditions, and to help agriculturalists plan and apply measures to control soil erosion.

The wind erosion equation, Eq. (7.1), is based on many years of wind tunnel and field experiments. The development of the equation is traced and the various factors involved are discussed by Skidmore (1976). A computer program was written to facilitate use of the equation by Skidmore and associates (1970) at the Wind Erosion Research Laboratory, Manhattan, Kansas. The wind erosion equation, which is used to determine average annual potential loss of soil by wind erosion, will probably continue to evolve so that it can eventually be utilized to obtain soil losses for more-specific climatic conditions and for periods of duration shorter than one year.

Field measurements of vertical and horizontal dust particle fluxes were carried out by Gillette (1976) and correlated to the erodibility index of different soils (Fig. 7.4). The total horizontal mass transport for soils 1–5 in Fig. 7.4, i.e., those with the highest erodibility index, closely follow the transport trend with friction speed predicted by Eq. (3.27). Soils with a lower erodibility index fall below the curve. The ratio of vertical fine particle flux, F_a', to horizontal flux, q', according to Gillette, depends to a great extent on the type of soil, including the percentage of fine particles. The vertical flux of fine particles for loamy soils increases very rapidly with increase in friction speed above threshold and is proportional to friction speed to an exponent of about 4. The fine particle vertical flux is the source for dust which is carried to high altitudes and for long distances.

7.2.4 *Dust devils*

Dust devils are interesting meteorological phenomena (Fig. 7.5).

Although they are responsible for injecting only a very small portion of the total dust contained in the atmosphere on Earth, they may be an important mechanism on Mars, as we discuss in Section 7.3. In Section 3.3.6 we showed that the threshold mechanism for dust devils is quite different from that involving the boundary-layer wind, and we discussed how small

Fig. 7.3. (*a*) Comparison of particle-size distributions resulting from sorting by the wind. (Junge, 1979.) (1) Libyan desert sand, (2) Libyan desert sand aerosol, (3) average of eight loess soils, and (4) Sahara dust over the Cape Verde Islands.

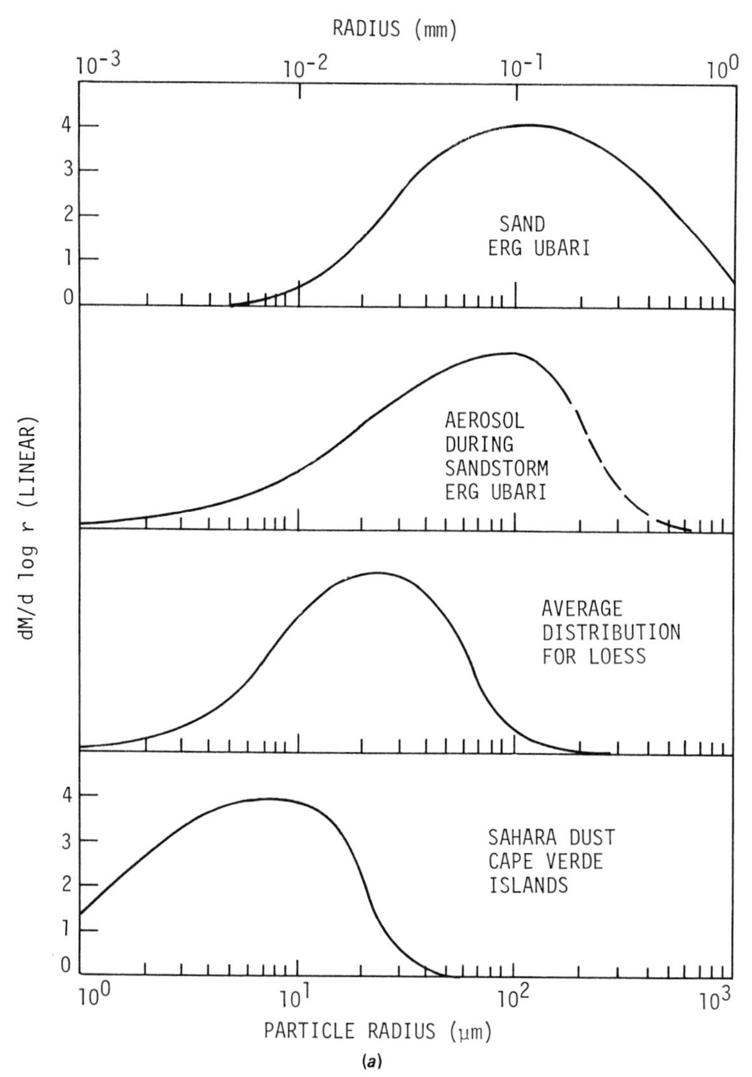

particles appear to be injected as easily as larger particles. In the dust devil shown in Fig. 7.5, the flared portion at the surface is probably due to larger particles which are thrown out of the vortex because of their relatively higher inertia. The dust-size particles within the vortex can be carried to great heights, as shown in the diagram. Centrifugal force keeps the particles out of the vortex core, which is seen as a clear region along the centerline of the vortex. Because of the high rotational speed within the vortex, the pressure inside it is lower than that outside the visible portion. This pressure gradient force on the dust particles, directed inward, balances the centrifugal force, directed outward, and leads to a sharply defined dust column (Fig. 7.5).

Although dust devils can probably be generated by shear in obstacle

Fig. 7.3. (*b*) Sand fractionation process by wind, showing mass (M) fraction distribution; the sum of curves 2, 3, 4 equals curve 1.

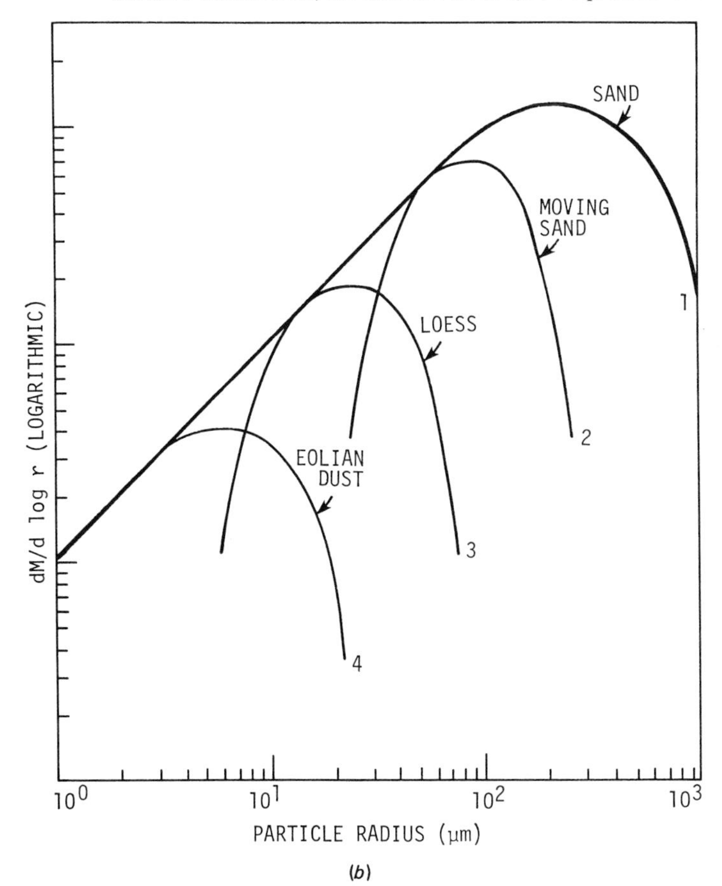

Fig. 7.4. Horizontal and vertical flux of soil. (From Gillette, 1981.) F_a' is the vertical flux of particles smaller than 20 μm and q is the horizontal flux of soil. Soils 1, 2, 4, 5 are from wind erosion group 1; soils 3, 7, 8 from group 2; soil 6 from group 3; and soil 9 from group 4. The soil groups are defined as follows (Lyles, 1977):

Group number	Group description	Soil erodibility I (Metric tons/(ha yr))
1	Very fine, fine, and medium dune sands	696
2	Loamy and loamy fine sands	301
3	Very fine and fine sandy loams, sandy loams	193
4	Clays, silty clays, silty clay loams	193

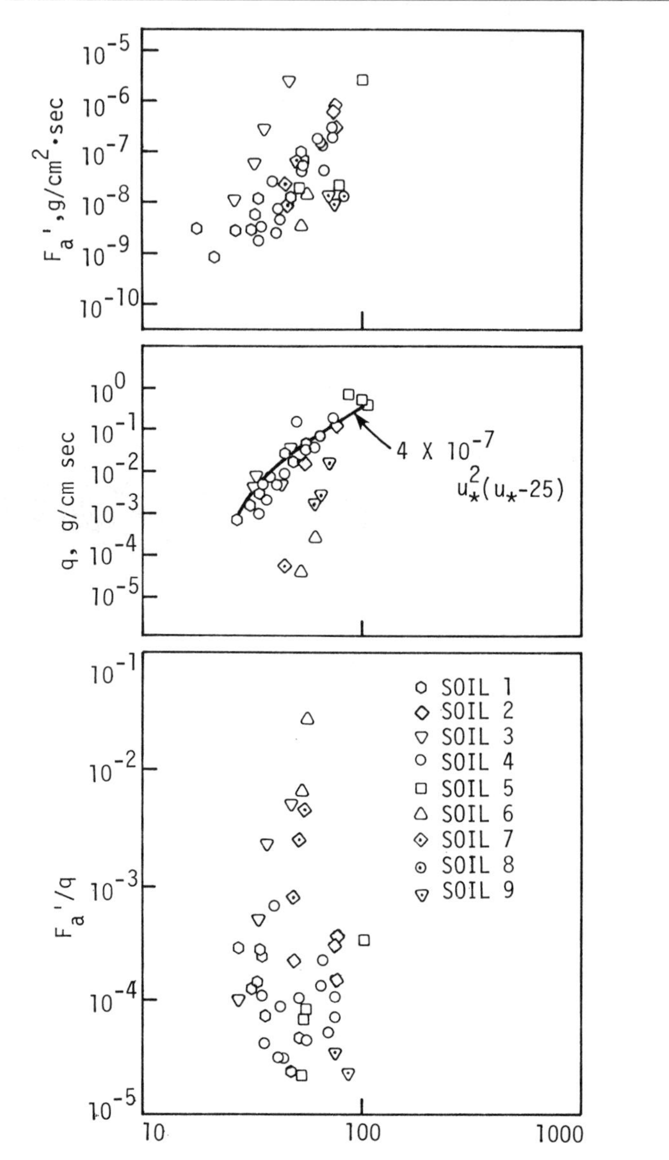

wakes, it appears that most are formed by convective heat transfer from warm surfaces in unstable atmospheres. The convective heat transfer from the ground consists of an unsteady system of updrafts accelerating away from the surface, accompanied by colder downdrafts toward the ground. These downdrafts must diverge horizontally near the surface and any

Fig. 7.5. Dust devil in southern Arizona. (Photograph by S. Idso.)

asymmetry in the flow pattern, caused by uneven heating, surface irregularities, or horizontal momentum convected from above, will result in a vertical component of vorticity from which the dust devil can evolve. The direction of rotation does not seem to be controlled by larger-scale vorticity, such as the Earth's rotation, or mesoscale atmospheric effects (Carroll & Ryan, 1970).

Sinclair (1966) made many measurements of dust devils in the Sonora Desert in Arizona. Size, location, rotational sense, and visible lifetimes were recorded for more than 2000 dust devils. Sinclair and his assistants also made measurements of pressure, temperature, wind velocity, and wind direction, with both a ground-based measurement system and with instruments mounted on a sailplane. The ground-based system was mounted on a jeep so that active dust devils could be 'chased'. Measurements within several dust devils were made at distances above the surface, of 2.1, 5.2, and 9.4 m. Measurements of vertical velocities and temperatures were obtained by penetration of the instrumented sailplane into a number of dust devils at altitudes of 1300–2900 m above the ground.

Sinclair's measurements near the surface show a strong inward radial component of velocity outside the visible dust column. This radial inflow diminishes in intensity as the dust column is approached. There is a strong updraft outside the core of the vortex, as expected, but there also is a downdraft inside the core, i.e., inside the radius for maximum swirl velocity component. Thus, the streamlines outside the core are helix-shaped with the radius of the helix near the surface decreasing rapidly with height.

The measurements are instantaneous and not time-averaged so the effect of turbulence makes it difficult to assess typical maximum values of the velocity components. From his plots, however, it appears that, for the three dust devil penetrations presented, maximum velocities for the tangential, vertical, and inward radial velocities were of the order of 10, 8, and 5 m/sec respectively. Maximum pressure deviations from the outside of the vortex to the center were of the order of -4 mb and temperature deviations were about $+5$°C. For the sailplane measurements far above the surface, vertical velocities of the order of 5 m/sec were measured with temperature increase within the vortex of about 0.5°C above the ambient. Some of these measurements were made far above the visible dust column.

The well-defined column of dust, as illustrated in Fig. 7.5, is somewhat difficult to explain. The central core is vacant because the dust particles, being heavier than air, are centrifuged out of the center of the column. Since the pressure decreases toward the center of the vortex, the force on each particle caused by the pressure gradient is in the inward direction but is much too small (by a factor of the order of the density ratio ρ/ρ_p) to keep the particle from being centrifuged further away from what appears to be an equilibrium radius for most of the visible dust column. There is apparently an additional factor necessary to keep particles from being centrifuged completely out of the column. That factor is the radial inflow which causes a relative inward air speed on the particle, and thus a drag force. A certain value of particle diameter can be calculated, which will be in equilibrium for a given combination of inward radial velocity, tangential speed, and radial position within the vortex. Too close to the core, the radial velocity is too small and tangential velocity too high to keep particles from being centrifuged further out. Too far outside the radius of maximum tangential speed, the centrifugal force on the particle becomes smaller than the inward drag force and the particles are forced inward. Thus a tubular column of dust is formed, as shown in Fig. 7.5. The conservation of mass requires that, for an axisymmetric vortex, the inward radial component must be balanced by an increase in vertical component of velocity with distance above the surface. Perhaps the upper limit of the visible portion of the vortex occurs at the altitude at which the vertical acceleration and, therefore, radial inflow, ceases.

Sinclair sampled the dust in a 3.5 m diameter dust devil at 2 m above the surface. The average particle diameter by mass was determined to be 90 μm. The large diameter sampled so far above the surface reflects the significant capability a dust devil has in raising particles to great heights by means of strong updrafts.

Sinclair's observations of large numbers of dust devils enabled him to generalize on their characteristics in several ways. As one would expect, the maximum frequency of occurrence is in the afternoon when the ground surface is warmest, the frequency decreasing with cloud cover. The probability of occurrence decreases rapidly with ambient wind speeds above 7 m/sec, with few dust devils forming at speeds greater than 9 m/sec because the higher wind speed causes greater turbulent mixing near the surface and lowers the unstable lapse rate toward the neutral value.

The duration of dust devils is a function of size. Those having diameters less than 3 m last only about 0.5 min, with those having diameters of 3–15, 15–30, and greater than 30 m, being of average 1.5, 3.5, and 6 min duration respectively. The ratio of duration to diameter is approximately constant, indicating from dimensional considerations that the maximum characteristic vertical and tangential speeds might be independent of diameter (a dimensionless duration time tu_θ/r_0 should be approximately constant). Large dust devils, however, can last considerably longer than the average values. Ives (1947) describes one that lasted 7 hr and traveled 65 km across the Bonneville salt desert in Utah.

7.2.5 *Fugitive dust*

Of importance from a human standpoint are the numerous problems associated with *fugitive dust*, i.e., dust caused to be airborne by mechanical means, such as traffic on unpaved roads, cultivation of agricultural land, mining operations, and coal handling at power plants and factories. Becker & Takle (1979) approach the unpaved road problem mathematically and compute the deposition which results downwind from the dust source. Their mathematical model predicts dust concentration and surface deposition for a given instantaneous two-dimensional fugitive dust source caused by passing vehicles. The model can account for differences in source strength, atmospheric stability, surface roughness, height of the source and deposition surfaces, and the wind speed. They show that the factors which have the greatest influence on the total deposition are the surface roughness and height of the deposition surface. The degree of atmospheric stability has the least effect of the parameters studied.

The source strengths for the study were obtained experimentally, in the atmosphere, by Becker & Takle. They made concentration measurements at a distance of 15 m from a gravel road centerline and collected samples of dust which averaged 3–5 μm in diameter. Calculations of effective area coverage of dust concentrations show that for some roads, only one automobile passage per day is sufficient to exceed US Environmental Protection Agency standards close to the roadway.

The environmental wind tunnel can be useful for the simulation of airborne dust problems. One of the authors has been involved with the physical modeling of fugitive dust blown by the wind from coal stockpiles (Iversen & Jensen, 1981). Both active (a pile being worked) and passive dust sources were simulated for a model power plant and associated coal yard. Deposition of simulated dust was measured downwind of the model stockpile for a variety of wind speeds and directions. A photograph of the model and example results are shown in Fig. 7.6. Although this kind of physical modeling for the fugitive dust problem is complex and difficult, feasibility of realistic qualitative and quantitative results from the wind tunnel model were demonstrated. Qualitative conclusions from the experiments, which would also be valid for stockpiles of other kinds of particles, such as these involved in mining operations and other storage situations, include:

(1) For isolated coal piles consisting of one lineal pile with a length much longer than its width, much more wind erosion will take place from the top of the pile if the wind direction is perpendicular to the long dimension of the pile than if it is parallel.
(2) For several parallel piles with the long direction oriented perpendicular to the wind, much more erosion takes place at the top edge of the windward pile than on those to the leeward.
(3) Vortices shed from the ends of a lineal pile or by buildings or other structures upwind of the pile, cause increased wind erosion on those portions of a downwind pile in the regions influenced by the vortices.
(4) Since high wind speeds are possible in the vortex interiors, these cause not only increased erosion on piles, but the particles can be carried much further downwind than would be possible in the absence of the vortex.

7.3 Dust storms on Mars

For many years, Earth-based telescopic observations of Mars revealed surface markings that changed size, shape, and position with time, seemingly in response to changes in the seasons on Mars. Although many possible origins have been attributed to these markings, including notions that they could be related to biological processes, the Mariner 9 and Viking missions have established that they are the result of atmospheric phenomena, primarily dust storm activity. At times, surface albedo features are totally obscured for several weeks as yellow clouds expand across the entire surface of Mars.

Transient yellow clouds have been recorded telescopically for more than 100 years. To coordinate Earth-based observations, a group of observatories formed a worldwide consortium in 1969, titled the International Planetary Patrol Program, whose purpose was to obtain uninterrupted observations and images of day-to-day and hour-to-hour changes in the atmospheric and surface features of the planets. Coordinated by W. A. Baum (1973) of the Lowell Observatory in Arizona, particular attention was given to Mars and the development and evolution of the dust storms. For example, in 1969, 7379 images of Mars were obtained (Baum *et al.*, 1970), which provided the first well-documented, seasonal patterns of clouds on a global scale.

The major dust storms of 1971 and 1973 were particularly well covered by International Planetary Patrol (Martin, 1974), as shown in Fig. 7.7.

Fig. 7.6 (a)

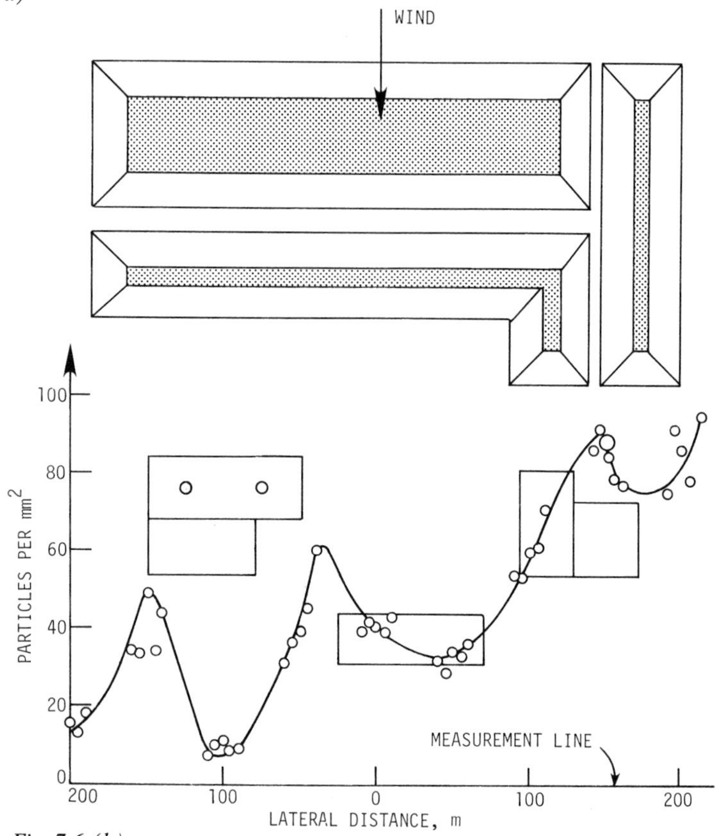

Fig. 7.6 (b)

Fig. 7.6. (*a*) Photograph of a wind tunnel model of a power plant and coal stockpile, built to a scale of 1–500 and wind tunnel tested by the wind engineering department at the Skibsteknisk Laboratorium (Danish Maritime Institute), Lyngby, Denmark. (*b*) Plan-view of the power plant and coal stockpile and the lateral distribution of dust deposited along the measurement line shown. The shaded area represents the surface of erodible dust. The dust was collected on oiled microscope slides, and the counts of collected lycopodium particles were made through a microscope. The wind tunnel was operated at a wind speed of 5 m/sec for 60 sec. (*c*) Same as (*b*) except that the power plant buildings were removed prior to the test. The buildings trap some dust but vortices generated within the boundary layer by the buildings concentrate dust to higher levels, as illustrated in (*b*), and also carry higher concentrations of particles further downwind. (From Iversen & Jensen, 1981.)

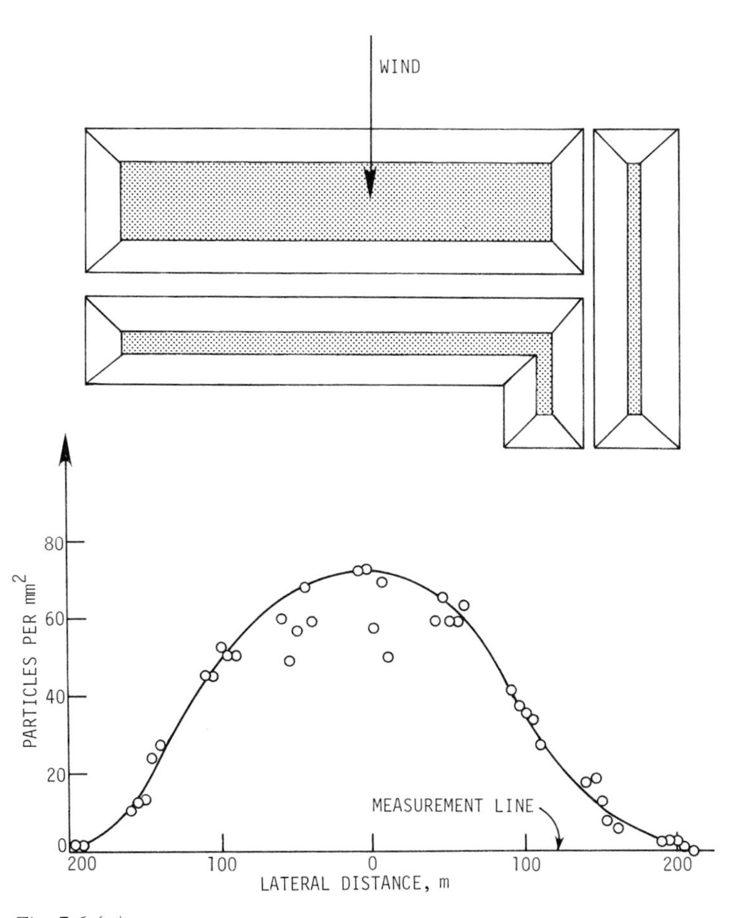

Fig. 7.6 (c)

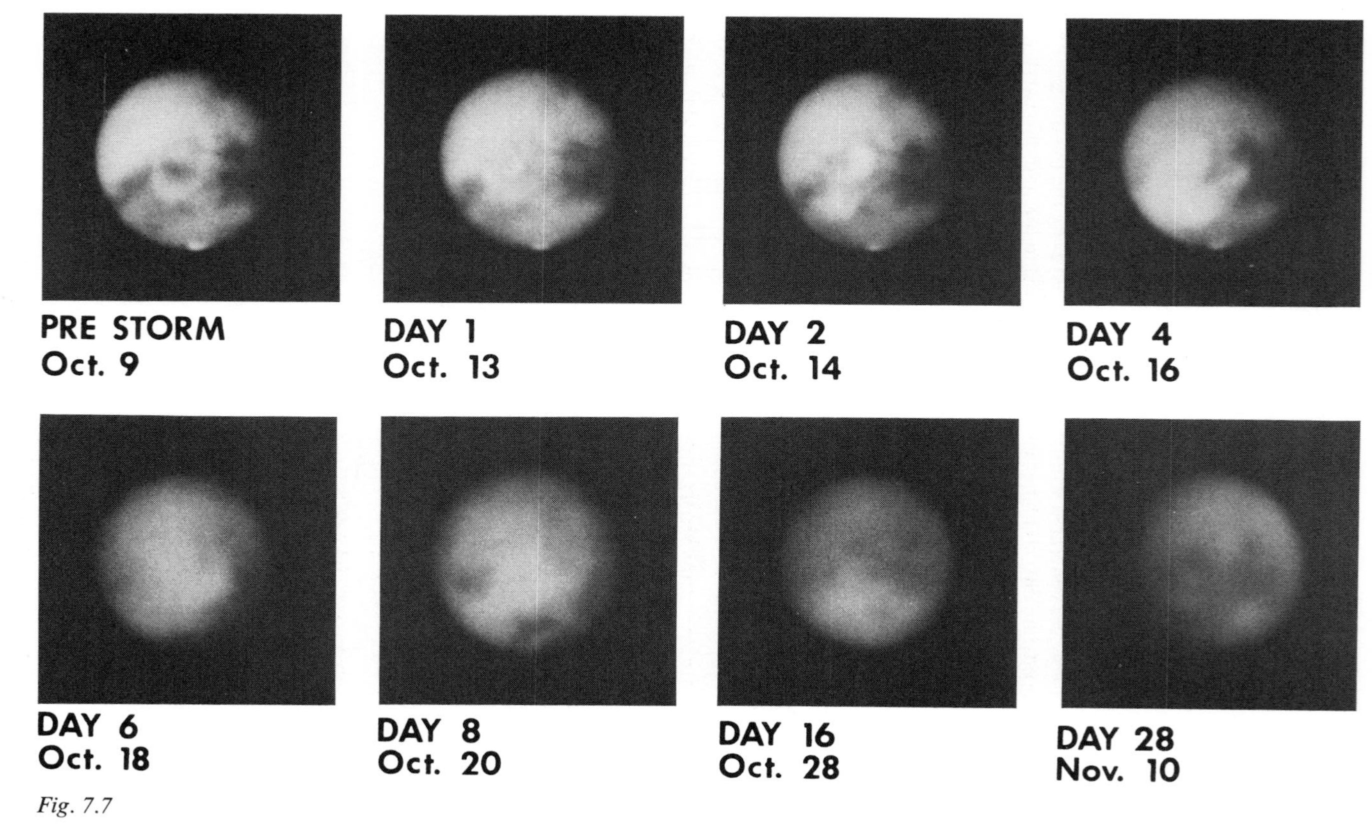
PRE STORM
Oct. 9
DAY 1
Oct. 13
DAY 2
Oct. 14
DAY 4
Oct. 16
DAY 6
Oct. 18
DAY 8
Oct. 20
DAY 16
Oct. 28
DAY 28
Nov. 10

Fig. 7.7

From the evidence of these and other studies, most global dust storms occur when Mars is closest to the sun, i.e., in late spring in the southern hemisphere. There is no evidence that dust is raised simultaneously on a global scale; rather, there are particularly favored geographical localities where the storms begin. Global obscuration can occur in as little as three weeks. In general, the storms are sandy yellow in color, but they can also be red or 'dirty' white.

The Mariner 9 spacecraft returned images and infrared and ultraviolet data which provided information on the fallout phase of global dust storms. Mariner 9 also recorded several instances of localized dust clouds. Additional data and greater detail on both local and global storms have been obtained by the Viking Orbiters and Landers. An excellent review of Earth-based observations and spacecraft results is provided by Briggs and his colleagues (1979), from which we summarize.

7.3.1 *Martian great dust storms*

Some years see the development of especially large dust storms, termed *martian great dust storms* or, more simply, *global dust storms*; however, the criteria for what constitutes a 'global' or 'great' dust storm are not well defined, as some storms so designated have not always led to global obscuration (Briggs *et al.*, 1979). Nonetheless, these are all big storms, which affect most of the planet, unlike the 'local' dust storms that we discuss later.

Gierasch & Goody (1973) and Gierasch (1974) described the great martian dust storm of 1971 – one of the best-documented storms – and presented an explanation for its origin. Their model is generally applied to all martian great dust storms. Although local dust storms occur frequently and are probably dynamically similar to dust storms on Earth, great martian dust storms are rarer and more spectacular.

The major dust storms originate in three general areas of Mars (Zurek, 1982): (1) on sloping plains between the northwest rim of the Hellas basin and the uplands of Noachis (sites for the 1956 and 1971 storms), (2) sloping plains to the west, south, and southeast of Claritas Fossae (sites of the 1973 and 1977 storms), and (3) on the low-lying region of Isidis Planitia, east of

Fig. 7.7. Earth-based telescopic images of the development of the great martian dust storm of 1973, showing normal clear view on 9 October, leading to global obscuration by 10 November. Photographs were obtained by Lowell, Mauna Kea, Perth, and the Republic Observatorie as part of the International Planetary Patrol Program. All have central meridians near Solis Lacus (near 80°W). (From Martin, 1974.)

Syrtis-Major. All three areas are in subtropical latitudes and constitute large east-facing slopes where one would expect differential heating to occur. In addition, thermal inertia data obtained by the Viking Orbiters indicate the presence of sand, which may aid in raising fine-grained dust (Peterfreund, 1981). Typically, the storms begin near or slightly before the southern hemisphere solstice, during the beginning of summer in the southern hemisphere (Fig. 7.8). In some years, however, storms do not develop at all.

Major storms appear to go through three phases. In phase I, the storms begin as bright spots, or cores, ≈ 400 km in diameter, or smaller. This first phase lasts ≈ 5 days. Phase II, lasting from 35 to 70 days, is the expansion phase which takes place by the development of secondary cores around the first phase primary core. Eventually, the entire planet is affected and the development of the storm path does not seem to be topographically controlled. For the largest storms, the entire planet may be totally obscured and, in the case of the 1971 storm, dust clouds may have reached a height of ≈ 60 km. Phase III marks the decay of the storm and lasts from 50 to 100 days.

The 1971 great dust storm started on 22 September 1971. The bright elongated core cloud first appeared on the northeast border of Noachis, covering $\approx 10^6$ km^2. It expanded slowly for five days, then more rapidly, and encircled the planet after 15 days with an average easterly speed of ≈ 65 m/sec. During October the obscuration of the planet spread but some features began to appear again in November. Improved visibility through the atmosphere began in early December with dramatic clearing occurring in the latter part of December. The significant fact about this dust storm is that the Mariner 9 spacecraft arrived right in the middle of it on 14 November. Prior to orbiting Mars, it returned images of the storm on 11 November, as well as orbital images during the entire decay phase of the storm. The only visible features from the first images were the row of three volcanic peaks in Tharsis and the gigantic volcano Olympus Mons (Fig. 1.13).

Gierasch & Goody (1973) hypothesize that great dust storms start on high-plateau areas during times of maximum insolation when winds are light and the atmosphere is unstable – conditions that are ideal for the creation of dust devils which lift dust particles to great heights above the surface (Sagan *et al.*, 1971). In this model, great dust storms can reach a mature state only when an initial source of dust exists over a sufficient area for a sufficient time. This implies fairly light winds – not strong enough to raise dust – and, therefore, a large number of dust devils, raising significant amounts of dust to high altitudes, seems to be necessary.

The initial area of dusty air, according to the Gierasch & Goody model, is warmed by solar radiation, causing updrafts within the dusty core and inward swirling flow within the boundary layer outside the core, similar to the growth of a hurricane on Earth. With sufficient ambient vertical vorticity, the swirl velocity can intensify to a sufficient level to reach boundary-layer threshold within a day or so. The time necessary for this to occur represents a constraint on the probability of occurrence of a great dust storm, because if a superimposed mean mesoscale wind of some strength exists it will ordinarily act to destroy the organization of such a storm right at the beginning of its development.

Fig. 7.8. Viking Orbiter image of a dust cloud in Thaumassia Fossae on Mars, preceding the first global dust storm of 1977. (From Briggs *et al.*, 1979.)

The representation by Gierasch & Goody of a great dust storm in its mature phase is shown in Fig. 7.9. This diagram illustrates the central rotating column of dusty air, with dust supplied by strong boundary-layer winds swirling inwards to the core and then rising within the central column to the stratosphere. Above the tropopause, at 2 scale heights, the dust column spreads rapidly. The time calculated for dust to fill the entire stratosphere is 20 days. The diagram shows to what heights the pumping action of the core will take particles of various sizes. Once the atmosphere is

Fig. 7.9. Representation of a martian great dust storm in its mature phase. (From Gierasch & Goody, 1973.)

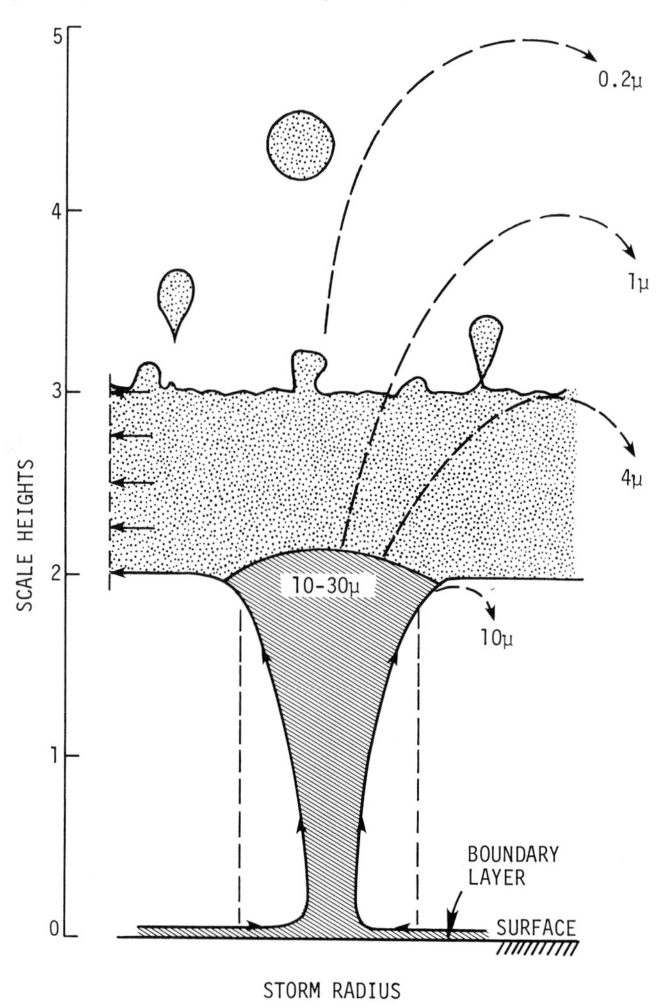

filled with dust, the central column should lose its buoyancy and should last until the 10 μm particles fall to the ground, which should take about ten days.

The final stage is decay of the dust storm. After 20 days the atmosphere is loaded with particles from 1 to 30 μm in diameter, with size decreasing with height. The time for 1 μm particles to fall from 4 scale heights is 130 days, giving an estimate of the decay phase lifetime.

The 1971 dust storm began as a bright yellow–white core $\approx 10^6$ km^2 in area, which expanded slowly until the seventh day. The core was seen until day 16, then an ochre-colored veil spread over the planet by about the 40th day and dust particles rose to heights above 30 km (3 scale heights). Dramatic clearing started after ≈ 50 days with some traces of the dust cloud left after 160 days. All of these data are in fair agreement with the Gierasch & Goody model.

Pollack *et al.* (1979) consider an alternative origin for global dust storms, since vortex motions within local dust storms have not been observed by Viking Orbiters. They list other types of winds which can possibly cause local dust storms. These include: (1) polar cap winds driven by strong temperature differences between frost-covered and frost-free terrain, (2) tidal winds created by diurnal heating, (3) topographical winds due to large elevation differences, (4) strong frontal winds due to weather systems, and (5) large-scale seasonal winds produced by latitudinal temperature gradients. They consider positive feedback mechanisms by which entrained dust can alter these winds by influencing temperature gradients and thus contribute to storm growth. They feel that for global dust storms, dust must be contributed from the surface of most of the planet rather than be fed from a relatively small source, as the 'hurricane' model might imply. The various great dust storm evolutionary models are reviewed by Leovy (1979), along with other aspects of martian meteorology.

Dust storms were also detected at the Viking Lander sites, by determining the *optical depth* of the atmosphere, which is a measure of its opacity. For vertical illumination, $I_l = I_o e^{-\tau}$, where I_l is the intensity of light at the surface, I_o is the intensity of light entering the upper atmosphere, and τ is the optical depth. Fig. 7.10 illustrates the contrast in view at Viking Lander 1 due to a dust storm and the change in measured optical depth due to a global storm.

The increase is dust loading is much more rapid than is the decay. The mass loading, m', is related to optical depth, τ, by

$$m' = D_p \, \rho_p \, \tau/3 \tag{7.2}$$

Fig. 7.10. (*a*) Viking Lander 1 images of seven sequential images. The sixth image (recorded 14 June 1981) is dark due to a great dust storm. (*b*) Optical depth at the Viking Lander 1 site as a function of time. The sol number is the number of martian days from the time of the spacecraft landing. The sudden increase in optical depth is due to the passage of the first great dust storm of 1977. (From Pollack *et al.*, 1979.)

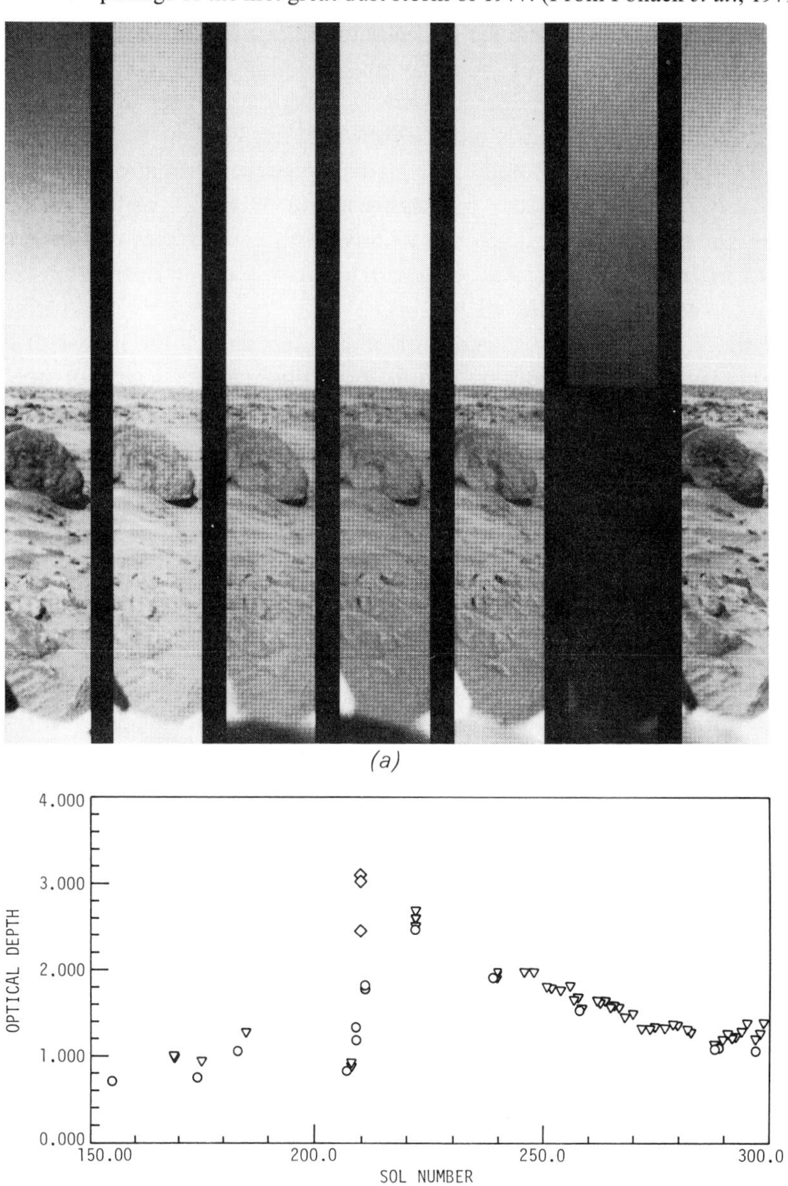

where D_p is mean diameter and ρ_p is particle density. For an optical depth, τ, of 1, particle diameter of 5 μm, and density of 3 g/cm^3, the mass loading, m', is $5(10)^{-4}$ g/cm^2. It appears from Fig. 7.10(*b*) that an optical depth of about 1 is necessary for initiation of a global storm, and that the value of τ increased to over 3 during the 1977 storm. Other meteorological characteristics measured at the Lander sites are described by Ryan & Henry (1979).

7.3.2 *Martian great dust storms clearing*

With the completion of the mature phase of great dust storms, the clearing phase begins. Partial clearing of the lower atmosphere should occur quickly because the particles are larger and have a higher fall speed. However, a thin haze remains, due to smaller dust particles settling from higher altitudes. Hartmann & Price (1974) have calculated dust particle sizes in the martian atmosphere during the 1971 storm by matching theoretical fallout curves with observed rate of clearing. Fig. 7.11 shows the clearing of the great 1971 dust storm as determined by measuring the clarity of Mariner 9 photographs at different surface elevations. The diagram shows a resurgence in dust storm activity in mid-December, followed by the significant clearing which started about 22 December. Note that the

Fig. 7.11. Identification of maximum particle sizes in the great martian 1971 dust storm by matching the rate of clearing to theoretical fallout curves. (From Hartmann & Price, 1974.)

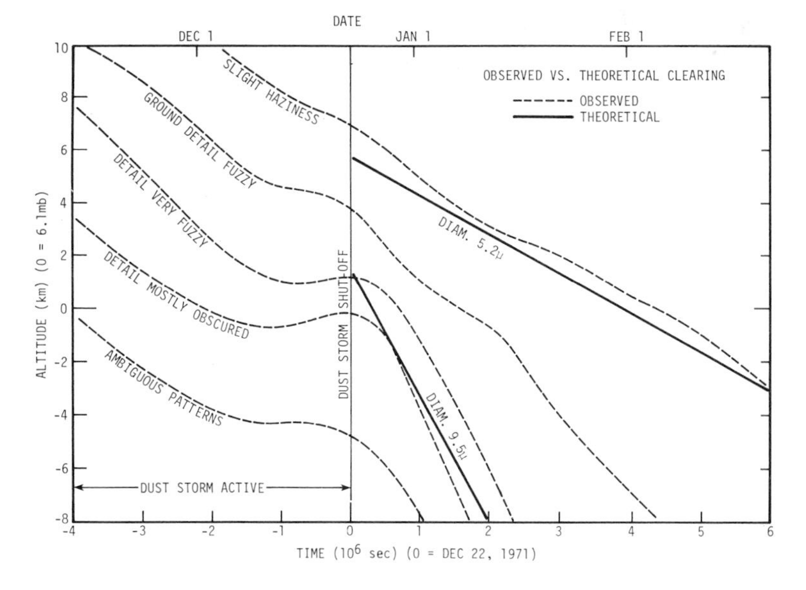

clearing rate at lower elevations corresponds to a particle diameter of ≈ 9.5 μm, and the slower rate at higher elevations corresponds to a 5.2 μm particle. These results are consistent with other estimates of particle size within the atmosphere.

Pollack *et al.* (1979) also discuss the clearing process. They indicate that the rate of clearing is controlled by: (1) gravitational settling, as discussed above, (2) degree of turbulence at the surface, and (3) growth by H_2O and CO_2 ice condensation followed by gravitational sedimentation. They suggest that because large amounts of dust are present in the atmosphere when seasonal deposits of CO_2 are formed at the north pole, dust is currently being preferentially deposited in the north polar region. Preferential deposition would occur at the south polar region in another 10^5 yr when axial precession leads to southern winter occurring near perihelion. Their calculations lead to the probability of a polar layered deposit 30 m thick being formed at alternate poles every 10^5 yr, to account for the so-called layered terrain (Fig. 7.12). The alternating preferential deposition at the

Fig. 7.12. Layered terrain in the north polar region of Mars, showing nearly horizontal deposits, presumed to be composed of windblown dust. (VO image 56B86.)

poles implies mass loss at equatorial and mid-latitudes, which, by their calculations, results in an erosion rate of 7 m per million years.

7.3.3 *Local dust storms*

Although the imaging experiment on board the Viking Orbiters observed only about 1% of the planet's surface each day, many local dust storms have been observed. Briggs *et al.* (1979) list 16 storms occurring in 1977, which means there were probably many more during that time. Using thermal inertia data to detect dust storms, Peterfreund (1982) notes at least 31 additional local dust storms. Most of the local dust storms observed by the Orbiters occurred during southern hemisphere spring and early summer, when Mars is near perihelion and insolation is near maximum. About one-half of the dust clouds appeared near the edge of the southern polar caps, and the other half occurred mainly in the southern hemisphere where topographical effects seem to be important.

Some of these local dust storms may result from slope winds related to the nocturnal cooling of a surface layer of atmosphere, as proposed by Megalhães & Gierasch (1982). Fig. 7.13 shows several local dust storms, photographed early in the Viking mission, over lava plains southeast of

Fig. 7.13. Local dust clouds rising from lava flow surfaces near Arsia Mons. (From Briggs *et al.*, 1979.)

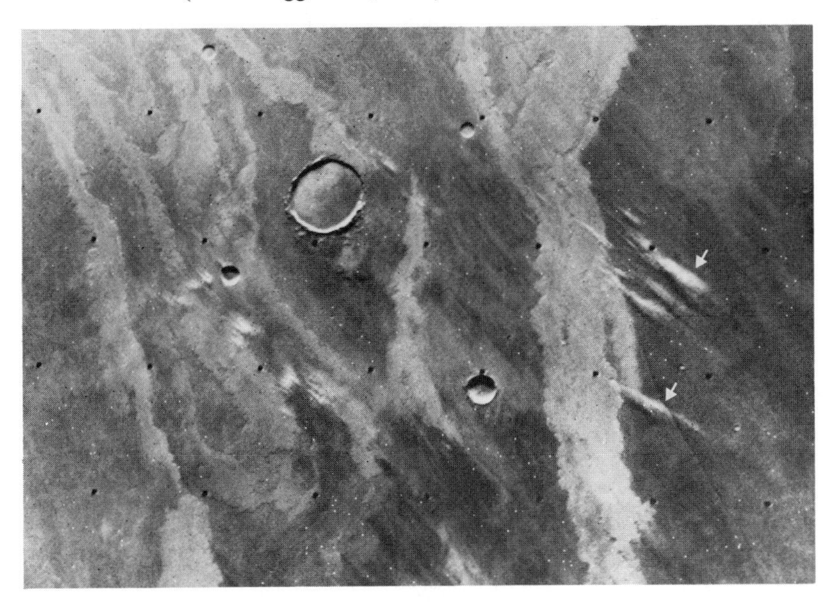

Arsia Mons. The dust clouds are estimated to rise about 1 km above the surface, based on measurements of the shadows cast by the clouds. These storms probably result from slope winds generated off the flanks of the volcanoes in the Tharsis region. For example, Lee and his colleagues (1982) have speculated that slope winds on the Tharsis volcanoes are effective in raising dust.

Similar slope winds are believed to be responsible for dust clouds seen in the Argyre basin (Fig. 1.14). The storms are shown in Fig. 1.17; each is $\approx 40\,000$ km^2 in extent and may be 4–6 km high. Other local dust storms were observed near Hellas, a 2000 km diameter basin where dust clouds 350 km long and 5 km high occurred.

The deep canyon systems of the equatorial zone are also sites of local dust storms and, although this region also has frequent condensate clouds, the dust clouds can usually be identified by their color and structural morphology.

In general, local dust storms appear to be random in their time of occurrence but to be regionally biased in their formation. Most appear to be associated with local slope winds and along the edges of the south polar cap where strong thermal gradients are found. Peterfreund (1982) found that local dust storms seldom occur in areas of high albedo and low thermal inertia (considered to be mostly dust deposit regions) but, rather, are usually confined to sandy regions. As Briggs *et al.* (1979) summarized, however, the record is very incomplete, and Viking probably recorded only a small fraction of those which actually occurred. Some of the Viking wind measurements are considered by Ryan and Lucich (1983) to have been vertically oriented vortices. A few of these could have been strong enough to raise dust and thus become dust devils.

7.4 Dust deposits

Deposits of windblown dust, commonly referred to as *loess*, cover significant parts of the surface of Earth. Pécsi (1968) estimates that one-tenth of the land's surface is covered by deposits of windblown dust ranging from 1 to 100 m thick. Mars, too, appears to have substantial areas blanketed with fine-grained particles.

From an agricultural point of view, we find that the rich grain-raising black soils of southern Russia, the Argentine Pampas, and the Great Plains of the US, owe their origins to windblown loess. Similarly, the soils of north China have produced grain crops for more than 4000 years without being depleted, primarily because of the annual influx of windblown dust which constantly renews the surface. In addition, loess and loess-like deposits have been used for housing. In some parts of the world, caves have been carved directly from loess to make dwellings. In some areas, loess is mixed with other materials to make durable adobe bricks, blocks, or plaster.

7.4.1 *Loess on Earth*

The term *loess* (from the German meaning 'loose') was first applied to the fine-grained, unconsolidated sediments of the Rhine Valley. It has since been applied to all relatively homogeneous silt deposits that are porous, permeable, and yellow–buff in color. For the most part, loess is considered to be aeolian in origin. Although this definition has been broadened by some investigators to include similar-appearing deposits of non-aeolian origin, we will confine our consideration to those deposits associated with aeolian processes.

The dominant size of loess grains is silt (10–50 μm in diameter and making up 40–50% of the total), but up to 30% clay (smaller than about 5 μm) and up to 10% fine sand (>250 μm) may also be present. Pécsi (1968) considers three possible explanations for this size distribution: (1) it results from size sorting during fluid transport (air or water), (2) it results from coagulation–aggregation of fine particles such as clay, and/or (3) the distribution results from the comminution of rocks, principally by processes involving insolation and frost heave.

Quartz is the primary mineral of the coarse grains in most loess deposits, averaging 65% in abundance; feldspars and various calcium and magnesium carbonates make up the remainder. Examination of loess grains via scanning electron microscopy shows the quartz grains to be angular; larger grains often have material (carbonates?) adhering to their surfaces.

The high porosity of loess has been attributed (Pécsi, 1968) to electrostatic effects when the grains were deposited, to biological activity (plant roots; burrowing organisms), to solution and redistribution of the carbonate components, or to frost-heaving processes.

Loess deposits occur primarily in semiarid and temperate zones between latitude 24° and 55°N and 30° and 40°S. Although occurring mostly as plains, loess deposits are also found on mountain slopes, in intermontane valleys, and as mantling deposits over hills and mountains. In general, loess deposits thicken toward the apparent source of the particles (Fig. 7.14). Loess deposits are not found in tropical areas, nor in areas formerly covered by ice during the last ice age.

Although there is considerable controversy over the origin of loess deposits – not only as to the basic source of the sediments and the mode of transport, but also as to the details of various proposals – most investigators consider loess to be windblown sediment associated with glacial activity (Flint, 1971). North American loess deposits are considered to have their sources as outwash plains from continental glaciers. The strong thermal winds found in association with glaciers are thought to have carried particles aloft where they were then transported easterly and deposited. Thus, the primary source of loess was glacio-fluviatile and, secondarily, aeolian. The vast loess deposits in China appear to owe their origin initially to winds which raised grains from the Gobi Desert (probably glacial outwash plains) and deposited them to the east where they were subjected to secondary reworking by fluvial processes.

Some loess also appears to form exclusive of glacial processes. Péwé (1981) notes that most desert dust is derived from *in situ* weathering of bedrock, where it can be windborne and later deposited as loess-like units.

7.4.2. *Dust deposits on Mars*

Analysis of the Mariner 9 images of Mars, beginning in 1971, revealed an unexpected distribution of impact craters smaller than ≈ 15 km in diameter. Soderblom *et al.* (1973*b*) noted that, from ≈40° latitude poleward (both north and south), there is a scarcity of small craters, in contrast to the Moon and Mercury where there is a logarithmic increase in the number of smaller craters. They proposed that the higher-latitude regions on Mars were mantled with a blanket of sedimentary deposits, which were inferred to be of aeolian origin. The presence of the blanket was defined on the presence of small (1–10 km in diameter), sharp-rimmed craters that have crater floors partly – or nearly completely – filled with sediment. Many of these craters are distinguished by ejecta fields which

Fig. 7.14. Thickness of loess deposit measured at primary- and secondary-divide ridge crests along the Rock Island Railroad line from the towns of Bentley in western Iowa (30 km from the Missouri River) to Adair in west central Iowa. The systematic decrease in thickness from the source (Missouri River flood plain) is of obvious aeolian origin. (Modified from Ruhe, 1969.)

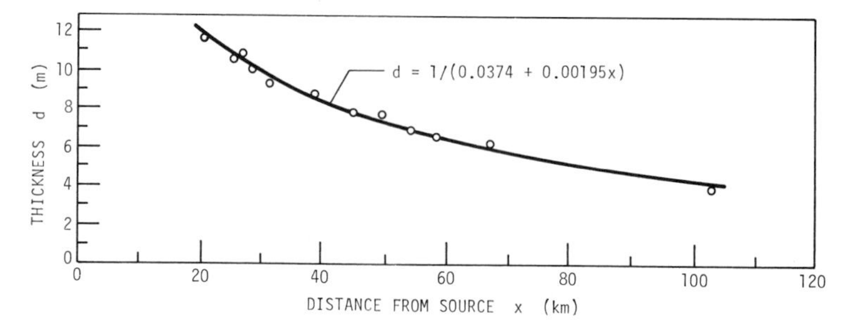

have distinctive scarps marking their outer margins, called *pedestal craters* (Fig. 7.15).

From the examination of high-resolution images and global mapping, Soderblom and his colleagues suggested that the mantled region has been subjected to repeated burial by aeolian sediments and wind erosion in response to planetwide changes in climate (Fig. 7.16). As discussed earlier,

Fig. 7.15. Pedestal craters on Mars. Unlike impact craters on the Moon and Mercury, the ejecta deposits terminate in a pronounced scarp (arrows), suggesting that the surrounding plains have been lowered by deflation. It has been proposed that blocks in the ejecta serve to shield the zone surrounding the crater and prevent deflation (VO image 60A53; from Carr, 1981.)

both Viking Lander sites appear to be undergoing deflation at present and many of the deposits of fine-grained material may be remnants of a much more extensive sedimentary mantle.

McCauley (1973) and, later, Arvidson *et al.* (1976) studied the pedestal craters in the high latitudes and suggested that the blocky ejecta deposits surrounding impact craters served as an 'armour' to prevent wind deflation. Thus, while the surrounding plains would be lowered by deflation, the crater and its zone of ejecta would remain high-standing. Although there are many uncertainties in the calculation, Carr (1981) estimates a rate of deflation of about 10^{-1} μm/yr of the plains surrounding the pedestal craters.

Analyses of Viking Orbiter data suggest that aeolian 'dust' mantles may be much more extensive than has been inferred from the Mariner 9 record. The infrared thermal mapping (IRTM) system provided data on the thermal inertia of the martian surface, which can be related to the particle

Fig. 7.16. Viking Orbiter view of Mars, showing smooth mantle deposits (lower right) in the southern highlands, which have been partly stripped – presumably by deflation – leaving yardangs. Stripped sediments may be a source of material forming the dunes in the lower left corner. Image centered at 11.8°S and 181.2°W; north is to the bottom of the photograph; the area shown is ≈60 by 85 km (VO frame 436S03.)

size. When combined with photogeology and radar data, Zimbelman & Greeley (1981, 1982) found that regions in the ancient cratered terrain at low latitudes appear to be mantled with dust-like material inferred to be of aeolian origin.

The polar terrains of Mars exhibit some of the best evidence for deposits of aeolian dust. Layered terrain (Fig. 7.12) is found in both polar regions within the 80° latitude circle. Careful analysis by Dzurisin & Blasius (1975) shows that individual layers are 10–50 m thick, and probably also exist as units which are thinner than can be resolved. They estimated a minimum total thickness of 1–2 km in the south and 4–6 km in the north. Cutts (1973) suggested that the layers represented cyclic deposition of dust, mixed with polar ice deposits. This generally accepted hypothesis has been expanded by many investigators, including Pollack and his colleagues (1979), who have estimated a deposition rate of $\approx 4 \times 10^{-2}$ cm/yr. In their model, Pollack *et al.* suggest that dust particles from global storms drift poleward and act as nuclei for CO_2 frost. As the CO_2 dust particles grow in size, they settle to the surface, gradually accumulating to form a layer.

Climatic variations on a global scale – partly related to the orbital and rotational characteristics of Mars – cause gaps in the deposits, leading to distinctive layering. Various valleys, sculpted hills, and irregular pits are attributed to periods of wind erosion and deflation.

APPENDIX A

Nomenclature and symbols

a	drag force moment arm (Eq. (3.3), Fig. 3.3)
a_p	pressure difference proportionality coefficient (Eq. (3.15))
a_1	drag force moment arm coefficient (Eq. (3.5))
A	threshold speed coefficient, $u_{*_t}(\rho/\rho_p g D_p)^{1/2}$
A_{cs}	longitudinal drift cross-section area
A_1	cohesionless threshold coefficient (Eq. (3.6))
A_c	plan area enclosed by crater rim
A_d	plan area of deposited or deflated material
b	lift force moment arm (Eq. (3.3), Fig. 3.3)
b_o	exponent in concentration distribution (Eq. (3.36), Fig. 3.22)
b_1	lift force moment arm coefficient (Eq. (3.5))
c	interparticle force moment arm (Eq. (3.3), Fig. 3.3)
c_r	bedform (dune of ripple) rate of advance (Eqs. (5.2)–(5.5))
c_1	interparticle force moment arm coefficient (Eq. (3.5))
C	mass transport coefficient (Eq. (3.20))
$\tilde{C}$	particle concentration
C_p	specific heat at constant pressure
C_τ	surface stress coefficient for a vortex
C_D	drag coefficient
$\tilde{C}_1$	particle concentration at reference altitude z_1 (Eq. (3.35))
C_1	coefficient (Eq. (5.4))
C_1	exponent (Eq. (3.32))
C_2	exponent (Eq. (3.32))
C_2	coefficient (Eq. (5.5))
C'	soil erosion climatic factor (Woodruff & Siddoway, 1965)
d	depth of deflated or deposited aeolian material
D'	saltation layer velocity profile factor (Eq. (3.39))
D	drag force
D_c	crater diameter

D_b distance associated with wind barriers (Woodruff & Siddoway, 1965)
D_f distance across an erodible soil surface (Woodruff & Siddoway, 1965)
D_p particle diameter
D_{po} reference particle diameter (Eq. (3.20))
e coefficient of restitution
f Coriolis parameter, $f = 2\omega \sin \phi$
F_a' vertical fine particle flux, mass per unit time per unit area
F_s surface crust stability (Woodruff & Siddoway, 1965)
g gravitational acceleration
h height of bedform (dune or ripple)
H_s scale height
H_r height of roughness elements (Fig. 3.11)
H characteristic vertical dimension
I, I' soil erodibility index (Woodruff & Siddoway, 1965)
I_l intensity of light at surface
I_o intensity of light entering upper atmosphere
I_p interparticle force
I_s Knoll erodibility index (Woodruff & Siddoway, 1965)
k von Karman's constant, approximately $= 0.4$
K_l interparticle force coefficient (Eq. (3.11))
K turbulent diffusivity
K_D drag parameter
K_L lift parameter
K_M moment parameter
K_o vegetative cover orientation (Woodruff & Siddoway, 1965)
K_r, K' soil ridge roughness (Woodruff & Siddoway, 1965)
L_i all other horizontal dimensions besides the reference length
L lift force
L_r characteristic horizontal dimension
L^* Monin–Obhukov stability length
L_p particle trajectory length in saltation
L_s areocentric longitude of the sun
L' field length (Woodruff & Siddoway, 1965)
L_x distance between roughness elements (Fig. 3.11)
m' mass loading due to dust in the atmosphere (Eq. (7.2))
m particle mass
M_o soil surface moisture (Woodruff & Siddoway, 1965)
M overturning moment
M_e ejected mass from hypervelocity impact

n interparticle force exponent (Eq. (3.11))
p pressure
Δp difference in pressure
p_t ratio of terminal to threshold friction speed, U_F/u_{*_t}
p_o atmospheric pressure at the surface
q mass-transport rate, mass per unit width per unit time
q_c mass transport rate at dune or ripple crest, mass per unit width per unit time
q_d deflation or deposition rate, mass per unit width per unit time
q_t mass transport rate at ripple trough, mass per unit width per unit time
r spherical radius
r_t threshold friction speed – friction speed ratio, u_{*_t}/u_*
r_0 characteristic vortex radius
r_u threshold wind speed ratio, U_t/U (Table 3.5)
R Reynolds number, (Eq. (2.7)) velocity times length/kinematic viscosity
$\tilde{R}$ correlation coefficient
R_F Reynolds number, $U_F D_p/v$
R_i Richardson number (Eq. (2.12))
R_* friction Reynolds number, $u_* D_p/v$
R_H crater-rim height
R' vegetative cover quantity (Woodruff & Siddoway, 1965)
R_g gas constant (Eq. (2.2))
R_{*_t} value of friction Reynolds number R_* at threshold, $u_{*_t} D_p/v$
S vegetative cover type (Woodruff & Siddoway, 1965)
S_a abrasion susceptibility
t time
T temperature
T_t Chepil's turbulence factor
T_o atmospheric temperature at surface
u horizontal particle speed
u_θ tangential wind speed component in a vortex
u_t wind speed at saltation focus height (Eq. 3.38)
u_* surface friction speed $(\tau/\rho)^{1/2}$
u_{*_t} threshold friction speed
$u_{*_{tR}}$ value of threshold friction speed for a rough surface
U wind speed in x direction
U_r reference undisturbed wind speed at reference height
U_∞ wind tunnel free stream speed (above boundary layer)
U_F particle terminal speed

U_g geostrophic wind speed component in x direction
U_o wind tunnel free-stream speed at threshold
U_{10} reference wind speed at 10 metres above the surface
U_p particle impact speed
U_t free stream or reference wind speed at initiation of motion (threshold)
V_D drift volume
V wind speed in y direction
V_g geostrophic wind speed component in y direction
w_o initial vertical particle speed
W vertical wind speed
W_t weight
x, y horizontal distance
z height above surface
z_o aerodynamic roughness height
z_o' aerodynamic roughness height in saltation
z_1 reference altitude (Eq. (3.35))
α_{op} particle lift-off angle in saltation (Fig. 3.19)
α_i particle impact angle
α_o angle between geostrophic and surface wind vectors
γ_p bulk density of sand grains
ζ ratio of height above surface to Monin–Obhukov stability length
δ planetary boundary layer thickness
η_p Chepil's packing factor (Eq. (3.9))
η all topographic vertical dimensions besides the reference length
ϕ potential temperature (Eq. (2.6))
λ ripple wave length
μ coefficient of absolute viscosity
v kinematic viscosity $= \mu/\rho$
ρ fluid density
ρ_p particle density
σ_I tensile stress due to interparticle cohesion
τ surface shear stress
τ optical depth
ϕ_o latitude, positive north of equator
ϕ parameter used to define particle diameter, $\phi = \ln D_p/\ln 2$, where D_p is in mm
ϕ_m dimensionless wind shear (Fig. 2.10)
ω_0 rotational angular velocity of a vortex generator
ω planetary rotation rate

APPENDIX B

Small-scale modeling of aeolian phenomena in wind tunnels

The similitude problem for modeling at small scale the effects of blowing particles of dust, sand, or snow is difficult because of the large number of variables. The important variables include:

d	linear deflation or deposition dimension (e.g., depth)
D_p	particle diameter
e	coefficient of restitution of particle
g	gravitational acceleration
H	reference vertical dimension
L_r	reference horizontal dimension
L_i	all other topographic horizontal dimensions
L^*	Monin–Obhukov atmospheric stability length
t	time
U	wind speed at reference height
u_*	surface friction speed
u_{*_t}	surface friction speed at threshold of motion
U_F	particle terminal speed
z_o	surface aerodynamic roughness height
η	all other topographical vertical dimensions
ν	kinematic viscosity
ρ	air density
ρ_p	particle density

According to the rules of dimensional analysis, the 18 preceding variables, with three basic dimensions, can be arranged in 15 dimensionless terms. These terms must be independent for proper similitude; for example, the dimensionless deposition or deflation depth, written below as parameter 1, would be considered as a function of all the rest in the following list:

(1) d/L_r	deposition or deflation depth
(2) D_p/L_r	particle diameter-characteristic length ratio
(3) U^2/gL_r	Froude number
(4) e	coefficient of restitution
(5) U/U_F	ratio of wind speed to particle terminal speed
(6) L_i/L_r, H/L_r, η/L_r	topographical geometric similarity
(7) z_0/L_r	roughness similarity
(8) L^*/L_r	boundary-layer stability similarity
(9) UL_r/ν	Reynolds number
(10) U/u_{*_t}, U/u_*	friction speed ratios
(11) ρ/ρ_p	density ratio
(12) Ut/L_r	time scale

For a true model to be attained, all of the 12 parameters above must have the same value in the model as in full scale. That is clearly impossible, as can be illustrated by considering parameters (2) and (3). For a small value of characteristic length, L_r, the particle diameter, D_p, would have to be so small that all particles would become suspended by turbulence upon becoming airborne. For the Froude number to be satisfied, the wind speed, U, would be far below threshold speed and no particle motion would take place. It is thus necessary to abandon the attempt at a true model. The only way it is possible to obtain realistic quantitative (and perhaps also qualitative) results for a distorted model is to vary the degree of distortion in order to facilitate extrapolation to full scale. The degree of distortion is varied by changing the experimental values of the dimensionless parameters as much as possible by changing particle density and diameter, wind speed and model scale. In addition, interpretation of results is aided by grouping the dimensionless parameters by theoretical means in order to reduce the number of variables.

Satisfaction of parameter (7), the roughness criterion, was first recognized by Jensen (1958) as necessary if appropriate modeling of the boundary layer is to be obtained (see Eq. (2.11)). It is desirable to have as long a test section as possible in order to simulate the atmospheric boundary layer with a naturally thick, fully turbulent layer in the wind tunnel. Others who have discussed the saltation modeling problem include Jensen (1959), Gerdel and Strom (1961), Odar (1962), Isyumov & Davenport (1974), Norem (1975), Kind (1976), and Tabler (1980a).

Transport rate similitude

The Mariner 9 spacecraft which began orbiting the planet Mars in 1971 revealed the presence of many craters on the surface which possess dark streaks extending in the leeward direction. Many of these streaks are probably caused by deflation of small particles resulting from increased shear stress in the crater wake.

Experimental correlation of gross erosional and depositional features near model craters in a wind tunnel was obtained (Iversen *et al.*, 1975*a*) by basing a similitude on rate of mass movement. The transport rate similitude is based on the theoretical particle mass transport rate,

$$q \propto \frac{\rho}{g} u_*^2 (u_* - u_{*_t}) \tag{3.27}$$

The rate at which an area is covered (or deflated) by windblown material can be expressed as

$$\frac{dA_d}{dt} = \frac{\text{area}}{\text{mass}} \times \frac{\text{mass}}{\text{time}} = \frac{1}{\rho_p d} \times q_d L_r \tag{B.1}$$

Similarly, the volume rate is

$$\frac{dV}{dt} = \frac{q_d L_r}{\rho_p} \tag{B.2}$$

and the cross-sectional area rate (perpendicular to wind direction) is

$$\frac{dA_c}{dt} = \frac{q_d}{\rho_p} \tag{B.3}$$

where q_d is the rate of deposition or deflation.

If these equations are combined with the equation for mass rate of movement, Eq. (3.27), the following equations result:

$$\frac{d\dfrac{A_d}{L_r^2}}{d\dfrac{u_* t}{L_r}}, \frac{d\dfrac{V}{L_r^2 \mathrm{H}}}{d\dfrac{u_* t}{L_r}}, \text{ or } \frac{d\dfrac{A_c}{L_r H}}{d\dfrac{u_* t}{L_r}} \propto \frac{\rho}{\rho_p}\left(\frac{u_*^2}{gH}\right)\left(1 - \frac{u_* t}{u_*}\right) \tag{B.4}$$

These equations provide a basis for analysis of saltation simulation by combining forms of several of the similitude parameters previously listed (parameters 3, 6, 10, 11, and 12).

Particle trajectory similitude

It has been shown (Odar, 1962; Iversen *et al.*, 1976*c*) that some of the dimensionless parameters can be combined through the dimensionless forms of the particle trajectory equations of motion. Two possible terms are

$$\frac{C_D L_r}{\rho_p D_p} \quad \text{and} \quad \frac{U^2 H}{g L_r^{\,2}}$$

which are combinations of parameters (2) and (11) for the first term and (3) and (6) for the second. These combined parameters in the model would be slightly closer to full-scale values than the original parameters if a low density particle and a vertical exaggerated scale were used. Ripple geometry would not be expected to scale correctly if the two parameters above were not satisfied. As long as only the gross features of depositional and deflational patterns are important, however, the two parameters can be ignored unless ripple wave lengths become long enough to distort the gross features.

Effect of Froude number and density ratio

The equivalent roughness height in saltation, z'_o, has not generally been determined. However, Owen (1964) shows that it is proportional to u_*^2/g, i.e.,

$$z'_o \propto \frac{u_*^{\,2}}{2g} \tag{B.5}$$

If the roughness height–characteristic length ratio z'_o/H is important, then the Froude number u_*^2/gH is also important, according to Eq. (B.5).

Consider a case in which deposition is governed by a topographical feature of height, H, and lateral width, L_r (such as a rock or leeward- or windward-facing escarpment). The volumetric rate at which material is deposited should be proportional to the product of wind speed (or friction speed), volumetric concentration, and frontal area. The maximum mass concentration possible in saltation is usually considered to be the fluid density, ρ, so that the volumetric concentration is ρ/ρ_p. Thus,

$$\mathrm{d}V/\mathrm{d}t \propto (\rho/\rho_p) H L_r u_* \tag{B.6}$$

and the mass deposition rate would be

$$q_d L_r = \rho_p\, \mathrm{d}V/\mathrm{d}t \propto \rho H L_r u_* \tag{B.7}$$

From Eq. (3.20), the ratio of deposition rate to total mass flow rate is thus

$$q_d/q \propto \rho H L_r u_* / (\rho u_*^{\,3} L_r/g) = gH/u_*^{\,2}$$

or

$$q_d \propto q/(u_*^{\,2}/gH) \tag{B.8}$$

so that a dimensionless deposition rate would be expected to decrease with increase in Froude number.

Experiments with different values of particle density have shown the deposition rate to be a function of density ratio as well (Iversen, 1980*a*). If the particle concentration within the deposition region is volume-limited rather than mass-limited, then the deposition rate becomes

$$q_d \propto q/(\rho u_*^2/\rho_p g H) \tag{B.9}$$

A series of wind tunnel experiments was performed in a manner similar to that employed in the model crater experiments to determine the snow drift deposition rate associated with an interstate highway grade separation structure (Iversen, 1980*a*). Values of friction speed were estimated from the wind speed measurements using Eq. (3.39) for velocity profile. The data for mass deposition rate from ten experiments were curve fit by linear regression to compare Eqs. (B.8) and (B.9). The results are as follows:

$$q_d g/\rho u_*^3 = 1.4\,(u_*^2/gH)^{-0.49},\ R = 0.74 \tag{B.10}$$

$$q_d g/\rho u_*^3 = 0.064\,(\rho u_*^2/\rho_p g H)^{-0.41},\ R = 0.90 \tag{B.11}$$

It is clear that the correlation (R=correlation coefficient) with densimetric Froude number (Eq. (B.11)) is better than with the ordinary Froude number (Eq. (B.10)). However, the dimensionless deposition rate is not as strong a function of Froude number or densimetric Froude number as predicted by Eqs. (B.8) or (B.9). By multiple regression, treating the Froude number and density ratio separately, the best fit was found to be

$$q_d g/\rho u_*^2\,(u_* - u_{*_t}) = 0.0696\,(\rho/\rho_p)^{-2/5}\,(u_*^2/gH)^{-7/9},\ R = 0.97 \tag{B.12}$$

The form of the best fit to any set of data is undoubtedly a strong function of the geometry of the boundary-layer obstruction. Since only a very few experiments of this type have yet been performed, much work remains to establish the mass transport similitude on a firmer foundation. Eq. (B.12) is used in plotting the data as shown in Fig. (B.1). Full-scale values of the density ratio–Froude number parameter are approached at the left side of the figure.

The similitude function

The mass transport function, Eq. (B.4) provides a basis for analysis of snow drifting simulation. Thus, considering parameters 3, 5, 6, 7, 9, 10, 11, 12 and Eq. (B.4), the similitude problem reduces to

$$\frac{d(A_d/L_r^2)}{d(u_* t/L_r)} = \frac{\rho u_*^2}{\rho_p g H}\left(1 - \frac{u_{*_t}}{u_*}\right) f\left(\frac{\rho}{\rho_p}, \frac{u_*^2}{gH}, \frac{u(H)L_r}{\nu}, \frac{U_F}{u_{*_t}}, \frac{\eta}{H}, \frac{L_i}{L_r}, \frac{z_0}{L_r}, \frac{H}{L_r}\right) \tag{B.13}$$

Eq. (B.13) allows for the possibility of vertical geometric distortion and uses Jensen's criterion for normal roughness modeling. It does not allow for the satisfaction of many of the original dimensionless terms. Thus, the saltation model is still highly distorted. By varying particle size and density, model speed, scale, and possibly vertical distortion, a range in degree of distortion can result so that full-scale predictions can be made. The effects of distortion become particularly serious, however, if small-scale features, such as surface ripples, become large enough to obscure or interfere with gross drift geometry associated with important topographical features of the model. Examples of model results and further discussion of the similitude parameters may be found in Greeley *et al.* (1974*a*), Iversen *et al.* (1973, 1975*a*, 1976*c*), and Iversen (1979, 1980*a*,*b*, 1981, 1982).

Fig. B.1. Rate of deposition of simulated snow on a model roadway, as a function of density ratio–Froude number parameter. (Iversen, 1982.)

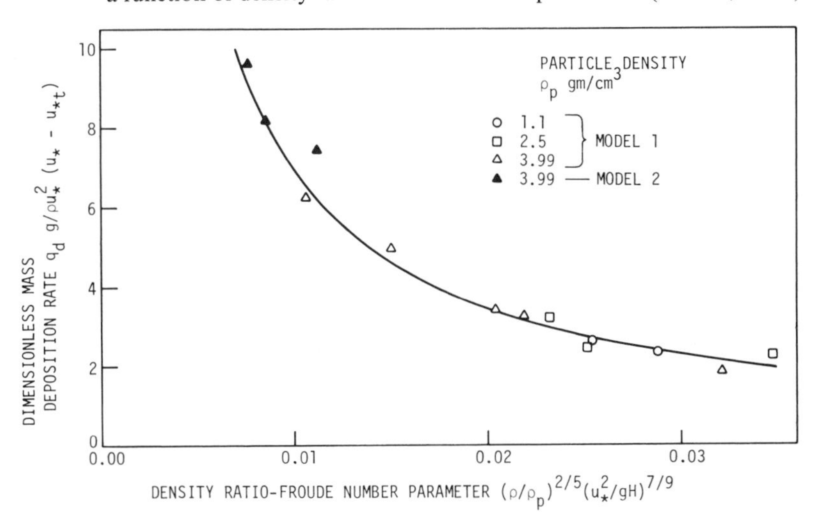

GLOSSARY

(countries or regions in parentheses indicate local usage or origin)

Adiabatic: negligible heat exchange between the atmosphere and the surroundings.
Adiabatic lapse rate: rate of decrease of temperature with altitude for a parcel of dry air lifted adiabatically (without heat transfer) through an atmosphere in hydrostatic equilibrium.
Aeolian: pertaining to wind; especially said of rocks, soils, and deposits (such as loess, dune sand, and some volcanic tuffs) whose constituents were transported (blown) and laid down by atmospheric currents, or of landforms produced or eroded by the wind, or of sedimentary structures (such as ripple marks) made by the wind, or of geologic processes (such as erosion and deposition) accomplished by the wind (Gary *et al.*, 1972).
Aerosol: a suspension in air of small particles (submicron size or of the order of microns) of a liquid and/or solid.
Aggregate: collection of grains held together by various mechanisms including electrostatic bonding, cementation, etc.
Albedo: that fraction of total incident light reflected in all directions.
Anchored dune: sand dune covered with vegetation whose movement is arrested. Also referred to as *Established dune* (Stone, 1967).
Angle of repose (rest): maximum angle at which loose material, such as talus or sand, is stable (Stone, 1967).
Anticyclone: high-pressure atmospheric circulation whose rotation direction is clockwise in the northern hemisphere.
Attrition: reduction of particle in mass.
Avalanche bedding: steeply inclined bedding in barchan and related dune forms produced by avalanche of sand down the slipface of the dune (Stone, 1967).
Bajada (Bahada): blanket deposit of alluvium at the base of desert mountain slopes, formed by the coalescing of alluvial fans or cones (Stone, 1967).
Barchan (or Barcan, Barchane, Barkan, Barkhan): crescent-shaped sand dune formed, generally, by moderate winds and with a moderate sand supply. Windward slope is gentle and lee slope or slipface is at the angle of repose of dry sand (Stone, 1967).
Bedform: a regularly repeated pattern which forms on a surface because of the shearing action of a fluid; includes features such as ripples and dunes (Wilson, 1972*a*).

Blowout: trough or saucer-shaped hollow formed by wind erosion in preexisting dune or sand deposits (Stone, 1967).

Blowout dune: sand accumulations derived from blowout troughs or depressions, especially where the accumulation is of large size and rises to considerable height above the source area (Stone, 1967).

Bora: cold, strong wind falling from mountain elevations to sea level, particularly in the Adriatic Sea area.

Boundary layer: layer of fluid flowing near a surface, which, because of the effects of viscosity, flows at reduced speed compared to the relatively inviscid flow outside the layer.

Brink (of dune): sharp break in slope to the slipface.

Canyon dune: sand dune formed in a box canyon (Stone, 1967).

Cavity flow: flow within a sharp-edged two- or three-dimensional concave pocket situated in an otherwise smooth surface.

Chevron dune: V-shaped dune formed by strong winds of constant direction where wind is in conflict with vegetation (Stone, 1967).

Chinook: warm, dry wind on the eastern side of the Rocky Mountains of North America.

Clay: particles 1/256 mm and smaller.

Clay dune: aeolian accumulation composed of clay fragments and particles (Stone, 1967).

Clay pan: dry lake or playa with a hard, sun-baked clay surface (Stone, 1967).

Climbing dune: sand piled up against a cliff or mountain slope by the wind (Stone, 1967).

Coastal desert: arid land bordering the sea as the Atacama, Sinai, and Namib deserts (Stone, 1967).

Coefficient of restitution: ratio of the relative speed of two spheres after direct impact to that before impact.

Cold desert: arid region in which the mean annual temperature is less than 18 °C, as in the Gobi Desert. Also, Arctic and Antarctic regions where plant and animal life are restricted by low temperatures (Stone, 1967).

Complex dune: sand accumulation formed by the intersection of two or more dune types.

Compound dune: sand accumulation formed by the intersection of two or more of the same dune type.

Concentration: mass per unit volume of particles in a saltating layer or in turbulent suspension.

Coppice dune: small sand hummocks or mounds that have accumulated around plants (McKee, 1982).

Coriolis force: pseudoforce in a reference coordinate system, which is rotating with respect to an inertial frame. The pseudoforce is proportional to the difference between the acceleration with respect to the rotating system and that with respect to the inertial system.

Creep: slow, forward motion of grains that are too large to be lifted by the wind, whose motion is produced by impact from saltating grains.

Crescentic dune: a barchan dune (Stone, 1967).

Crest (of dune): highest part of dune.

Cyclone: (1) low-pressure atmospheric circulation whose rotation direction is counterclockwise in the northern hemisphere; (2) counterpart to the hurricane in the Indian Ocean.

Deflation: removal of loose, granular particles by the wind, caused by less material coming upwind than leaving downwind.

Deflation armor: surface mosaic of pebbles; a *Desert pavement* (Stone, 1967).

Deflation basin: hollow formed by removal of sand and dust by aeolian action and commonly with a rim of resistant material surrounding depression (Stone, 1967).

Deflation ripple: unusually large wind ripples composed in part of material approaching, or attaining, granule size (2–4 mm). Refer to *Granule ripple* (Stone, 1967).

Deposition: piling up of loose, granular particles by the wind, caused by more material coming upwind than leaving downwind.

Desertification: process of conversion of land to desert.

Desert lacquer: surface stain or crust of manganese or iron oxide on gravel and rock surfaces. See *Desert varnish* and *Desert patina* (Stone, 1967).

Desert mosaic: a desert pavement, particularly applied to those with tightly interlocking fragments. Refer to *Desert pavement* (Stone, 1967).

Desert patina: dark brown and black surface coating of manganese or iron oxide on rock surfaces in desert regions. See *Desert varnish* (Stone, 1967).

Desert pavement: coarse, angular-to-subround rock fragments on undissected portions of alluvial fans, terraces, and the valley floors, arranged in such a way as to make a smooth and interlocking surface. Formed by the continual removal of fine material by the wind and by rain splash and sheetwash leaving a mosaic of fragments. Also see, *Reg* and *Desert mosaic* (Stone, 1967).

Desert polish: smooth and polished surfaces on rocks, formed by the action of windblown sand and dust (Stone, 1967).

Desert varnish: surface stain or crust of manganese or iron oxide which characterizes many exposed rock surfaces in deserts. It coats ledges or rocks, as well as boulders and pebbles, imparting a brown or black color and a glisteny luster to these surfaces. See *Desert lacquer* and *Desert patina* (Stone, 1967).

Draa: very large sand dune.

Drag: aerodynamic force component parallel to wind direction.

Dreikanter: wind faceted and polished pebble exhibiting three distinct edges. A dreikanter resembles a Brazil nut (Stone, 1967).

Dry playa (or Dry type playa): dry lake with a hard, dense, sun-baked surface located in a basin in which subsurface water can escape and hence the water table is at a considerable distance beneath the surface. Refer to *Clay pan* (Stone, 1967).

Dry quicksand: surface of a sand accumulation which suddenly sinks when loaded, due to the alternation of firmly compacted sand with loose sand (Stone, 1967).

Dry valley: one originally carved by running water but which no longer has a running stream (Stone, 1967).

Dune: any accumulation of sand-size, windblown material (Stone, 1967).

Dune lake: lake formed, as a result of the blocking of mouths of streams by wind or wave transport (Stone, 1967).

Dune valley: area with furrows or swales between dune ridges (Stone, 1967).

Duricrust: Case-hardened crust of soil in semiarid climates, formed by precipitation of aluminous, siliceous, and/or calcareous salts at the surface as rising groundwater evaporates (Stone, 1967).

Dust devil: swirling, vertical updraft of air developed by local heating of the air above the flat desert floor. They are most common on calm summer and autumn days around dry lakes or on the valley floor (Stone, 1967).

Dust storm: strong wind carrying large clouds of dust or silt and larger than a *Dust devil* (Stone, 1967).

Dust whirl: *(see Dust devil)*.

Einkanter: wind faceted and polished pebble on which only one face or facet has been carved (Stone, 1967).

Ekman layer: layer of transition in the atmospheric boundary layer from the surface where friction forces predominate to the relatively inviscid flow above where pressure gradient and Coriolis forces predominate.

Environmental wind tunnel: a wind tunnel with an unusually long test section for the purpose of small-scale simulation of the mean wind and turbulence characteristics of the atmospheric boundary layer.

Eolation: action of the wind on the land in the transportation of dust and sand and the formation of dunes (Stone, 1967).

Eolian: *(see Aeolian)*.

Eolianite: all consolidated sedimentary rocks derived from wind-deposited sediments.

Eolian Phase: that portion of the geomorphic dune cycle marked by diminished vegetal control and active dune growth (Stone, 1967).

Equivalent roughness height: the value of height in the logarithmic boundary-layer velocity profile at which the wind speed is zero; the height depends upon the size, shape, and spacing of surface roughness elements and, for quiescent closely packed sand grains, is approximately equal to 1/30 the average grain diameter.

Erg (or Ergh): vast region covered with sand (sand 'sea').

Established dune: sand dune covered with vegetation, whose movement is arrested. Also referred to as *Anchored dune* (Stone, 1967).

Exfoliation: process by which successive sheets or layers of rocks are split or peeled off from the parent mass and caused by removal of overburden, chemical changes or other causes (Stone, 1967).

Falling dune: sand accumulation sloping at the angle of repose of dry sand at the base of a cliff, on a mountain slope, or in a valley, which is formed as sand is blown off a mesa top or over a cliff face or steep slope (Stone, 1967).

Fish hook dune: dune consisting of a long sinuous sigmoidal ridge forming the shaft and a well-defined crescent forming the hook. A *Hooked dune* (Stone, 1967).

Fixed dune: non-migratory dune fixed by vegetation or by cementation (Stone, 1967).

Flow separation: when the primary flow near the surface no longer continues to follow the surface contour because of sudden change in contour or because of insufficient flow momentum to allow flow to proceed into a region of increasing pressure.

Foehn: warm, dry wind on the lee side of mountain range, due to adiabatic compression as air descends the slopes.

Forset beds (aeolian): deposits formed on the slipface of a dune.

Froude number: dimensionless number in fluid mechanics; important when gravitational effects are critical; proportional to the ratio of inertial to gravitational forces.

Fugitive dust: dust generated by mechanical activities, such as from passing vehicles or livestock.

Fulgurite: long tube of fused sand resulting from lightening strikes into sand dunes.

Fulji (or Falj or Fulje): deep horseshoe depression between dunes; interdune hollow formed when a barchan encroaches on another barchan in such a way that the horns of one barchan join with the foot of the leeward dune (North Africa) (Stone, 1967).

Geostrophic wind: wind speed at the top of the Ekman layer, at which the Coriolis and pressure–gradient forces are in balance.

Ghourd: sand dune (Egypt) (Stone, 1967).

Gibber: residual fragment or lag gravel (Australia) (Stone, 1967).

Gibber plain: desert pavement; a rock-floored plain (Australia) (Stone, 1967).

Gobi: flat desert floor; rock-floored desert plain; the Great Desert (Central Asia) (Stone, 1967).

Goz: a long sand ridge (Egypt) (Stone, 1967).

Granule ripple: large wind ripples built in part of granule-size grains (2–4 mm). Forms where local lag concentrations of grains mostly larger than 1 mm in diameter develop. They can have wavelengths of 3 m and heights of up to 25 cm (Stone, 1967).

Grus: the fragmental products of *in situ* granular disintegration of granite and granitic rocks (Bates & Jackson, 1980).

Haboob: severe dust storm; dust-laden winds of great intensity; result from density currents generated by cool, descending air masses.

Hairpin dune: sand accumulation with deep U-shape in plan-view, formed where wind is in conflict with vegetation, and where winds are strong and of constant direction (Stone, 1967).

Hammada (or Hamada): widespread area of flat, bare rock floor swept almost clear of sand; desert surface developed on bedrock or bedrock covered with a thin veneer of pebbles, the sand having been carried away by the wind (Stone, 1967).

Hertzian fracture: circular crack on rock surface, resulting from impact of rounded grains.

Hooked dune: dune consisting of a long sinuous sigmoidal ridge forming a shaft with a hook consisting of a well-defined crescent. Refer to *Fish hook dune* (Stone, 1967).

Horseshoe vortex: the U-shaped vortex wrapped around a boundary-layer obstruction. The arms of the vortex trail downwind and can persist for many obstruction diameters. The vortex is created from the upwind laterally oriented vorticity being forced to 'pile-up' on the windward side of the obstruction and being stretched and reoriented in the downwind direction in the trailing arms of the vortex.

Hurricane: a tropical cyclone spawned in the Atlantic Ocean and of great intensity and breadth, with strong winds near the center of the vortex. Wind speeds greater than 32.5 m/sec are said to be of hurricane force.

Impact creep: movement of material along the surface by being struck by the impact of grains.

Impact threshold: lowest wind speed or friction speed at which continuous grain motion is possible when the motion is triggered by airborne grains introduced from upwind.

Isobar: line drawn through all points of equal pressure; in the atmosphere referenced to a given altitude, such as sea level.

Jardang: *(see Yardang)*.

Knudsen number: ratio of molecular mean free path in the atmosphere to characteristic length, such as particle diameter.

Lag deposit: residual cover of pebbles and gravel remaining after finer material has been removed by the wind (Stone, 1967).

Lag gravel: residual accumulation of coarser particles from which the fine material has been deflated. Similar to *Desert pavement* but more restricted in terms of its extent and continuity (Stone, 1967).

Lee: sheltered, or the part of an object turned away from the wind.

Lee dune: general term for a dune formed to the lee of an obstruction (Stone, 1967).

Lee waves: wave disturbance caused by, and stationary with respect to, a barrier to the flow.

Lift: aerodynamic force component perpendicular to wind direction.

Loam: soil composed of about equal amounts of clay, silt, and sand.

Loess: unstratified deposits of loosely arranged, angular grains of silt considered to be deposited by the wind.

Longitudinal dune: any linear dune ridge, usually more or less symmetrical in cross-profile, which extends parallel to the direction of the dominant wind (Stone, 1967).

Lunette: crescent-shaped dune ridge or crescentic mound of loamy material on the lee side of lakes and rarely more than 6–10 m high (Australia) (Stone, 1967).

Magnus force: aerodynamic force component on a rotating body perpendicular to wind direction and axis of rotation, caused by the rotation.

Megabarchan: giant barchan dunes that may be 100 m high (Stone, 1967).

Megaripple: large ripples with wavelengths of 10 m or more.

Migration: movement of a dune due to transfer of sand from the windward to leeward side (Stone, 1967).

Migratory dune: any dune which moves or is displaced by wind activity (Stone, 1967).

Monin-Obhukov stability length: length scale characterizing degree of stability in the atmospheric boundary layer, defined as $L^* = C_p \theta_o u_*^3 / kg\bar{H}$ where θ_o is surface potential temperature, k, is von Karman's constant, and $\bar{H}$ is heat flux at the surface.

Monsoon: large-scale wind system which strongly influences climate of large regions, and in which flow direction reverses from winter to summer.

Mushroom rock: mushroom-shaped rock formed by differential weathering and erosion. The thin base of the mushroom is composed of soft sedimentary material which cannot resist the effects of disintegration and decomposition as well as the more resistant upper layers of the rock mass. Also, refer to *Pedestal rock* (Stone, 1967).

Musical sand: *(see Singing sand)*.

Oghurd: more-massive summits above the general sand dune level; a massive mountainous dune formed by some underlying rocky topographical feature (North Africa) (Stone, 1967).

Optical depth: in calculations of transfer of radiant energy, a parameter related to the mass of a given absorbing or emitting material, lying in a vertical column of unit-section area.

Orographic desert: arid land on the lee side of a mountain or a mountain range, which intercepts moisture-laden air. Refer to *Rain shadow desert* (Stone, 1967).

Parabolic dune: long scoop-shaped hollow of sand with points tapering to the windward and whose groundplan approximates the form of a parabola (Stone, 1967).

Parna: aeolian clay occurring in sheets (Butler, 1956).

Pedestal rock (or Boulder): isolated mass or rock resting or balanced on a smaller rock or pedestal (Stone, 1967).

Pediment: slightly inclined rock plain thinly veneered with fluvial gravels; a rock-carved plain formed as desert mountains retreat under the influence of plantation by streams, sheetwash and rillwash and backweathering (Stone, 1967).

Phi-units: measurement of particle diameter, D_p (in mm) $=(0.5)^{\phi}$.

Playa: from Spanish '*beach*'. An extremely flat, vegetation-free area of clay and silt in the lower-most portions of basins of interior drainage in arid and semiarid regions. Formed by temporary lakes which quickly evaporate leaving behind fine sediment. Also referred to as *Dry lake*, *Clay pan*, *Salt flat*, or *Salt pan* (Stone, 1967).

Plinth: lower and outer portion of a seif dune beyond the slipface boundaries, which have never been subjected to sand avalanches (Stone, 1967).

Polar desert: a high-latitude desert where water is frozen in ice sheets and is not available for plant growth (Stone, 1967).

Pyroclastic: *(see Tephra)*.

Rain shadow: region of diminished rainfall on the lee side of a mountain or mountain range (Stone, 1967).

Rain shadow desert: arid land on the lee side of a mountain or mountain range. See *Orographic desert* (Stone, 1967).

Reg: desert pavement. Usually small rounded pebbles, tightly packed together. See *Desert mosaic* and *Pebble armor* (Stone, 1967).

Reg desert: extensive areas of desert pavement in an arid region (Stone, 1967).

Regolith: general term for the zone of erosion products covering bedrock; includes alluvium, weathering residues, glacial aeolian deposits, and impact cratering deposits.

Reynolds number: a dimensionless number which is proportional to the ratio of inertial force to viscous force in a fluid flow and which is, therefore, a measure of the relative importance of viscosity.

Rhourd: dune of pyramidal or roughly pyramidal form (Stone, 1967).

Richardson number: a dimensionless number which is a measure of stability or instability of the atmospheric boundary layer. The flux Richardson number is the ratio of turbulent kinetic energy produced by buoyancy (instability) to that produced by wind shear.

Ripple: small-scale aeolian or aqueous bedform of regularly repeated pattern.

Ripple index: ratio of wavelength to wave height.

Rising dune: sand accumulation blown up against the face of a steep slope, mountain flank or cliff. Refer to *Climbing dune* (Stone, 1967).

Roaring sand: dune sand which when disturbed sets up a low roaring sound that sometimes can be heard for a distance of half a kilometer. See *Singing sand* (Stone, 1967).

Rossby number: a dimensionless number which is the ratio of inertial force to Coriolis force for a rotating fluid system.

Sabkha: salt flat; a playa or salt-encrusted dry lake (Arabia) (Stone, 1967).

Saltation: bouncing or leaping movement of rock particles, usually sand-size, carried by wind or water currents (Stone, 1967).

Sand: particles 1/16–2 mm in diameter.

Sand avalanche: movement of large masses of sand down a dune face when the angle of repose is exceeded, or when the dune is disturbed (Stone, 1967).

Sand drift: general term for all windblown sand. Also sand accumulation on steep slopes or in valleys; sand deposits formed to the lee of a gap between two obstructions (Stone, 1967).

Sand dune: ridge or pile of sand resulting from aeolian action (Stone, 1967).

Sand flood: vast body of sand migrating or borne along a desert, especially in the Arabian deserts (Stone, 1967).

Sand glacier: sand accumulations formed as broad, fan-shaped plains that develop where sand is blown up the sides of mountains and through passes and saddles to spread out on the opposite side (Stone, 1967).

Sand sea: extensive sand deposits consisting of several composite dune types and with absence of travel lines or directional indicators (Stone, 1967).

Sand shadow: wherever wind velocity is checked by a fixed obstruction, sand may strike the obstruction and then fall at its base or be swept into the lee of the obstruction and accumulate as a streamlined mound (Bloom, 1978).

Sand sheet: sand area marked by an extremely flat surface broken only by small sand ripples (Stone, 1967).

Sandstorm: transportation of sand particles through the atmosphere, the preponderance of which are carried close to the ground (Stone, 1967).

Sand streak: patches of sand lacking slipfaces and having sharply defined edges; often associated with topographical obstacles.

Sand stream: sand occurrence along beds of small creeks or spread out at the mouths of gullies. Formed by run-off from torrential rains which are overloaded with a large quantity of sand (Stone, 1967).

Sand strip: long, narrow strip of sand which extends for a long distance downwind from each horn (Stone, 1967).

Sastrugi: wind-eroded formations of a moist or cohesive granular surface – usually refers to formation in snow and often characterized by irregularly sculpted patterns showing effects of stratified layers of particles.

Scale height: altitude in an isothermal atmosphere for which the pressure has decreased from the surface value by a factor 1/e (where e is the base of the natural logarithm).

Sebkha: *(see Sabkha)*.

Seif (or Sif): variety of longitudinal dune or long dune chain oriented in the direction of wind movement. Term originated in North Africa but is applied in North America and other areas to similar dunes (Stone, 1967).

Serir: desert pavement comprised of coarse, rounded pebbles and stones rather than gravel. Analogous to *Reg* but much older (Stone, 1967).

Shear stress: *(see Surface shear stress)*.

Shrub-coppice dune: *(see Coppice dune)*.

Sigmoidal dune: S-shaped, steep-sided, sharp-crested dune formed under the influence of alternating and opposing winds of roughly equal velocities. It is a transitional form between the crescentic dune and some of the dune complexes (Stone, 1967).

Silt: particle 1/256–1/16 mm in diameter.

Similitude: study of comparative characteristics between large-scale and small-scale physical systems.

Singing sand: sand and sand dune which when disturbed if walked upon or when sand slides down the slipface, sets up a musical tone or humming or booming sound. Also known as *Acoustical*, *Barking*, *Booming*, *Musical*, *Roaring*, *Sonorous*, *Sounding*, *Whispering*, and *Whistling sands* (Stone, 1967).

Sirocco (or Scirocco): hot desert wind in the Sahara and particularly in Syria and Israel. Wind blows from the desert over the Mediterranean (North Africa) (Stone, 1967).

Slipface: steep face on the lee side of a dune which stands at the angle of repose of sand (Stone, 1967).

Soil erodibility: potential loss of soil, due to wind erosion; expressed as mass per unit area per unit time.

Source-bordering lee dune: dune ridge developed on the lee side of an area of loose sand and where the wind may be gentle or strong; not necessarily of unvarying direction, and the sand supply is constantly replenished. Also referred to as a *Leesource dune*. See *Umbrafon dune* (Stone, 1967).

Star dune: stationary dune, with radial buttresses extending in many directions, which develops in an area where the wind blows from multiple directions (Stone, 1967).

Static threshold: lowest wind speed or friction speed at which continuous grain motion is possible when motion is due only to aerodynamic forces.

Stoss: windward side of an object such as a dune.

Surface friction speed: a quantity having dimensions of length/unit time and which is equal to the square root of surface shear stress divided by fluid density.

Surface shear stress: tangential force per unit area, produced by viscous effect of fluid flowing parallel to a surface.

Suspension: mixture of fine particles within a liquid or gas for which the terminal speed of the particles is small compared to characteristic turbulent fluctuation speed.

Tephra: rock fragment erupted in solid state from volcano.

Terminal speed: speed with which a particle moves relative to a fluid when the resultant force is zero (drag = weight).

Threshold friction speed: value of surface friction speed at initiation of particle motion.

Threshold wind speed: value of wind speed at a given reference height, at initiation of grain motion.

Tornado: an intense atmospheric vortex which can contain wind speeds sufficiently great to cause extensive destruction; the diameters of tornados range from meters to several kilometers.

Traction: mode of transport in which material rolls, slides, or is pushed along the surface.

Transverse dune: asymmetrical dune ridge transverse to the dominant direction of sand transporting winds, in which the leeward slope is at the angle of repose of sand and the windward slope is comparatively gentle (Stone, 1967).

Turbulent diffusion: diffusion of heat or material through a fluid, due to turbulence – a much faster process than can occur by molecular diffusion.

Typhoon: western Pacific equivalent of the hurricane.

Umbracer dune: dune formed under constant wind direction on the lee side of an obstacle that tapers to a point downwind. They may form behind prominent bedrock obstacles or more commonly behind clumps of bushes. For the latter, see *Coppice dune* (Stone, 1967).

Umbrafon dune: dune ridge which develops to the lee side of an area of loose sand. Refer to *Source-bordering lee dune* (Stone, 1967).

U-shaped dune: *(see Parabolic dune)*.

Variable features: albedo markings on planetary surfaces, which change size, shape, or position with time; originally applied to features on Mars.

Ventifact: wind-modified object.

Viscosity: that property of a fluid by which resistance occurs due to shearing motion. For a Newtonian fluid, the shear stress is linearly proportional to the time rate of strain. The coefficient of proportionality is termed the absolute viscosity. The kinematic viscosity is the absolute viscosity divided by fluid density.

Vortex: any flow with closed streamlines, usually such that fluid motion takes place in circular or near-circular paths about some common axis.

Vorticity: mathematically, a vector equal to the curl of the flow velocity vector. A fluid parcel may translate, deform, and rotate. Physically, the vorticity at a point in the fluid is equal to twice the rate of rotation at the point.

Wadi (or Wady): dry river course in the desert (Egypt).

Whaleback: coarse-grained residue or platform built up and left behind by the passage of a succession of seif dunes along the same path. A whaleback may be 1–3 km wide, perhaps 50 m in height and run in a straight line for 300 km (Stone, 1967).

Whirlwind: swirling column of heated air and dust rising above the desert floor. See *Dust devil* (Stone, 1967).

Willy-Willy: a whirlwind or dust devil (Australia) (Stone, 1967).

Winddrift dune (or Windrift dune): linear dune deposited on the lee side of blowout areas (Stone, 1967).

Windkanter: irregularly shaped, wind-polished, grooved or faceted stone (Stone, 1967).

Wind shadow: that portion of a slope, cliff or escarpment which is protected from the direct action of the wind blowing over it (Stone, 1967).

Wind shadow dune: longitudinal dune formed under constant wind direction in the lee of an obstacle, commonly clumps of vegetation and more rarely permanent bedrock obstacles (Stone, 1967).

Wind streaks: lineal surface pattern of aeolian origin.

Yardang (or Yarding): streamlined, aerodynamically shaped elongate hill oriented parallel to the wind; term is derived from the Turkistani word *yar* meaning ridge or steep bank from which material is being removed (McCauley *et al.*, 1977*b*).

Zastrugi: *(see Sastrugi)*.

Zibar: rolling transverse dune ridge of low relief without vegetation cover, or a slipface that is formed between dune complexes and especially in low areas between linear dune belts (Arabia) (Stone, 1967).

REFERENCES

Allen, J.R.L., 1965. 'Scour marks in snow'. *Journal of Sedimentary Petrology*, **35,** 331–8.

Allen, J.R.L., 1968. *Current Ripples*. North-Holland, Amsterdam, 433 pp.

Arvidson, R.E., 1974. 'Wind blown streaks, splotches and associated craters on Mars: statistical analysis of Mariner 9 photographs'. *Icarus*, **21,** 12–27.

Arvidson, R.E., A. Carusi, A. Coradini, M. Coradini, M. Fulchignoni, C. Federico, R. Funicello & M. Salomone, 1976. 'Latitudinal variation of wind erosion of crater ejecta deposits on Mars'. *Icarus*, **27,** 503–16.

Arvidson, R.E., E.A. Guinness & S.W. Lee, 1979. 'Differential aeolian redistribution rates on Mars'. *Nature*, **278,** 533–5.

Arvidson, R.E., T.A. Mutch & K.L. Jones, 1974. 'Craters and associated aeolian features on Mariner 9 photographs: An automated data gathering and handling system and some preliminary results'. *The Moon*, **9,** 105–14.

Aufrere, L., 1930. 'L'orientation des dunes continentales'. *Report Proceedings 12th International Geog. Conference*. Cambridge, July 1928, pp. 220–31.

Bagnold, R.A., 1941. *The Physics of Blown Sand and Desert Dunes*. Methuen, London, 265 pp.

Bagnold, R.A., 1953. 'The surface movement of blown sand in relation to meteorology'. *In: Desert Research, Proceedings International Symposium, Jerusalem. Research Council Israel Special Publication*, **2,** 89–93.

Bagnold, R.A., 1956. 'The flow of cohesionless grains in fluids'. *Philosophical Transactions Royal Society, Ser. A.*, **249,** 239–97.

Bagnold, R.A., 1960. 'The re-entrainment of settled dusts'. *International Journal of Air Pollution*, **2,** 357–63.

Bagnold, R.A., 1973. 'The nature of saltation and of 'bedload' transport in water'. *Proceedings Royal Society London, Ser. A.*, **332,** 473–504.

Baker, V.R., 1982. *The Channels of Mars*. University of Texas Press, Austin, 198 pp.

Barndorff-Nielsen, O., K. Dalsgaard, C. Halgreen, H. Kuhlman, J. T. Møller & G. Schou, 1982. 'Variation in particle size distribution over a small dune'. *Sedimentology*, **29,** 53–65.

Barsukov, V.L., C.P. Florensky & O.V. Nikolaeva, 1981. 'On the rock types on Venus'. *In: Abstracts, International Conference on the Venus Environment*, NASA-Ames Research Center, p. 5.

Bates, R.L. & J.A. Jackson, editors, 1980. *Glossary of Geology*. American Geological Institute, Falls Church, 751 pp.

Baum, W.A., 1973. 'The International Planetary Patrol Program: an assessment of the first three years'. *Planetary and Space Science*, **21,** 1511.

Baum, W.A., R.L. Millis, S.E. Jones & L.J. Martin, 1970. 'The International Planetary Patrol Program'. *Icarus*, **12,** 435–9.

Bayliss, P., 1976. 'NOAA4 satellite picture showing cloud vortices'. *Weather*, **31,** 346.

Beadnell, H.J.L., 1910. 'The sand dunes of the Libyan Desert'. *Geographical Journal*, **35,** 379–95.

Beavers, A.H., 1957. 'Source and deposition of clay minerals in Peorian loess'. *Science*, **126,** 1285.

Becker, D.L. & E.S. Takle, 1979. 'Particulate deposition from dry unpaved roadways'. *Atmospheric Environment*, **13,** 661–8.

Belcher, D., J. Veverka & C. Sagan, 1971. 'Mariner photography of Mars and aerial photography of Earth: some analogies'. *Icarus*, **15,** 241–52.

Belly, P.Y., 1964. *Sand Movement by Wind.* TM No. 1, US Army Coastal Engineering Research Center, 80 pp. (Available from US Army Coastal Engineering Research Center, 5201 Little Falls Road, NW, Washington, DC 20016.)

Binder, A.B., R.E. Arvidson, E.A. Guinness, K.L. Jones, E.C. Morris, T.A. Mutch, D.C. Pieri & C. Sagan, 1977. 'The geology of the Viking Lander 1 site'. *Journal Geophysical Research*, **82,** 4439–51.

Blackwelder, E., 1934. 'Yardangs'. *Geological Society of America Bulletin*, **45,** 159–66.

Blake, W.P., 1855. 'On the grooving and polishing of hard rocks and minerals by dry sand'. *American Journal of Science*, 2nd Series, **20,** 178–81.

Blom, J. & L. Wartena, 1969. 'The influence of changes in surface roughness on the development of the turbulent boundary layer in the lower layers of the atmosphere'. *Journal of the Atmospheric Sciences*, **26,** 255–65.

Bloom, A.L., 1978. *Geomorphology: A Systematic Analysis of Late Cenozoic Landforms*. Prentice-Hall, Englewood Cliffs, New Jersey, 510 pp.

Boning, P., 1927. 'Dust electricity'. *Zeitschrift für Technische Physik*, **8,** 385.

Bosworth, T., 1922. *Geology of the Tertiary and Quaternary Periods in the Northwest Part of Peru.* Macmillan, London, pp. 269–309.

Bowler, J.M., 1973. 'Clay dunes: their occurrence, formation and environmental significance'. *Earth Science Reviews*, **9,** 315–38.

Breed, C.S., 1977. 'Terrestrial analogs of the Hellespontus dunes, Mars'. *Icarus*, **30** (no. 2), 326–40.

Breed, C.S. & W.J. Breed, 1979. 'Dunes and other windforms of Central Australia and a comparison with linear dunes on the Moenkopi Plateau, Arizona'. *In:* Farouk El-Baz & D.M. Warner, editors, Apollo-Soyuz Test Project Summary Science Report, Vol. 11, *Earth Observations and Photography*, NASA SP-412, pp. 319–58.

Breed, C.S., M.J. Grolier & J.F. McCauley, 1979. 'Morphology and distribution of common 'sand' dunes on Mars. Comparison with the Earth'. *Journal of Geophysical Research*, **84,** 8183–204.

Breed, C.S. & T. Grow, 1979. 'Morphology and distribution of dunes in sand seas observed by remote sensing'. *In:* E.D. McKee, editor, *A Study of Global Sand Seas.* US Geological Survey Professional Paper 1052, pp. 253–308.

Breed, C. & A.W. Ward, 1979. *Longitudinal Dunes on Mars* (abstract). NASA Conference Publication 2072, p. 10.

Briggs, G.A., W.A. Baum & J. Barnes, 1979. 'Viking orbiter imaging observations of dust in the martian atmosphere'. *Journal of Geophysical Research*, **84,** 2795–820.

Bryan, K., 1931. 'Wind-worn stones or ventifacts – a discussion and bibliography'. Report Committee Sedimentation 1929–30, *National Research Council Circular*, **98,** pp. 29–50.

Businger, J.A., J.C. Wyngaard, Y. Izumi & E.F. Bradley, 1971. 'Flux profile relationships in the atmospheric surface layer'. *Journal of the Atmospheric Sciences*, **28,** 181–9.

Butler, B.E., 1956. 'Parna – an aeolian clay'. *Australian Journal of Science*, **18,** 145–51.

Butler, B.E., 1974. 'A contribution towards the better specification of parna and some other aeolian clays in Australia'. *Zeitschrift Geomorphologie N. F.*, **20,** 106–16.

Carr, M.H., 1981. *The Surface of Mars*. Yale University Press, New Haven, 232 pp.

Carroll, J.J. & J.A. Ryan, 1970. 'Atmospheric vorticity and dust devil rotation'. *Journal of Geophysical Research*, **75,** 5179–84.

Chamberlain, J.W., 1978. *Theory of Planetary Atmospheres*. Academic Press, New York, 330 pp.

Chepil, W.S., 1945. 'Dynamics of wind erosion: II. Initiation of soil movement'. *Soil Science*, **60** (no. 5), 397–411.

Chepil, W.S., 1950. 'Properties of soil which influence wind erosion: I. The governing principle of surface roughness'. *Soil Science*, **69,** 149–62.

Chepil, W.S., 1958. 'The use of evenly spaced hemispheres to evaluate aerodynamic forces on a soil surface'. *EOS, American Geophysical Union*, **39** (no. 3), 397–403.

Chepil, W.S., 1959. 'Equilibrium of soil grains at the threshold of movement by wind'. *Soil Science Society of America Proceedings, 1959,* 422–8.

Chepil, W.S. & N.P. Woodruff, 1957. 'Sedimentary characteristics of dust storms: II. Visibility and dust concentration'. *American Journal of Science*, **255,** 104–14.

Chepil, W.S. & N.P. Woodruff, 1963. 'The physics of wind erosion and its control'. *Advances in Agronomy*. US Department of Agriculture, **15,** 211–302.

Chorley, R. & B.A. Kennedy, 1971. *Physical Geography, A Systems Approach*. Prentice-Hall, Englewood Cliffs, 370 pp.

Clark, B.C. & A.K. Baird, 1979. 'Chemical analyses of martian surface materials'. *Lunar and Planetary Science*, Vol. X, Planetary Science Institute, Houston, pp. 215–17.

Coffey, G.N., 1909. 'Clay dunes'. *Journal of Geology*. **17,** 754–5.

Coleman, N.L., 1967. 'A theoretical and experimental study of drag and lift forces acting on a sphere resting on a hypothetical streambed. *Proceedings 12th Congress International Association Hydraulic Research*, Ft. Collins, Colorado, **3,** 185–92.

Coleman, N.L., 1972. 'The drag coefficient of a stationary sphere on a boundary of similar spheres'. *La Houille Blanche*, **1,** 17–21.

Coleman, N.L. & W.M. Ellis, 1976*a*. 'Model study of the drag coefficient of a streambed particle'. *Proceedings 3rd Federal Interagency Sedimentation Conference*, Denver, Colorado, pp. 4–12.

Coleman, N.L. & W.M. Ellis, 1976*b*. *A Streambed-particle Model-study Facility using Hydroxyethylcellulose Solutions as a Fluid*. US Department of Agriculture, Agriculture Research Service, ARS-S-147, 8 pp.

Cooke, R.U. & I.J. Smalley, 1968. 'Salt weathering in deserts'. *Nature*, **220,** 1226–7.

Cooke, R.U. & A. Warren, 1973. *Geomorphology in Deserts*. University of California Press, Berkeley, 374 pp.

Cornish, V., 1902. 'On snow-waves and snow-drifts in Canada'. *Geographical Journal*, **20,** 137–73.

Cornish, V., 1914. *Waves of Sand and Snow*. Fisher-Unwin, London, 383 pp.

Counselman, C.C., S.A. Gourevitch, R.W. King, G.B. Loriot & R.G. Prinn, 1979. 'Venus winds and zonal and retrograde below the clouds'. *Science*, **205,** 85–7.

Csanady, G.T., 1972. 'Geostrophic drag, heat and mass transfer coefficients for the diabatic Ekman layer'. *Journal of the Atmospheric Sciences*, **29,** 488–96.

Cutts, J.A., 1973. 'Nature and origin of layered deposits in martian polar regions'. *Journal of Geophysical Research*, **78,** 4231–49.

Cutts, J.A., K.R. Blasius, G.A. Briggs, M.H. Carr, R. Greeley & H. Masursky, 1976. 'North polar region of Mars: Imaging results from Viking 2'. *Science*, **194,** 1329–37.

Cutts, J.A. & R.S.U. Smith, 1973. 'Eolian deposits and dunes on Mars'. *Journal of Geophysical Research*, **78,** 4139–54.

Dietrich, R.V., 1977*a*. 'Impact abrasion of harder by softer materials'. *Journal of Geology*, **85,** 242–6.

Dietrich, R.V., 1977*b*. 'Wind erosion by snow'. *Journal of Glaciology*, **18,** 148–9.

Donahue, T.M., 1979. 'Pioneer Venus results: An overview'. *Science*, **205,** 41–4.

Donahue, T.M., J.G. Hoffman, R.R. Hodges & A.J. Watson, 1982. 'Venus was wet: A measurement of the ratio of deuterium to hydrogen'. *Science*, **216,** 630–4.

Dregne, H.E., 1980. 'Desertification of arid lands'. *Workshop on Desertification*, International Centre for Theoretical Physics, Trieste, Italy, November 1980.

Dyunin, A.K., 1954*a*. 'O mekhanicheskikh usloviiakh erozii snega'. *Trudy Transportino Energetischesgo Instituta* ('The mechanical conditions of snow erosion'). National Research Council of Canada, Technical Translation 1101–963, **4,** 59–69.

Dyunin, A.K., 1954*b*. *Trudy Transportino Energicheskogo Instituta* ('Solid flux of snow-bearing air flow'). National Research Council of Canada, Technical Translation 1102–63, **4,** 71–88.

Dyunin, A.K., 1959. 'Fundamentals of the theory of snowdrifting'. *Isvest. Sibirsk. Otdel. Akad. Nauk SSSR*. National Research Council of Canada, translated in Technical Translation 952–1961, **12,** 11–24.

Dyunin, A.K. & A.A. Komarov, 1954. 'On the construction of snow fences'. *Trudy Transportino Energeticheskogo Instituta*. English Translation in National Research Council of Canada, Technical Translation 1103, 1963, **4**, 111–18.

Dzurisin, D. & K.R. Blasius, 1975. 'Topography of the polar layered deposits of Mars'. *Journal of Geophysical Research*, **82,** 4225–48.

Eckholm, E. & L.R. Brown, 1977. *Spreading Deserts – The Hand of Man.* Worldwatch paper 13, Worldwatch Institute, 40 pp.

Einstein, H.A. & E. El-Samni, 1949. 'Hydrodynamic forces on a rough wall'. *Reviews of Modern Physics*, **21** (no. 3), 520–4.

Ekman, V.W., 1902. *Nyt Magazin for Naturvidenskaberne*, **40** (no. 1).

Elachi, C., W.E. Brown, J.B. Cimino, T. Dixon, D.L. Evans, J.P. Ford, R.S. Saunders, C. Breed, H. Masursky, J.F. McCauley, G. Schaber, L. Dellwig, A. England, H. MacDonald, P. Martin-Kaye & F. Sabins, 1982. 'Shuttle imaging radar experiment'. *Science*, **218,** 996–1003.

El Baz, F., C.S. Breed, M.J. Grolier & J.F. McCauley, 1979. 'Eolian features in the western desert of Egypt and some applications to Mars'. *Journal of Geophysical Research*, **84,** 8205–21.

Ellwood, J.M., P.D. Evans & I.G. Wilson, 1975. 'Small scale aeolian bedforms'. *Journal of Sedimentary Petrology*, **45,** 554–61.

Embabi, N.S., 1982. 'Barchans of the Kharga Depression'. *In:* F. El-Baz & T.A. Maxwell, editors, *Desert Landforms of Southwest Egypt: A Basis for Comparison with Mars*, NASA CR-3611, pp. 141–55.

Evans, J.W. 1911. 'Dreikanter'. *Geological Magazine*, **8,** 334–5.

Finkel, H.J., 1959. 'The barchans of southern Peru'. *Journal of Geology*, **67,** 614–47.

Fisher, C., 1964. 'Something Missing'. *Aerospace Safety*, **20,** 4–7.

Flint, R.F., 1971. *Glacial and Quaternary Geology*. John Wiley, New York, 892 pp.

Florensky, C.P., L.B. Ronca, A.T. Basilevsky, G.A. Burba, O.V. Nikolaeva, A.A. Pronin, A.M. Trakhtman, V.P. Volkov & V.V. Zuzetsky, 1977. 'The surface of Venus as revealed by Soviet Venera 9 and 10'. *Geological Society of America, Bulletin*, **88,** 1537–45.

Florensky, C.P., A.T. Basilevsky, V.P. Kryuchkov, R.D. Kusmin, O.V. Nikolaeva, A.A. Pronin, I.M. Chernaya, Yu.S. Tyuflin, A.S. Selivanov, M.K. Naraeva & L.B. Ronca, 1983. 'Venera 13 and Venera 14: sedimentary rocks on Venus?' *Science*, **221,** 57–9.

Folk, R.L., 1971*a*. 'Longitudinal dunes of the northwestern edge of the Simpson Desert, Northern Territory, Australia: I. Geomorphology and grain size relationships'. *Sedimentology*, **16,** 5–54.

Folk, R.L., 1971*b*. 'Genesis of longitudinal and oghurd dunes elucidated by rolling upon grease'. *Geological Society of America Bulletin*, **82,** 3461–8.

Fox, R. & A.T. McDonald, 1973. *Fluid Mechanics*. Wiley, New York, 630 pp.

French, R.G., P.J. Gierasch, B.D. Popp & R.J. Yerdon, 1981. 'Global patterns in cloud forms on Mars'. *Icarus*, **45,** 468–93.

Frey, H., B.L. Lowry & S.A. Chase, 1979. 'Pseudocraters on Mars'. *Journal of Geophysical Research*, **84,** 8075–86.

Fryberger, S.G. & C. Schenk, 1981. 'Wind sedimentation tunnel experiments on the origins of aeolian strata'. *Sedimentology*, **28,** 805–21.

Gabriel, A., 1938. 'The southern Lut and Iranian Baluchistan'. *Geographical Journal*, **92,** 193–210.

Garvin, J.B., P.J. Mouginis-Mark & J.W. Head, 1981. 'Characterization of rock populations on planetary surfaces: Techniques and a preliminary analysis of Mars and Venus'. *The Moon and Planets*, in press.

Gary, M., R. McAfee, Jr & C.L. Wolf, editors, 1972. *Glossary of Geology*. American Geological Institute, Washington, DC, 805 pp.

Gault, D.E., E.M. Shoemaker & H.J. Moore, 1963. *Spray Ejected from the Lunar Surface by Meteoroid Impact*. NASA TN D-1767, 39 pp.

Gerdel, R.W. & G.H. Strom, 1961. 'Scale simulation of a blowing snow environment'. *Proceedings of the Institute of Environmental Sciences*, **53,** 53–63.

Gierasch, P.J., 1974. 'Martian dust storms'. *Reviews of Geophysics and Space Physics*, **12,** 730–4.

Gierasch, P.J. & R.M. Goody, 1973. 'A model of a Martian great dust storm'. *Journal of the Atmospheric Sciences*, **30,** 169–79.

Gifford, F.A., R.P. Hosker, Jr & K.S. Rao, 1978. 'Diffusion–deposition patterns on Mars'. *Icarus*, **36,** 133–46.

Gillette, D.A., 1977. 'Fine particulate emissions due to wind erosion'. *Transactions American Society for Agricultural Engineers*, **20,** 890–7.

Gillette, D.A., 1981. 'Production of dust that may be carried great distances'. *In:* T.J. Péwé, editor, *Desert Dust: Origin, Characteristics, and Effect on Man*, Geological Society of America Special Paper 186, pp. 11–26.

Gillette, D.A., J.A. Adams, D. Muhs & R. Kihl, 1982. 'Threshold friction velocities and rupture moduli for crusted desert soils for the input of soil particles into the air'. *Journal of Geophysical Research*, **87,** 9003–15.

Gillette, D.A. & P.A. Goodwin, 1974. 'Microscale transport of sand-sized soil aggregates eroded by wind'. *Journal of Geophysical Research*, **79,** 4080–4.

Glennie, K.W., 1970. *Desert Sedimentary Environments*. Elsevier, Amsterdam, 222 pp.

Goldman, A.R., R.G. Cox & H. Brenner, 1967. 'Slow viscous motion of a sphere parallel to a plane wall'. *Chemical Engineering Science*, **22,** 653–60.

Gooding, J.L., 1983. 'Planetary surface weathering'. *In:* M.G. Kivelson, editor, *The Solar System: Observations and Interpretations*, Prentice-Hall, Englewood Cliffs.

Goody, R.M. & J.C.G. Walker, 1972. *Atmospheres*. Prentice-Hall, Englewood Cliffs, 150 pp.

Goudie, A.S., 1978. 'Dust storms and their geomorphological implications'. *Journal of Arid Environments*, **1,** 291–311.

Goudie, A.S., R.U. Cooke & J.C. Doornkamp, 1979. 'The formation of silt from quartz dune sand by salt-weathering processes in deserts'. *Journal of Arid Environments*, **2,** 105–12.

Government Printing Office, 1962. *US Standard Atmosphere*. Washington, DC, 278 pp.

Greeley, R., 1977. 'Aerial guide to the geology of the central and eastern Snake River Plain'. *In:* Greeley *et al.*, editors, *Volcanism of the Eastern Snake River Plain, Idaho*. NASA CR-154621, pp. 59–111.

Greeley, R., 1979. 'Silt-clay aggregates on Mars'. *Journal of Geophysical Research*, **84,** 6248–54.

Greeley, R., 1982. 'Aeolian modification of planetary surfaces'. *In:* A. Coradini & M. Fulchignoni, editors, *The Comparative Study of the Planets*, D. Reidel, Dordrecht, pp. 419–34.

Greeley, R. & J.D. Iversen, 1978. 'Field guide to Amboy lava flow'. San Bernardino

County, California. *In: Aeolian Features of Southern California: A Comparative Planetary Geology Guidebook*, NASA, pp. 23–52.

Greeley, R. & J.D. Iversen, 1983. 'Aeolian processes and features at Amboy lava field California'. *Proceedings UNESCO Workshop on Physics of Desertification*, Trieste, Italy (in press).

Greeley, R., J. Iversen, R. Leach, J. Marshall, B. White & S. Williams, 1984. 'Windblown sand on Venus: preliminary results of laboratory simulations'. *Icarus*, **57,** 112–24.

Greeley, R., J.D. Iversen, J.B. Pollack, N. Udovich & B.R. White, 1974*a*. 'Wind tunnel studies of Martian eolian processes'. *Proceedings Royal Society London*, **Ser. A,** 331–50.

Greeley, R., J.D. Iversen, J.B. Pollack, N. Udovich & B. White, 1974*b*. 'Wind tunnel simulations of light and dark streaks on Mars'. *Science*, **183,** 847–9.

Greeley, R., J.D. Iversen, B.R. White & J.B. Pollack, 1974*c*. 'Aeolian erosion on Mars. Part II: Estimated thickness of surface dust in the Daedalia region of Mars, 1971'. *Geological Society of America, Abstracts with Programs*, **6,** 765–6.

Greeley, R. & R. Leach, 1978. *A Preliminary Assessment of the Effects of Electrostatics on Aeolian Processes.* Reports Planetary Geology Program 1977–78. NASA TM 79729, pp. 236–7.

Greeley, R., R. Leach, B.R. White, J.D. Iversen & J.B. Pollack, 1980*a*. 'Threshold windspeeds for sand on Mars: Wind tunnel simulations'. *Geophysical Research Letters*, **7,** 121–4.

Greeley, R., R.N. Leach, S.H. Williams, B.R. White, J.B. Pollack, D.H. Krinsley & J.R. Marshall, 1982. 'Rate of wind abrasion on Mars'. *Journal of Geophysical Research*, **87,** 10009–24.

Greeley, R., K. Malone, R. Leach, R. Leonard & B.R. White, 1980*b*. *Flux of Windblown Particles on Mars: Preliminary Wind Tunnel Determination.* Reports of Planetary Geology Program – 1980. NASA TM 82385, pp. 278–9.

Greeley, R., R. Papson & J. Veverka, 1978. 'Crater streaks in the Chryse Planitia Region of Mars: early Viking results'. *Icarus*, **34,** 556–7.

Greeley, R., S.H. Williams & J.R. Marshall, 1983. 'Velocities of windblown particles in saltation: preliminary laboratory and field measurements'. *In*: M.E. Brookfield & T.S. Ahlbrandt, editors, *Eolian Sediments and Processes*, Elsevier, Amsterdam, pp. 133–48.

Greeley, R. & A.R. Peterfreund, 1981. 'Aeolian 'megaripples': examples from Mono Craters, California and northern Iceland'. *Geological Society of America Abstracts with Programs*, **13** (no. 7), 463.

Greeley, R. & P.D. Spudis, 1981. 'Volcanism on Mars'. *Reviews of Geophysics and Space Physics*, **19,** 13–41.

Greeley, R., B.R. White, R.N. Leach, J.D. Iversen & J.B. Pollack, 1976. 'Mars: Wind friction speeds for particle movement'. *Geophysical Research Letters*, **3,** 417–20.

Greeley, R., B.R. White, R. Leach, R. Leonard, J. Pollack & J.D. Iversen, 1980*c*. *Venus Aeolian Processes: Saltation Studies and the Venusian Wind Tunnel.* NASA TM–82385, pp. 275–7.

Greeley, R., B.R. White, J.B. Pollack, J.D. Iversen & R.N. Leach, 1981. 'Dust storms on Mars: Considerations and simulations'. In T. Péwé, editor, *Desert*

Dust: Origin, Characteristics, and Effect on Man. Geological Society of America Special Paper 186, pp. 101–21.

Grolier, M., G.E. Ericksen, J.F. McCauley & E.C. Morris, 1974. 'The desert landforms of Peru: A preliminary photographic atlas': US Geological Survey, Interagency Report, *Astrogeology*, **57,** 146 pp.

Hack, J.T., 1941. 'Dunes of the western Navajo country'. *Geographical Review*, **31,** 240–63.

Hagen, L.H. & E.L. Skidmore, 1977. 'Wind erosion and visibility problems'. *Transactions American Society for Aeronautical Engineers*, **20,** 898–903.

Hanna, S.R., 1969. 'The formation of longitudinal sand dunes by large helical eddies in the atmosphere'. *Journal of Applied Meterology*, **8,** 874–83.

Hartmann, W.K. & M.J. Price, 1974. 'Mars: Clearing of the 1971 dust storm'. *Icarus*, **21,** 28–34.

Hastenrath, S.L., 1978. 'Mapping and surveying – dune shape and multiannual displacement'. *In* H.H. Lettau & K. Lettau, editors, *Exploring the World's Driest Climate*, University of Wisconsin-Madison, Institute for Environmental Studies, IES Report 101, pp. 74–88.

Head, J.W. & L. Wilson, 1981. 'Volcanic processes on Venus'. *Lunar and Planetary Science* XIII, Lunar and Planetary Institute, Houston, pp. 312–13.

Hedin, Sven, 1903. *Central Asia and Tibet*. Vols. 1 and 2, Charles Scribners and Sons, New York, 608 pp.

Hedin, Sven, 1905. *Journey in Central Asia 1899–1902*. Lithographic Institute, General Staff Swedish Army, Stockholm, 241 pp.

Hertzler, R.G., E.S.J. Wang & O.J. Wilbers, 1967. 'Martian sand and dust storm experimentation'. *Journal Spacecraft Rockets*, **4,** 284–6.

Hess, S.L., 1973. 'Martian winds and dust clouds'. *Planetary and Space Science*, **21,** 1549–57.

Hess, S.L., 1975. 'Dust on Venus'. *Journal of the Atmospheric Sciences*, **32,** 1076–87.

Hickox, C.F., 1959. 'Formation of ventifacts in a moist, temperate climate'. *Geological Society of America Bulletin*, **79,** 1489–90.

Higgins, C.G., 1956. 'Formation of small ventifacts'. *Journal of Geology*, **64,** 506–16.

Hills, E.S., 1940. 'The Lunette, a new land form of aeolian origin'. *Australian Geographer*, **3,** 15–21.

Hodges, C.A. & H.J. Moore, 1979. 'The subglacial birth of Olympus Mons and its Aureoles'. *Journal of Geophysical Research*, **84,** 8061–74.

Holm, D.A., 1960. 'Desert geomorphology in the Arabian Peninsula'. *Science*, **132,** 1369–79.

Hovde, M.R., 1934. 'The great duststorm of November 12, 1933'. *Monthly Weather Review*, **62,** 12–13.

Howard, A.D., 1977, 'Effect of slope on the threshold of motion and its application to orientation of wind ripples'. *Geological Society of America Bulletin*, **88,** 853–6.

Howard, A.D., J.A. Cutts & K.R. Blasius, 1982. 'Stratigraphic relationships within Martian polar cap deposits'. *Icarus*, **50,** pp. 161–215.

Hsu, C.T. & B. Fattahi, 1976. 'Mechanism of tornado funnel formation'. *Physics of Fluids*, **19,** 1853–7.

Huffman, G.G. & W.A. Price, 1949. 'Clay dune formation near Corpus Christi, Texas'. *Journal of Sedimentary Petrology*, **19,** 118–27.

Hunter, R.E., 1977. 'Basic types of stratification in small eolian dunes'. *Sedimentology*, **24,** 361–87.

Hunter, W.A., 1979. 'Ventifacts on Garnet Hill'. Unpubl. senior thesis, Department of Earth Sciences, California State Polytechnic University, Pomona, 223 pp.

Idso, S.B., 1976. 'Dust storms'. *Scientific American*, **235,** 108–14.

Idso, S.B., 1981. 'Climatic change: The role of atmospheric dust'. *In:* T. Péwé, editor, *Desert Dust: Origin, Characteristics, and Effect on Man.* Geological Society of America Special Paper 186, pp. 207–15.

Isyumov, N. & A.G. Davenport, 1974. 'A probabilistic approach to the prediction of snow loads'. *Canadian Journal of Civil Engineering*, **1,** 28–49.

Iversen, J.D., 1979. 'Drifting snow similitude'. *Journal of the Hydraulics Division, Proceedings, American Society of Civil Engineering*, **105,** 737–53.

Iversen, J.D., 1980*a*. 'Wind tunnel modeling of snow fences and natural snow fence controls'. *Proceedings Eastern Snow Conference*, pp. 106–24, published by Eastern Snow Conference, Downsview, Ontario, Canada, Barry Goodison, editor.

Iversen, J.D., 1980*b*. 'Drifting snow similitude – transport rate and roughness modeling'. *Journal of Glaciology*, **26,** 393–403.

Iversen, J.D., 1981. 'Comparison of wind-tunnel model and full-scale fence drifts'. *Journal Wind Engineering Industrial Aeronautics*, **8,** 231–49.

Iversen, J.D., 1982. 'Small-scale modeling of snow drift phenomena'. *Wind Tunnel Modeling for Civil Engineering Applications*, US National Bureau of Standards, pp. 522–45.

Iversen, J.D. & R. Greeley, 1978. *Atmospheric and Wind Tunnel Experiments of the Amboy Crater Sand-covered Lava Flow.* Iowa State University Engineering Research Institute Report ERI-78235, 79 pp.

Iversen, J.D. & R. Greeley, 1984. 'Martian crater dark streak lengths – explanation from wind tunnel experiments'. *Icarus*, in press.

Iversen, J.D., R. Greeley & J.B. Pollack, 1976*b*. 'Windblown dust on Earth, Mars, and Venus'. *Journal of the Atmospheric Sciences*, **33,** 2425–9.

Iversen, J.D., R. Greeley, J.B. Pollack & B.R. White, 1973. 'Simulation of martian eolian phenomena in the atmospheric wind tunnel'. *Space Simulation, NASA Special Publication*, **36,** 191–213.

Iversen, J.D., R. Greeley, B.R. White & J.B. Pollack, 1975*a*. 'Eolian erosion of the martian surface, part I: erosion rate similitude', *Icarus*, **26,** 321–31.

Iversen, J.D., R. Greeley, B.R. White & J.B. Pollack, 1975*b*. *Estimates of Saltation Threshold and Erosion Rates on Mars.* AIAA paper 75–1144, 5 pp.

Iversen, J.D., R. Greeley, B.R. White & J.B. Pollack, 1976*c*. 'The effect of vertical distortion in the modeling of sedimentation phenomena'. *Journal of Geophysical Research*, **81,** 4846–56.

Iversen, J.D. & V. Jensen, 1981. *'Wind Transportation of Dust from Coal Piles'.* Skibsteknisk Laboratorium Report SL 81054, Copenhagen, Denmark, 82 pp.

Iversen, J.D., J.B. Pollack, R. Greeley & B.R. White, 1976*a*. 'Saltation threshold on

Mars: the effect of interparticle force, surface roughness, and low atmospheric density'. *Icarus*, **29** (no. 3), 381–93.

Iversen, J.D. & B.R. White, 1982. 'Saltation threshold on Earth, Mars, and Venus'. *Sedimentology*, **29,** 111–19.

Ives, R.L., 1947. 'Behavior of dust devils'. *Bulletin American Meteorological Society*, **28,** 168–74.

Jackson, M.L., D.A. Gillette, E.F. Danielsen, J.H. Blifford, R.A. Bryson & S.K. Syers, 1973. 'Global dustfall during the quaternary as related to environments'. *Soil Science*, **116,** 135–45.

Jensen, M., 1958. 'The model-law for phenomena in natural wind'. *Ingeniøren*, **2,** 121–8.

Jensen, M., 1959. *Aerodynamik i den Naturlige Vind.* Teknisk Forlag, Kobenhavn, Denmark, pp. 219–27.

Jensen, A.M., 1976. 'A review of some dune afforestation procedures'. *In: Conservation in Arid and Semi-Arid Zones.* Food and Agriculture Organization of the United Nations, Rome, Italy, pp. 85–94.

Johnson, K.L., J.J. O'Connor & A.C. Woodward, 1973. 'The effect of the indenter elasticity on the Hertzian fracture of brittle materials'. *Proceedings, Royal Society London*, **Ser. A, 334,** 95–117.

Junge, C.E., 1971. 'The nature and residence times of tropospheric aerosols'. *In:* W.H. Matthews, W.W. Kellogg & G.D. Robinson, editors, *Man's Impact on the Climate:* MIT Press, Cambridge, Massachusetts, pp. 302–9.

Junge, C.E., 1979. 'The importance of mineral dust as an atmospheric constituent in the atmosphere' *In:* C.Morales, editor, *Saharan Dust*, John Wiley and Sons, New York, pp. 49–60.

Jungerius, P.D., A.J.T. Verheggen & A.J. Wiggers, 1981. 'The development of blowouts in 'De Blink' a coastal dune near Noordwijkerhout, the Netherlands'. *Earth Surface Processes and Landforms*, **6,** 375–96.

Kawamura, R., 1951. *Study on Sand Movement by Wind.* Institute of Science and Technology, Tokyo, Report 5, pp. 95–112.

Kimberlin, L.W., A.R. Hidlebaugh & A.R. Grunewald, 1977. 'The potential wind erosion problem in the United States'. *Transactions of the American Society for Agricultural Engineers*, **20,** 873–79.

Kind, R.J., 1976. 'A critical examination of the requirements for model simulation of wind-induced erosion/deposition phenomena such as snow drifting'. *Atmospheric Environment*, **10,** 219–27.

King, D., 1960. 'The sand ridge deserts of South Australia and related aeolian landforms'. *Transactions Royal Society S. Australia*, **83,** 99–108.

King, L.C., 1936. 'Wind-faceted stones from Marlborough, New Zealand'. *Journal of Geology*, **44,** 201–13.

King, W.J.H., 1916. 'The nature and formation of sand ripples and dunes'. *Geographical Journal*, **47,** 189–209.

Knollenberg, R.G. & D.M. Hunter, 1980. 'The microphysics of the clouds of Venus: Results of the Pioneer Venus Particle Size Spectrometer Experiment'. *Journal of Geophysical Research*, **85,** 8039–57.

Kocurek, G. & R.H. Dott, 1981. 'Distinctions and uses of stratification types in the interpretation of eolian sand'. *Journal of Sedimentary Petrology*, **51,** 579–95.

Koscielniak, D.E., 1973. 'Aeolian deposits on a volcanic terrain near St. Anthony, Idaho'. Unpubl. MS Thesis, State University of New York at Buffalo, 28 pp.

Krinsley, D.H. & J.C. Doornkamp, 1973. *Atlas of Quartz and Surface Textures*. Cambridge University Press, 91 pp.

Krinsley, D.H., R. Greeley & J.B. Pollack, 1979. 'Abrasion of windblown particles on Mars – erosion of quartz and basaltic sand under simulated martian conditions'. *Icarus*, **39,** 364–84.

Krinsley, D.H. & I.J. Smalley, 1972. 'Sand'. *American Scientist*, **60,** 286–91.

Krumbein, W.C. & L.L. Sloss, 1963. *Stratigraphy and Sedimentation*. W.H. Freeman, San Francisco, 660 pp.

Kuenen, P.H., 1928. 'Experiments on the formation of wind-worn pebbles'. *Leidsche Geologische Medellinger*, **3,** 17–38.

Kuenen, P.H., 1960, 'Sand'. *Scientific American*, **202,** 94–110.

Kuenen, P.H. & W.G. Perdok, 1962. 'Experimental abrasion 5. Frosting and defrosting of quartz grains'. *Journal of Geology*, **70,** 648–58.

Kuhlman, H., 1958. 'Quantitative measurements of aeolian sand transport'. *Geografisk Tidsskrift*, **57,** 51–74.

Kuzmin, R.O., 1978. 'The dark and bright streaks of Mars'. *Modern Geology*, **6,** 139–46.

Lawn, B. & R. Wilshaw, 1975. 'Review – indentation fracture: principles and applications'. *Journal Material Science*, **10,** 1049–81.

Lee, S.W., P.C. Thomas & J. Veverka, 1982. 'Windstreaks in Tharsis and Elysium: Implications for sediment transport by slope winds'. *Journal of Geophysical Research*, **87,** 10025–41.

Leovy, C.B., 1979. 'Martian meteorology'. *Annual Review of Astronomy and Astrophysics*, **17,** 387–413.

Leovy, C.B. & Y. Mintz, 1969. 'Numerical simulations of the atmospheric circulation and climate of Mars'. *Journal of the Atmospheric Sciences*, **26,** 1167–90.

Lettau, K. & H. Lettau, 1969. 'Bulk transport of sand by the barchans of the Pampa La Joya in southern Peru.' *Zeitschrift fur Geomorphologie*, **13,** 182–95.

Lettau, K. & H.H. Lettau, 1978. 'Experimental and micro-meteorological field studies of dune migration'. *In:* H.H. Lettau & K. Lettau, editors, *Exploring the World's Driest Climate*, University of Wisconsin-Madison, Institute for Environmental Studies, IES Report 101, pp. 110–47.

Lindsay, J.F., 1973. 'Ventifact evolution in Wright Valley, Antarctica'. *Geological Society of America Bulletin*, **84,** 1791–8.

Lindsay, J.F., 1976. *Lunar Stratigraphy and Sedimentology*. Elsevier, Amsterdam, 302 pp.

Long, J.T. & R.P. Sharp, 1964. 'Barchan-dune movement in the Imperial Valley, California'. *Geological Society of America Bulletin*, **75,** 149–56.

Lyles, L., 1977. 'Wind erosion: Processes and effect on soil productivity'. *Transactions, American Society Agricultural Engineering*, **20,** 880–4.

Lyles, L., L.F. Hagen & E.L. Skidmore, 1979. 'Theory and principles of wind erosion'. Unpubl. manuscript.

Lyles, L. & R.K. Krauss, 1971. 'Threshold velocities and initial particle motion as influenced by air turbulence'. *Transactions American Society Agricultural Engineering*, **14,** 563–6.

Lyles, L., R.L. Schrandt & N.F. Schneidler, 1974. 'How aerodynamic roughness elements control sand movement'. *Transactions, American Society Agricultural Engineering,* **17,** 134–9.

Mabbutt, J.A., 1977. *Desert Landforms.* The MIT Press, Cambridge, Massachusetts, 340 pp.

Mabbutt, J.A., R.A. Wooding & J.N. Jennings, 1969. 'The asymmetry of Australian desert sand ridges'. *Australian Journal of Science,* **32,** 159–60.

Madigan, C.T., 1946. 'The sand formations, Simpson Desert Expedition, 1939, Scientific Report 6'. *Geology, Transactions Royal Society S. Australia,* **70,** 45–63.

Maegley, W.J., 1976. 'Saltation and Martian sandstorms'. *Reviews of Geophysics and Space Physics,* **14,** 135–42.

Maeno, N., K. Araoka, K. Nishimuma & Y. Kaneda, 1979. 'Physical aspects of the wind-snow interaction in blowing snow'. *Journal Faculty Science,* **6,** 127–41.

Mainguet, M., 1968. 'Le Borkou, aspects d'un modelé éolien'. *Ann. de Géog.,* **77,** 296–322.

Mainguet, M., 1972. *Le Modelé des Grès.* L'Institut Géographique National, Paris.

Malin, M.C., 1974. 'Salt weathering on Mars'. *Journal of Geophysical Research,* **79,** 3888–94.

Malin, M.C., & R.S. Saunders, 1977. 'Surface of Venus: Evidence of diverse landforms from radar observations'. *Science,* **196,** 987–90.

Marshall, J.R., 1979. 'Experimental abrasion of natural materials'. Unpubl. Ph.D. Thesis, University College, London, 301 pp.

Martin, L.J., 1974. 'The major martian dust storms of 1971 and 1973'. *Icarus,* **23,** 108–15.

Maull, D.J. & L.F. East, 1963. 'Three-dimensional flow in cavities'. *Journal of Fluid Mechanics,* **16,** 620–32.

McCall, G.J.H., 1965. 'Possible meteorite craters – Wolf Creek, Australia and analogs'. *Annals New York Academy of Science,* **123,** 970–98.

McCauley, J.F., 1973. 'Mariner 9 evidence for wind erosion in the equatorial and midlatitude regions of Mars'. *Journal of Geophysical Research,* **78,** 4123–37.

McCauley, J.F., C.S. Breed, F. El-Baz, M.I. Whitney, M.J. Grolier & A.W. Ward, 1979. 'Pitted and fluted rocks in the Western Desert of Egypt: Viking comparisons'. *Journal of Geophysical Research,* **84,** 8222–31.

McCauley, J.F. & M.F. Grolier, 1976. 'Terrestrial yardangs'. *In: Reports of Accomplishments of Planetology Programs 1975–1976,* NASA TMX-3364, pp. 110–14.

McCauley, J.F., M.J. Grolier & C.S. Breed, 1977*a. Yardangs of Peru and Other Desert Regions.* US Geological Survey, Interagency Report, *Astrogeology,* **81,** 177.

McCauley, J.R., M.J. Grolier & C.S. Breed, 1977*b*. 'Yardangs'. *In:* D.O. Doehring, editor, *Geomorphology in Arid Regions.* Proceedings 8th Geomorphology Symposium, State University of New York, Binghamton, pp. 233–69.

McCauley, J., A.W. Ward, C.S. Breed, M.J. Grolier & R. Greeley, 1977*c*. 'Experimental modeling of wind erosion forms'. *In: Reports Accomplishments of Planetology Programs 1976–1977,* NASA TMX-3511, pp. 150–2.

McGill, G.E., J.L. Warner, M.C. Malin, R.E. Arvidson, E. Eliason, S. Nozette &

R.D. Reasenberg, 1983. 'Topography, surface properties and tectonic evolution'. *In:* D.M. Hunten, L. Colin, T.M. Donahue & V.I. Moroz, editors, *Venus*, University of Arizona Press, Tucson (in press).

McKee, E.D., 1966. 'Structures of dunes at White Sands National Monument, New Mexico (and a comparison with structure of dunes from other selected areas).' *Sedimentology*, **7,** 1–69.

McKee, E.D., editor, 1979*a*. *A Study of Global Sand Seas.* US Geological Survey Professional Paper 1052, 429 pp.

McKee, E.D., 1979b. 'Sedimentary structures in dunes'. *In:* E.D. McKee, editor, *A Study of Global Sand Seas.* US Geological Survey Professional Paper 1052, pp. 83–113.

McKee, E.D., 1982. 'Sedimentary structures in dunes of the Namib Desert, Southwest Africa'. *Geological Society of America Special Paper*, **188,** 177.

McKee, E.D. & G.C. Tibbitts, 1964. 'Primary structures of a seif dune and associated deposits in Libya'. *Journal of Sedimentary Petrology*, **34,** 5–17.

McLaughlin, D., 1954*a*. 'Volcanism and aeolian deposition on Mars'. *Bulletin Geological Society of America*, **65,** 715–17.

McLaughlin, D., 1954*b*. 'Wind patterns and volcanoes on Mars'. *Observatory*, **74,** 166–8.

Megalhães, J. & P. Gierasch, 1982. 'A model of martian slope winds: implications for eolian transport'. *Journal of Geophysical Research*, **87,** 9975–84.

Meigs, P., 1953. 'World distribution of arid and semi-arid homoclimates'. *In: Reviews of Research on Arid Zone Hydrology* (UNESCO, Paris), pp. 203–9.

Mellor, M., 1965. *Blowing snow: Cold Regions Science and Engineering*, Part III, Section A3c. Cold Regions Research and Engineering Laboratory, Hanover, New Hampshire, 79 pp.

Miller, E.R., 1934. 'The dustfall of November 12–13, 1933'. *Monthly Weather Review*, **62,** 14–15.

Mills, A.A., 1977. 'Dust clouds and frictional generation of glow discharges on Mars'. *Nature*, **268,** 614.

Møller, Jens, 1980. 'Desertification problems in a humid region'. *Workshop on Physics of Desertification*, International Centre for Theoretical Physics, Trieste, Italy.

Monin, A.S. & A.M. Yaglom, 1971. *Statistical Fluid Mechanics: Mechanics of Turbulence:* MIT Press, Cambridge, Massachusetts (translation from 1965 Russian edition), 769 pp.

Moore, H.J., R.E. Hutton, R.F. Scott, C.R. Spitzer & R.W. Shorthill, 1977. 'Surface materials of the Viking landing sites'. *Journal of Geophysical Research*, **82,** 4497–523.

Moore, H.J., C.R., Spitzer, K.Z. Bradford, P.M. Cates, R.E. Hutton & R.W. Shorthill, 1979. 'Sample fields of the Viking landers, physical properties, and aeolian processes'. *Journal of Geophysical Research*, **84,** 8365–77.

Morales, C., 1979. *Saharan Dust.* John Wiley, New York, pp. 3–20.

Murray, B.C., M.J.S. Belton, G.E. Danielson, M.E. Davies, D. Gault, B. Hapke, B. O'Leary, R.G. Strom, V. Soumi & N. Trask, 1974. 'Venus: Atmospheric motion and structure from Mariner 10 pictures'. *Science*, **183,** 21–9.

Mutch, T.A., R.E. Arvidson, A.B. Binder, E.A. Guinness & E.C. Morris, 1977.

'The geology of the Viking Lander 2 site'. *Journal of Geophysical Research*, **82,** 4452–67.

Mutch, T.A., A.B. Binder, F.O. Huck, E.C. Levinthal, S. Liebes, E.C. Morris, W.R. Patterson, J.B. Pollack, C. Sagan & G.R. Taylor, 1976. 'The surface of Mars: The view from the Viking 1 Lander'. *Science*, **193,** 791–801.

Nickling, W.G., 1978. 'Eolian sediment transport during dust storms: Slims River Valley, Yukon Territory'. *Canadian Journal of Earth Science*, **15,** 1069–84.

Nickling, W.G. & M. Ecclestone, 1981. 'The effects of soluble salts on the threshold shear velocity of fine sand'. *Sedimentology*, **28,** 505–10.

Norem, H., 1975. *Designing Highways Situated in Areas of Drifting Snow*. Cold Regions Research and Engineering Laboratory, Hanover, New Hampshire, Draft Translation 503, 141 pp.

Norem, H., 1979. 'Vurdering av forskjellige sneskjermtyper'. *Veglaboratoriet*, Oslo, Norway, Intern rapport nummer 905, 30 pp.

Norris, R.M., 1966. 'Barchan dunes of Imperial Valley, California'. *Journal of Geology*, **74,** 292–306.

Nozette, S. & J.S. Lewis, 1982. 'Venus: chemical weathering of igneous rocks and buffering of atmospheric composition'. *Science*, **216,** 181–3.

Odar, F., 1962. 'Scale factors for simulation of drifting snow'. *Journal of the Engineering Mechanics Division, American Society of Civil Engineers*, **88, EM2,** 1–16.

Oh, H.L., K.P.L. Oh, S. Vaidyanathan & I. Finnie, 1972. 'On the shaping of brittle solids by erosion and ultrasonic cutting'. *Proceedings Symposium on the Science of Ceramic Machining and Surface Finishing, U.S. National Bureau of Standards Special Publication*, **348,** 119–32.

O'Neil, M.E., 1968. 'A sphere in contact with a plane wall in a slow linear shear flow'. *Chemical Engineering Science*, **23,** 1293–8.

Owen, P.R., 1964. 'Saltation of uniform grains in air'. *Journal of Fluid Mechanics*, **20,** 225–42.

Owen, P.R., 1980. 'Sand movement mechanism'. *Workshop on Physics of Desertification*, International Centre for Theoretical Physics, Trieste, Italy.

Pécsi, Marton, 1968. 'Loess'. *In:* R.W. Fairbridge, editor, *Encyclopedia of Geomorphology*. Reinhold, New York, pp. 674–8.

Peel, R.F., 1970. 'Landscape sculpture by wind'. *International Geographical Congress* (Calcutta), *21st Proceedings*, pp. 99–104.

Peterfreund, A.R., 1981. 'Visual and infrared observations of wind streaks on Mars'. *Icarus*, **45,** 447–67.

Peterfreund, A.R., 1982. 'A search for local martian dust storms (abstract)'. *Bulletin of American Astronomical Society*, **14,** 757.

Peterfreund, A.R. & H.H. Kieffer, 1979. 'Thermal infrared properties of the martian atmosphere 3, local dust clouds'. *Journal of Geophysical Research*, **84,** 2853–64.

Peterson, S.T. & C.E. Junge, 1971. 'Sources of particulate matter in the atmosphere'. *In:* W.H. Matthews, W.W. Kellogg & G.D. Robinson, Editors, *Man's Impact on the Climate*, MIT Press, Cambridge, Massachusetts, pp. 310–20.

Pettijohn, F.J., P.E. Potter & R. Siever, 1972. *Sand and Sandstone*. Springer-Verlag, New York, 618 pp.

Péwé, T.L., 1981. 'Desert dust: an overview'. *In:* T.L. Péwé, editor, *Desert Dust: Origin, Characteristics, and Effect on Man*. Geological Society of America Special Paper 186, pp. 1–10.

Pollack, J.B., 1979. 'Climatic change on the terrestrial planets'. *Icarus*, **37,** 479–553.

Pollack, J.B., 1981. 'Atmospheres of the terrestrial planets'. *In:* J.K. Beatty, B. O'Leary & A. Chaikin, editors, *The New Solar System*, Cambridge University Press, pp. 57–70.

Pollack, J.B., D.S. Colburn, F.M. Flasar, R. Kahn, C.E. Carlston & D. Pidek, 1979. 'Properties and effects of dust particles suspended in the martian atmosphere'. *Journal of Geophysical Research*, **84,** 2929–45.

Pollack, J.B., D. Colburn, R. Kahn, J. Hunter, W. Van Camp, C.E. Carlston & M.R. Wolf, 1977. 'Properties of aerosols in the martian atmosphere, as inferred from Viking lander imaging data'. *Journal of Geophysical Research*, **82,** 4479–95.

Pollack, J.B., R. Haberle, R. Greeley & J.D. Iversen, 1976. 'Estimates of the wind speeds required for particle motion on Mars'. *Icarus*, **29,** 395–417.

Pollack, J.B. & Y.L. Yung, 1980. 'Origin and evolution of planetary atmospheres'. *Annual Review of Earth and Planetary Sciences*, **8,** 425–87.

Powers, W.E., 1936. 'The evidences of wind abrasion'. *Journal of Geology*, **44,** 214–19.

Radok, U., 1977. 'Snow Drift'. *Journal of Glaciology*, **19,** 123–39.

Reineck, H.E. & I.B. Singh, 1980. *Depositional Sedimentary Environments*. Springer-Verlag, Berlin, 549 pp.

Ring, S., J.D. Iversen, J.B. Sinatra & J.D. Benson, 1979. *Wind Tunnel Analysis of the Effects of Planting at Highway Grade Separation Structures*. Iowa Highway Department of Transportation. Iowa Highway Research Board HR-202, 210 pp.

Rogers, J.J.W., W.C., Krueger & M. Krog, 1963. 'Sizes of naturally abraded materials'. *Journal of Sedimentary Petrology*, **33,** 628–32.

Rubin, D.M., & R.E. Hunter, 1982. Bedform climbing in theory and nature'. *Sedimentology*, **29,** 121–38.

Rubin, D.M. & R.E. Hunter, 1983. 'Reconstructing bedform assemblages from compound crossbedding'. *In*: M.E. Brookfield & T.S. Ahlbrandt, editors, *Eolian Sediments and Processes*, Elsevier, Amsterdam, pp. 407–427.

Ruhe, R.V., 1969. *Quaternary Landscapes in Iowa*. Iowa State University Press, Ames, pp. 33–5.

Ryan, J.A., 1964. 'Notes on the martian yellow clouds'. *Journal of Geophysical Research*, **69,** 3759–70.

Ryan, J.A. & R.M. Henry, 1979. 'Mars' atmospheric phenomena during major dust storms, as measured at surface'. *Journal of Geophysical Research*, **84,** 2821–9.

Ryan, J.A. & R.D. Lucich, 1983. 'Possible dust devils, vortices on Mars'. *Journal of Geophysical Research*, **88,** 11005–11.

Saffman, P.G., 1965. 'The lift on a small sphere in a slow shear flow'. *Journal of Fluid Mechanics*, **22,** 385–400.

Saffman, P.G., 1968. 'Corrigendum'. *Journal of Fluid Mechanics*, **31,** 624.

Sagan, C., 1975. 'Windblown dust on Venus'. *Journal of the Atmospheric Sciences*, **32,** 1079–83.

Sagan, C. & R.A. Bagnold, 1975. 'Fluid transport on Earth and aeolian transport on Mars'. *Icarus*, **26,** 209–18.

Sagan, C., D. Pieri, P. Fox, R.E. Arvidson & E.A. Guinness, 1977. 'Particle motion on Mars inferred from the Viking Lander cameras'. *Journal of Geophysical Research*, **82,** 4430–8.

Sagan, C. & J.B. Pollack, 1969. 'Windblown dust on Mars'. *Nature*, **223,** 791–4.

Sagan, C., J. Veverka, P. Fox, R. Dubisch, R. French, P. Gierasch, L. Quam, J. Lederberg, E. Levinthal, R. Tucker, B. Eross & J.B. Pollack, 1973. 'Variable features on Mars. II. Mariner 9 global results'. *Journal of Geophysical Research*, **78,** 4163–96.

Sagan, C., J. Veverka, P. Fox, R. Dubisch, J. Lederberg, E. Levinthal, L. Quam, R. Tucker, J.B. Pollack & B.A. Smith, 1972. 'Variable features on Mars: preliminary Mariner 9 television results'. *Icarus*, **17,** 346–72.

Sagan, C.A., J. Veverka & P. Gierasch, 1971. 'Observational consequences of martian wind regions'. *Icarus*, **15,** 253–78.

Sagan, C., J. Veverka, R. Steinbacher, L. Quam, R. Tucker & B. Eross, 1974. 'Variable features on Mars. IV. Pavonis Mons'. *Icarus*, **22,** 24–47.

Sakamoto-Arnold, C.M., 1981. 'Eolian features produced by the December, 1977 windstorm, southern San Joaquin Valley, California'. *Journal of Geology*, **89,** 129–37.

Schmidt, R.A., 1980, 'Threshold wind-speeds and elastic impact in snow transport'. *Journal of Glaciology*, **26,** 453–67.

Schmidt, R.A., 1982. 'Properties of blowing snow'. *Reviews of Geophysics and Space Physics*, **20,** 39–44.

Schoewe, W.H., 1932. 'Experiments on the formation of wind-faceted pebbles'. *American Journal of Science*, 5th series, **224,** 111–34.

Secretariat of the United Nations, 1977. *Desertification: Its Causes and Consequences:* Conference on Desertification, Pergamon Press, Oxford, 448 pp.

Seiff, A. & D.B. Kirk, 1977. 'Structure of the atmosphere of Mars in summer at mid-latitudes'. *Journal of Geophysical Research*, **82,** 4364–78.

Seiff, A., D.B. Kirk, R.E. Young, S.C. Sommer, R.C. Blanchard, J.T. Findlay & G.M. Kelly, 1979. 'Thermal contract in the atmosphere of Venus: Initial appraisal from Pioneer Venus probe data'. *Science*, **205,** 46–8.

Seppala, M. & K. Linde, 1978. 'Wind tunnel studies of ripple formation'. *Geografiska Annaler*, **60,** 29–40.

Sharp, R.P., 1949. 'Pleistocene ventifacts east of the Big Horn Mountains, Wyoming'. *Journal of Geology*, **57,** 175–95.

Sharp, R.P., 1963. 'Wind ripples'. *Journal of Geology*, **71,** 617–36.

Sharp, R.P., 1964. 'Wind-driven sand in Coachella Valley, California'. *Geological Society of America Bulletin*, **74,** 785–804.

Sharp, R.P., 1966. 'Kelso dunes, Mojave Desert, California'. *Geological Society of America Bulletin*, **77,** 1045–74.

Sharp, R.P., 1973. 'Mars: South polar pits and etched terrain'. *Journal of Geophysical Research*, **78,** 4222–30.

Sharp, R.P., 1980. 'Wind-driven sand in Coachella Valley, California: Further data'. *Geological Society of America Bulletin*, Part I, **91,** 724–30.

Sinclair, P.C., 1966. 'A quantitative analysis of the dust devil. Unpubl. Ph.D. dissertation, University of Arizona, Tucson, 292 pp.

Skidmore, E.L., 1976. 'A wind erosion equation: Development, application, and limitations'. *Atmosphere-Surface Exchange of Particulate and Gaseous Pollutants. EROA Symposium Series*, **38,** 452–65.

Skidmore, E.L., P.S. Fisher & N.P. Woodruff, 1970. 'Wind erosion equation: Computer solution and application'. *Soil Science Society of America Proceedings*, **34,** 931–5.

Smalley, I.J., 1966. 'Formation of quartz sand'. *Nature*, **211**, 476–9.

Smalley, I.J. & D.H. Krinsley, 1979. 'Eolian sedimentation on Earth and Mars: Some comparisons'. *Icarus*, **40,** 276–88.

Smith, H.T.U., 1972. 'Aeolian deposition in martian craters'. *Nature Physical Science*, **238** (no. 83), 72–4.

Smith, R.S.U., 1982. 'Sand dunes in the North American Deserts': *In:* G.L. Bender, editor, *Reference Handbook on the Deserts of North America*, Greenwood Press, Westport, pp. 481–524.

Soderblom, L.A., T.J., Kreidler & H. Masursky, 1973*a*. 'Latitudinal distribution of a debris mantle on the martian surface'. *Journal of Geophysical Research*, **78,** 4117–22.

Soderblom, L.A., M.C. Malin, J.A. Cutts & B.C. Murray, 1973*b*. 'Mariner 9 observations of the surface of Mars in the north polar region'. *Journal of Geophysical Research*, **78,** 4197–210.

Stöffler, D., D.E. Gault, J. Wedekind & G. Polkowski, 1975. 'Experimental hypervelocity impact into quartz sand: Distribution and shock metamorphism of ejecta'. *Journal of Geophysical Research*, **80,** 4062–77.

Stone, E.C., & E.D. Miner, 1981. 'Voyager 1 encounter with the Saturnian system'. *Science*, **212,** 159–63.

Stone, R.O., 1967. 'A desert glossary'. *Earth Science Review*, **3,** 211–68.

Stone, R.O. & H.J. Summers, 1972. *Study of Subaqueous and Subaerial Sand Ripples*. US Office Naval Research, Report. No. USC Geology 72–1, Arlington, Virginia, 274 pp.

Sugden, W., 1968. 'Ventifacts'. *In:* R.W. Fairbridge, editor, *Encyclopedia of Geomorphology*. Reinhold, New York, pp. 1192–3.

Suzuki, T. & K. Takahashi, 1981. 'An experimental study of wind abrasion'. *Journal of Geology*, **89,** 23–36.

Svasek, J.N. & J.H.J. Terwindt, 1974. 'Measurements of sand transport by wind on a natural beach'. *Sedimentology*, **21,** 311–22.

Tabler, R.D., 1975. *Estimating the Transport and Evaporation of Blowing Snow*. Great Plains Agricultural Council, Research Committee, Publication No. 73, pp. 85–104.

Tabler, R.D., 1980*a*. 'Geometry and density of drifts formed by snow fences'. *Journal of Glaciology*, **26,** 405–19.

Tabler, R.D., 1980*b*. 'Self-similarity of wind profiles in blowing snow allows outdoor modeling'. *Journal of Glaciology*, **26,** 421–34.

Tabler, R.D., & R.L. Jairell, 1980. 'Studying snowdrifting problems with small-scale models outdoors'. *Western Snow Conference*, April 15–17, Laramie, Wyoming, 13 pp.

Takeuchi, M., 1980. 'Vertical profile and horizontal increase of drift-snow transport'. *Journal of Glaciology*, **26,** 481–92.

Tennekes, H., 1973. 'Similarity laws and scale relations in planetary boundary layers'. *In: Workshop on Micrometeorology*, American Meteorological Society, Boston, Massachusetts, pp. 177–216.

Thomas, P., 1981. 'North–south asymmetry of eolian features, in martian polar regions: Analysis based on crater-related wind markers'. *Icarus*, **48,** 76–90.

Thomas, P., 1982. 'Present wind activity on Mars: relation to large latitudinally zoned sediment deposits'. *Journal of Geophysical Research*, **87,** 9999–10008.

Thomas, P. & J. Veverka, 1979. 'Seasonal and secular variation of wind streaks on Mars: An analysis of Mariner 9 and Viking data'. *Journal of Geophysical Research*, **84,** 8131–46.

Thomas, P., J. Veverka & R. Campos-Marquetti, 1979. 'Frost streaks in the south polar cap of Mars'. *Journal of Geophysical Research*, **84,** 4621–33.

Thomas, P., J. Veverka, S. Lee & A. Bloom, 1981. 'Classification of wind streaks on Mars'. *Icarus*, **45,** 124–53.

Tillman, J.E., R.M. Henry & S.L. Hess, 1979. 'Frontal systems during passage of the martian north polar hood over the Viking lander 2 site prior to the first 1977 dust storm'. *Journal of Geophysical Research*, **84,** 2947–55.

Tsoar, H., 1974. 'Desert dunes morphology and dynamics'. *Al-Arish, Zeitschrift fur Geomorphologie Supplement*, **20,** 41–61.

Tsoar, H., 1978. 'The dynamics of longitudinal dunes'. Unpubl. Ph.D. Thesis, Ben-Gurion University of the Negev, 171 pp.

Tsoar, H., 1982. 'Internal structure and surface geometry of longitudinal (seif) dunes'. *Journal of Sedimentary Petrology*, **52,** 823–31.

Tsoar, H., 1983. 'Wind tunnel modeling of echo and climbing dunes'. *In*: M.E. Brookfield & T.S. Ahlbrandt, editors, *Eolian Sediments and Processes*, Elsevier. Amsterdam, pp. 247–59.

Tsoar, H., R. Greeley & A.R. Peterfreund, 1979. 'Mars: The north polar sand sea and related wind patterns'. *Journal of Geophysical Research*, **84,** 8167–80.

Twidale, C.R., 1972. 'Evolutions of sand dunes in the Simpson Desert, Central Australia'. *Transactions Institute British Geographers*, **56,** 77–109.

Tyler, G.L., V.R., Esheman, J.D. Anderson, G.S. Levy, G.F. Lindal, G.E. Wood & T.A. Croft, 1981. 'Radio science, investigations of the Saturn system with Voyager 1: Preliminary results'. *Science*, **212,** 201–6.

Verstappen, H.T., 1968. 'On the origin of longitudinal (seif) dunes'. *Zeitschrift fur Geomorphologie*, **12,** 200–12.

Veverka, J., K. Cook & J. Goguen, 1978*a*. 'A statistical study of crater-associated wind streaks in the north equatorial zone of Mars'. *Icarus*, **33,** 466–82.

Veverka, J., P. Gierasch & P. Thomas, 1981. 'Wind streaks on Mars: Meteorological control of occurrence and mode of formation'. *Icarus*, **45,** 154–66.

Veverka, J. & C. Sagan, 1974. 'McLaughlin and Mars'. *American Scientist*, **62,** 44–53.

Veverka, J., C. Sagan, L. Quam, R. Tucker & B. Eross, 1974. 'Variable features on Mars III: Comparison of Mariner 1969 and Mariner 1971 photography'. *Icarus*, **21,** 317–68.

Veverka, J., P. Thomas & C. Sagan, 1978*b*. 'On the nature and visibility of crater-associated streaks on Mars'. *Icarus*, **36,** 147–52.

Ward, A., 1978. 'Windforms and wind trends on Mars: An evaluation of martian surficial geology from Mariner 9 and Viking spacecraft television images'. Unpubl. Ph.D. Thesis, University of Washington, Seattle, 201 pp.

Ward, A.W., 1979. 'Yardangs on Mars: Evidence of recent wind erosion'. *Journal of Geophysical Research*, **84,** 8147–66.

Ward, A.W. & R. Greeley, 1984. 'The yardangs at Rogers Lake, California', in press, *Geological Society of America Bulletin.*

Warren, A., 1969. 'A bibliography of desert dunes and associated phenomena'. *In:* W.G. McGinnies & B.J. Goldman, editors, *Arid Lands in Perspective*, University Arizona Press, Tucson, pp. 75–99.

Wellman, H.W. & A.T. Wilson, 1965. 'Salt weathering, a neglected geological erosive agent in coastal and arid environments'. *Nature*, **205,** 1097.

White, B.R., 1979. 'Soil transport by winds on Mars.' *Journal of Geophysical Research*, **84,** 4643–51.

White, B.R., 1981*a*. 'Low-Reynolds-number turbulent boundary layers'. *Journal of Fluids Engineering*, **103,** 624–30.

White, B.R., 1981*b*. 'Venusian saltation'. *Icarus*, **46,** 226–32.

White, B.R., R. Greeley, J.D. Iversen, & J.B. Pollack, 1976. 'Estimated grain saltation in a martian atmosphere'. *Journal of Geophysical Research*, **81,** 5643–50.

White, B.R., & J.C. Schulz, 1977. 'Magnus effect on saltation'. *Journal of Fluid Mechanics*, **81,** 497–512.

White, S.J., 1970. 'Plane bed thresholds of fine grained solids'. *Nature*, **228,** 162–3.

Whitney, M.I., 1978. 'The role of vorticity in developing lineation by wind erosion'. *Geological Society of America Bulletin*, Part I, **89,** 1–18.

Whitney, M.I., 1979. 'Electron micrography of mineral surfaces subject to wind-blast erosion'. *Geological Society of America Bulletin*, Part I, **90,** 917–34.

Whitney, M.I. & R.V. Dietrich, 1973. 'Ventifact sculpture by windlbown dust'. *Geological Society of America Bulletin*, **84,** 2561–82.

Whitney, M.I. & J.F. Splettstoesser, 1982. 'Ventifacts and their formation: Darwin Mountains, Antarctica'. *In:* D.H. Yaalon, editor, *Aridic Soils and Geomorphic Processes*, *Catena*, Suppl. 1, pp. 175–94.

Williams, G., 1964. 'Some aspects of the eolian saltation load'. *Sedimentology*, **3,** 257–87.

Wilshire, H.G., J.K. Nakata & B. Hallet, 1981. 'Field observations of the December, 1977 wind storm, San Joaquim Valley, California'. *In:* T.J. Péwé, editor, *Desert Dust: Origin, Characteristics, and Effect on Man*, Geological Society of America Special Paper 186, pp. 233–51.

Wilson, I.G., 1971. 'Desert sandflow basins and a model for the development of ergs'. *Geographical Journal*, **137,** 180–99.

Wilson, I.G., 1972*a*. 'Aeolian bedforms – their development and origins'. *Sedimentology*, **19,** 173–210.

Wilson, I.G., 1972*b*. 'Sand waves'. *New Scientist*, **53,** 634–7.

Wilson, I.G., 1972*c*. 'Universal discontinuities in bedforms produced by the wind'. *Journal of Sedimentary Petrology*, **42,** 667–9.

Wilson, I.G., 1973. 'Ergs'. *Sedimentary Geology*, **10,** 77–106.

Wilson, L. & J.W. Head, 1981. 'Ascent and eruption of basaltic magma on the Earth and Moon'. *Journal of Geophysical Research*, **86,** 2971–3001.

Wood, C.D. & P.W. Espenschade, 1965. 'Mechanisms of dust erosion'. *Society of Automotive Engineers*, **73,** 515–23.

Woodruff, N.P. & F.H. Siddoway, 1965. 'A wind erosion equation'. *Soil Science Society Proceedings*, **29,** 602–8.

Zhonglong, W. & C. Yuan, 1980. 'Research on prevention of snow-drifts by blower fences'. *Journal of Glaciology*, **26,** 435–45.

Zimbleman, J. & R. Greeley, 1981. 'High resolution visual, thermal, and radar observations in the northern Syrtis Major region of Mars'. *Proceedings, Lunar Planetary Science*, **12,** 1419–29.

Zimbleman, J. & R. Greeley, 1982. 'Surface properties of ancient cratered terrain in the Northern Hemisphere of Mars'. *Journal of Geophysical Research*, **87,** 10181–9.

Zingg, A.W., 1953. 'Wind tunnel studies of the movement of sedimentary material'. *Proceedings 5th Hydraulic Conference Bulletin*, **34,** 111–35.

Zingg, A.W. & W.S. Chepil, 1950. 'Aerodynamics of wind erosion'. *Agricultural Engineering*, **31,** 279–82.

Zurek, R.W., 1982. 'Martian great dust storms: An update'. *Icarus*, **50,** 288–310.

Index

Page references in *italics* refer to figures and tables.